Pan Pacific Microelectronics Symposium 2013

(PAN PAC 2013)

Maui, Hawaii, USA
22-24 January 2013

ISBN: 978-1-62276-898-1

Printed from e-media with permission by:

Curran Associates, Inc.
57 Morehouse Lane
Red Hook, NY 12571

Some format issues inherent in the e-media version may also appear in this print version.

Copyright© (2013) by Surface Mount Technology Association (SMTA)
All rights reserved.

Printed by Curran Associates, Inc. (2013)

For permission requests, please contact Surface Mount Technology Association (SMTA)
at the address below.

Surface Mount Technology Association (SMTA)
5200 Wilson Road
Suite 215
Edina, MN 55424

Phone: (952) 920-4682
Fax: (952) 926-1819

www.smta.org

Additional copies of this publication are available from:

Curran Associates, Inc.
57 Morehouse Lane
Red Hook, NY 12571 USA
Phone: 845-758-0400
Fax: 845-758-2634
Email: curran@proceedings.com
Web: www.proceedings.com

TABLE OF CONTENTS

2013 Pan Pacific Symposium Technical Papers

Session TA1: Advanced Manufacturing
Chair: Keith Sweatman, Nihon Superior, Co. Ltd.

Cleaning High Reliability Assemblies with Tight Gaps, a Detailed Analysis   1
Thomas M. Forsythe, Kyzen Corporation

Requirements on a Class "0" EPA – Basics, Standards, ESD Equipments and Measurements   9
Hartmut Berndt, B.E.STAT European ESD Competence Center

Design and Fabrication of Ultra-Thin Flexible Substrate   15
Ming-Kun Chen, Yi-Lung Lin, Advanced Semiconductor Engineering Test RD; Yu-Jung Huang, Shen-Li Fu,
I-Shou University

Session TP1 Failure Analysis Tools and Techniques
Chair: Hartmut Berndt, B.E.STAT European ESD Competence Center

Side Wall Wetting Induced Void Formation Due To Small Solder Volume in Microbumps of Ni/SnAg/Ni
Upon Reflow   20
Y. C. Liang, Chih Chen, National Chiao Tung University; K. N. Tu, University of California
Los Angeles

A Mechanistically Justified Model for Life of SnAgCu Solder Joints in Thermal Cycling   25
P. Borgesen, L. Yang, A. Qasaimeh, L. Yin, Binghamton University; M. Anselm, Universal Instruments Corporation

TP2 Plenary Session: Strategic Directions
Chair: Bernard Courtois, Ph.D., CMP

A New Manufacturing Model for Successfully Competing in High Labor Rate Markets: How to Minimize
Labor and Material, the Controllable Contributions to a High-Tech Electronic Product's Cost, and Assess a
Manufacturing Region's Business Climate   31
Tom Borkes, The Jefferson Project

Tamper Proof, Tamper Evident Encryption Technology   54
Phil Isaacs, Thomas Morris, Jr., Michael J. Fisher, IBM Corporation; Keith Cuthbert, W. L. Gore & Associates

Impact of Lead-Free Components and Technology Scaling for High Reliability Applications   63
Chris Bailey, Ph.D., University of Greenwich

Keynote Presentation I
Chair: Soren Norlyng, MICRONSULT

Panel Level Packaging – A Manufacturing Solution for Cost Effective Systems   68
R. Aschenbrenner, Ph.D., K.-F. Becker, T. Braun, A. Ostmann, Fraunhofer Institute for Reliability and
Microintegration

Session WA1 3D Structures
Chair: Hajime Tomokage, Ph.D., Fukuoka University

3D-TSV Vertical Interconnection Using Cu/SnAg Double Bumps and Non-Conductive Films (NCFs)   74
Kyung-Wook Paik, Yongwon Choi, Jiwon Shin, KAIST

Microstructure Control of Uni-Directional Growth of η-Cu_6Sn_5 in Microbumps on (111) Oriented and Nanotwinned Cu   78
Han-Wen Lin, Jia-Ling Lu, Chen-Min Liu, Chih Chen, National Chiao Tung University; King-ning Tu, University of California Los Angeles; Delphic Chen, Jui-Chao Kuo, National Cheng Kung University

3D Integration A Thermal-Electrical-Mechanical-Reliability Study   84
K. Weide-Zaage, J. Schlobohm, A. Farajzadeh, J. Kludt, Leibniz University Hannover; H. Frémont, University Bordeaux I

Low-cost and High Performance Silicon Interposers and Packages (LSIP) – A New Georgia Tech PRC Industry Consortium   93
Rao R. Tummala, Ph.D., Venkatesh Sundaram, Ph.D., Qiao Chen, Hao Lu, Gokul Kumar, Georgia Institute of Technology

BVA: Solution for Next Generation Very Fine-Pitch Package-on-Package (PoP) Applications   98
Vern Solberg, Ilyas Mohammed, Invensas Corporation

Session WP1 Advanced Materials
Chair: Kirsten Weide-Zaage, Leibniz University Hannover

Jetting Fine Lines for High Viscosity Fluids onto 2D and 3D Electronic Packages   103
Horatio Quinones, Ph.D., Nordson ASYMTEK

Aerosol Jet® Printing of Conductive Epoxy for 3D Packaging   107
Michael J. Renn, Ph.D., Kurt K. Christenson, Ph.D., Optomec, Inc.; Donald Giroux, Resin Designs, LLC; Daniel Blazej, Ph.D., Assembly Answers, LLC

Dielectrics for Embedding Active and Passive Components   114
J. Kress, R. Park, A. Bruderer, N. Galster, Atotech Deutschland GmbH; SH Cho, Dongyang Mirae University

A Nano Silver Replacement for High Lead Solders in Semiconductor Junctions   120
Keith Sweatman, Tetsuro Nishimura, Nihon Superior Co. Ltd.; Teruo Komatsu, Applied Nanoparticle Laboratory Co., Ltd.

Keynote Presentation II
Chair: Dock Brown, Medtronic (Retired)

Technical Communication: Strategies for Success   127
Chrys Shea, Shea Engineering Services

Session THA1 State of the Art Tools and Techniques
Chair: Chris Bailey, Ph.D., University of Greenwich

Tools and Techniques for Material Assessment in Advanced Technologies   132
Martin Anselm, Ph.D., Wayne Jones, Universal Instruments Corporation

Computed Tomography on Electronic Components-Better Ways to Do Failure Analysis Plus 4D CT The New Frontier   139
Wesley F. Wren, North Star Imaging

Acoustic Micro Imaging Analysis Methods for 3D Packages   145
Janet E. Semmens, Sonoscan, Inc.

Silicon V-Groove Alignment Bench for Optical Component Assembly   151
Terry Bowen, TE Connectivity

Keynote Presentation III
Chair: Charles Bauer, Ph.D., TechLead Corporation

Three Dimensional Integration Research Focusing on Device Embedded Substrate   156
Hajime Tomokage, Ph.D., Fukuoka University

THP1 Trends and Roadmaps
Chair: Rolf Aschenbrenner, Ph.D., Fraunhofer Institute IZM

The Challenges of LGA Server Socket Trends   160
Jackson Chang, Michael Hung, Bono Liao, Nick Lin, Andrew Gattuso, Bob McHugh, Foxconn Electronics, Inc.

The Quest for Reliability Standards   167
Dieter W. Bergman, IPC Inc.

Alternatives to Solder in Interconnect, Packaging, and Assembly   177
Herbert J. Neuhaus, Ph.D., Charles E. Bauer, Ph.D., TechLead Corporation

2013 iNEMI Technology Roadmap Overview   183
Bill Bader, Chuck Richardson, iNEMI

3D IC Integration Technology Development in China   186
Wei Koh, Ph.D., Pacrim Technology

Technical Committee
2013 Pan Pacific Symposium

North American Liaison:
Chuck Bauer, Ph.D., *TechLead Corporation*

Asian Liaison:
Phil Isaacs, *IBM Corporation*

European Liaison:
Soren Norlyng, *MICRONSULT*

Committee:
Bill Bader, *iNEMI*
Tom Chung, Ph.D., *ASTRI*
Bernard Courtois, Ph.D., *CMP*
Krista Crotty, *Alberi EcoTech*
Joseph Fjelstad, *Verdant Electronics*
Yu-Jung Huang, Ph.D., *I-Shou University*
Ricky Lee, *HKUST-Center for Advanced Microsystems Packaging*
Charles Lin, Ph.D., *Bridge Semiconductor*
Teng Hoon Ng, *Celestica (Thailand) Co., Ltd.*
Tetsuro Nishimura, *Nihon Superior Company Ltd.*
Kyung Paik, Ph.D., *KAIST*
Michael Pecht, Ph.D., *University of Maryland*
Peter Pooh, Ph.D.
Horatio Quinones, Ph.D., *Nordson ASYMTEK*
Alan Rae, Ph.D., *TPF Enterprises LLC*
Vern Solberg, *STC Madison / Invensas Corporation*
Keith Sweatman, *Nihon Superior Company l td.*
Rao Tummala, Ph.D., *Georgia Institute of Technology*
Henry Utsunomiya, *Interconnection Technologies, Inc.*
M. Juergen Wolf, *Fraunhofer Institute*

Thanks to all of the 2013 Pan Pacific Symposium Sponsors!

Corporate Package Sponsors:

Conference Sponsor:

Luau and Reception Sponsor:

Refreshment Break Sponsors:

Media Sponsors:

CLEANING HIGH RELIABILITY ASSEMBLIES WITH TIGHT GAPS, A DETAILED ANALYSIS

Thomas M. Forsythe
Kyzen Corporation
Nashville, TN, USA
tom_forsythe@kyzen.com

ABSTRACT

As electronic assemblies have grown ever more capable over recent years, their form factors had decreased at an impressive rate of their won. These small gaps and design challenges have emerged in concert with a renewed requirement for entire assembly cleaning driven by an array of requirements led by increasing reliability requirements. Of course, while cleaning has always been mission critical for a number of segments, such as medical and military assemblies, today it is being adopted broadly throughout the industry.

This presents both advanced technology groups and manufacturing engineers with a new process to implement with little tribal knowledge within their organization to base their evaluation on. This paper will study these small gaps and evaluate ionic residues post cleaning by a variety of cleaning agents versus a pair of commonly encountered water soluble fluxing materials. This will allow users to understand the challenges presented by low gap height and the risks associated with various cleaning approaches to remove those residues.

INTRODUCTION

State of the art electronic devices continue to advance at a rapid rate delivering new capabilities to consumers throughout the world. Mobile phones alone account for over 400 million units per quarter, a solid 35% of which are smart phones. Smart phone production volumes now eclipse PS shipments, even with the generous inclusion of tablet sales in the PC statistics.

As even the casual observer is aware, these smart phones are smart indeed and their capabilities are steady improving. This enhanced performance is a key element driving demand for these devices, not surprisingly as performance improves so does the user's expectations of quality and reliability; if one has their "life on one's phone", we certainly are not happy to see it go up in smoke in any way.

Electro Chemical Migration (ECM) is a critical risk factor in any electronic reliability analysis. Since every electronic device is powered up to function, and virtually every device does so in the presence of humidity, the sure way to prevent ECM is the absence of ionic residues.

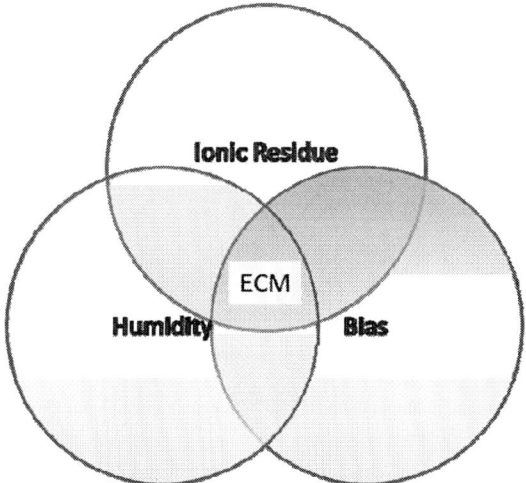

Figure 1. Factors contributing to ECM

This paper will detail conduct a thorough review of these residues detected during the DOE developed for this paper.

Key words: electronics cleaning, POP cleaning, flip chip cleaning

Reducing or eliminating residues starts with a well-designed, validated, well run cleaning process. Such a process has two major building blocks: the equipment delivering the mechanical energy, and the cleaning agent delivering a well matched chemical solution that together remove all undesired contaminants not only from readily accessible surface areas but difficult to reach gaps beneath components and other devices.

The balance between chemical and mechanical elements in the process is critical to robust process design, equally important to a detailed understanding of the assemblies or packages which are to be cleaned.

Those schooled in the art of cleaning know that board density can increase the cleaning challenge, but the critical driver in today's complex designs is the "gap". The gap, also known as the stand-off height, is the distance between the bottom of a device and the board surface; the shorter this distance, generally referred to as the smaller that gap the more difficult the cleaning challenge. Not surprisingly, truly flush mounted components present the greatest challenge.

With a sound understanding of the challenges presented by the assembly design, next we turn to the cleaning process itself. This evaluation begins with certain fundamentals developed during decades of research into cleaning technology which can act as a guide during the process.

1. Increased temperature generally enhances processes results. However, the results provided by slightly elevated temperature are often not bettered at very high temperatures. More is not always better, and our data set will guide us to defining the point of diminishing returns.
2. Likewise, higher concentrations of the cleaning agents often enhance performance. As with temperature, there is routinely an inflection point of diminishing returns that should be understood in any process design. Operating concentrations have a linear effect on operating costs and always receive close scrutiny.
3. The mechanical energy delivery system: pressure, spray patterns, exposure gaps.
4. Exposure time to the cleaning agent and mechanical energy. Time is always a precious commodity, and frequently subject to arbitrary limits determined prior to the device evaluation. When considering tight gaps or low standoff height device cleaning, a fifth element comes into play: cleaning agent surface tension and propensity for capillary action. In conjunction with the driving force of mechanical impingement lower surface tension improves capillary action. Together these forces enhance wetting and penetration of the fluid into tight gaps beneath components.

The purpose of this designed experiment is to evaluate the effectiveness of a variety of cleaning agents under an array of process conditions. As such, mechanical energy was limited to allow for full understanding of the chemical driving forces at work as evaluated by ion chromatography.

EXPERIMENTAL DESIGN
This DOE focused on cleaning effectiveness of a selected low gap chip scale package.

Two commonly used water soluble fluxes typical for this type of package were selected for comparative purposes, referred to in the paper as WS#1 and WS#2.

Three different cleaning solutions were included plus the commonly used water alone baseline. These materials are referred to throughout the paper as Agents A, B, & C which were evaluated at 2%, 4%, 6%, & 10%. An un-cleaned control samples were evaluated with each soldering material as well as samples cleaned with 100% water which were evaluated at each temperature condition.

Three temperatures of 20C, 40C & 60C were evaluated all with minimal agitated soak via mild shaking agitation. This approach was taken to fully evaluate the chemical driving forces of the cleaning agents. Follow on testing is planned to evaluate various mechanical energy options and their impact on the results.

Response variable included vision inspection at 100x and both anion and cation evaluation via ion chromatography (IC).

LITERATURE REVIEW AND DATA ANALYSIS METHODOLOGY
Dozens of papers have been presented over the past 10 years, evaluating various aspects of new and novel cleaning processes. These evaluations provide a range of perspectives:

1. One mechanical approach compared to various chemical options.
2. One or two chemical options across a variety of equipment platforms.
3. User driven papers walking through their DOE, often employing sophisticated test cards to simulate the wide variety of designs encountered in their operation.

Each of these approaches has their benefits and contributes to the industries body of knowledge, and indeed this paper follows point one employing one, limited mechanical action approach contrasted with a number of temperature, concentration and agent variations. This DOE attempted to bring another facet to the discussion. That being a large data set, 82 different points each with IC results to compare and contrast an unusually large body of IC data we will attempt to analyze thoroughly.

DATA ANALYSIS
We begin the data review with our control sample. What is the state of the substrates prior to any cleaning step at all? Figure 2 provides the anions detail. We have chlorides, nitrates and weak organic acid present. WS#2 has generally lower levels of WOA than WS#1.

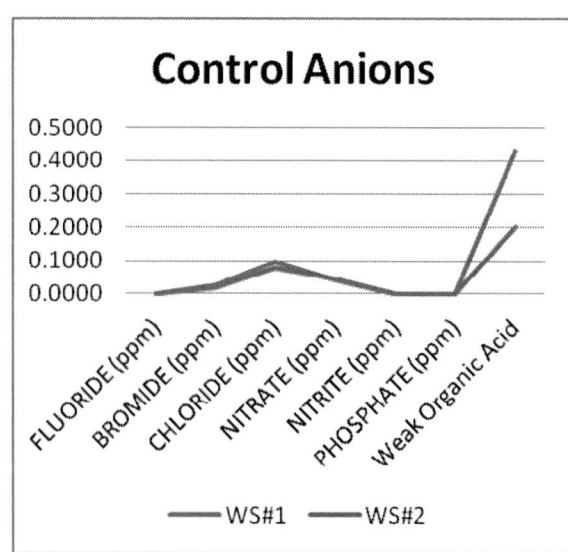

Figure 2. Control Anions

One challenge with this DOE is it is a point source analysis. We did not evaluate full assemblies. The reason is surface cleaning is generally not very challenging these days. It can be, but cleaning in these tight gaps is the critical success criteria. For this reason the results are a bit different from other recent studies.

The challenge comes with interpreting the data, current industry standards are logically focused on the acceptability of a full assembly not a single challenging device. The proper approach for scaling down these full assembly acceptable standards is also work that will be addressed in the future.

Cations detail is in Figure 3. We have sodium, lithium, potassium present. WS#2 also generally has lower levels of cations than WS#1.

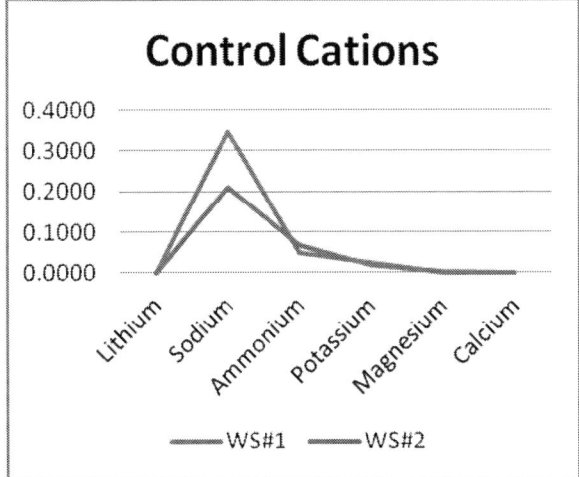

Figure 3. Control Cations

Water alone was included in the evaluation for one reason. It is the most common cleaning agent used to clean water soluble fluxes throughout the world. The key question is how does it measure up versus the control and the various cleaning agents evaluated.

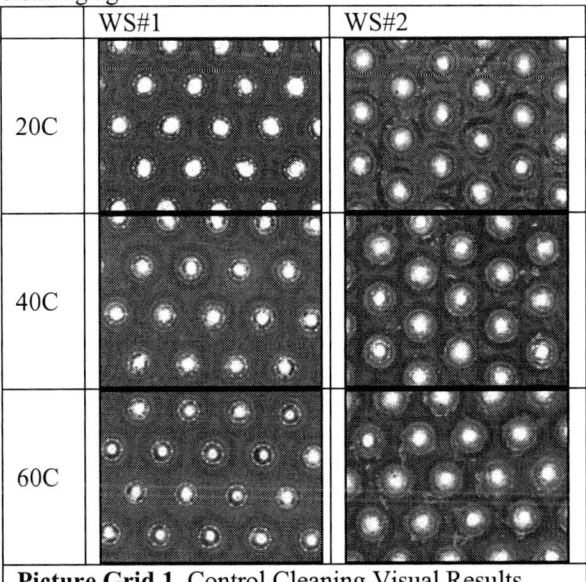

Picture Grid 1. Control Cleaning Visual Results

As shown in Picture Grid 1, WS#1 visually has less residue than WS#2 at each temperature point.

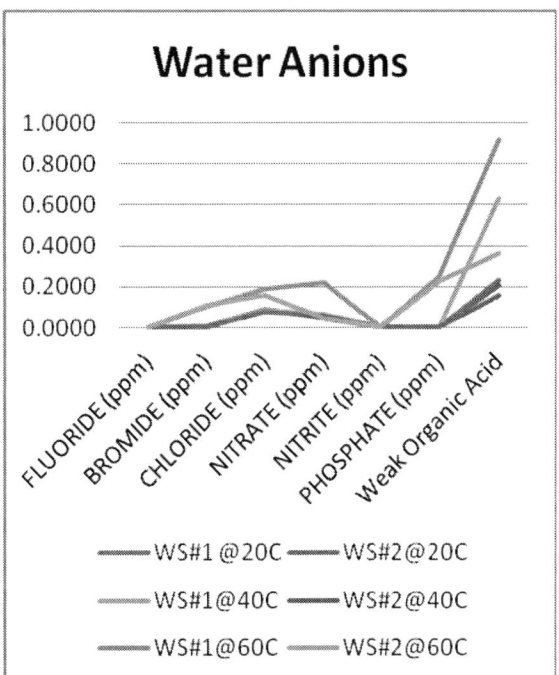

Figure 4. Water Only Anion Results

Evaluating the water only anion results we see that chlorides, bromides, nitrates and weak organic acids are all present while WS#2 has lower levels of WOA.

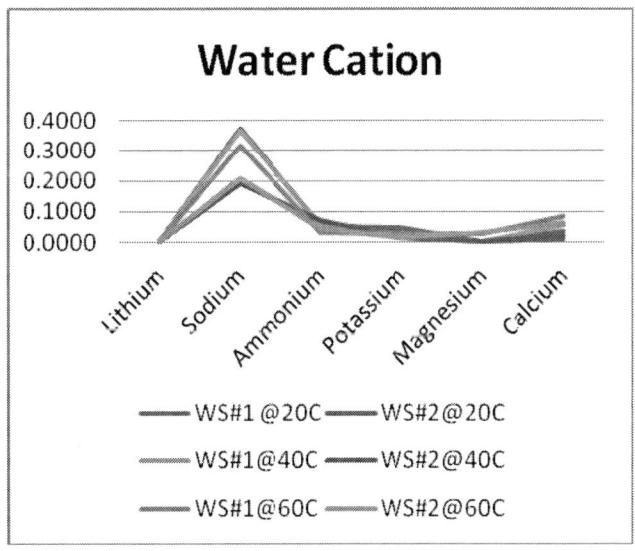

Figure 5. Water Only Cation Results

Reviewing the water only cation results, we see sodium, ammonium, potassium, magnesium, and calcium present. WS#2 displays lower levels of cations.

Next we will review the results from Agent A.

		WS#1	WS#2
10%	40C		
10%	60C		
6%	40C		
6%	60C		
4%	40C		
4%	60C		
2%	40C		
2%	60C		

Picture Grid 2. Agent A Visual Cleaning Results

The visual results for Agent A are comparable for both soldering materials with much less difference between the significant visual contrast displayed by water alone.

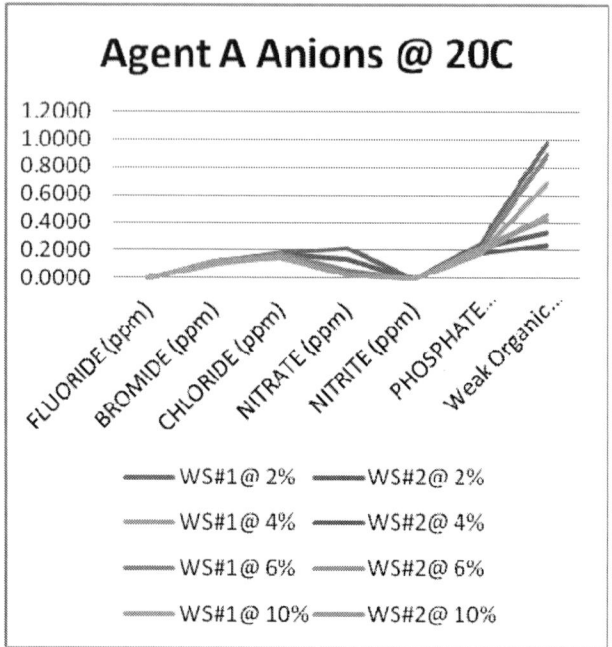

Figure 6. Agent A Anions @ 20C

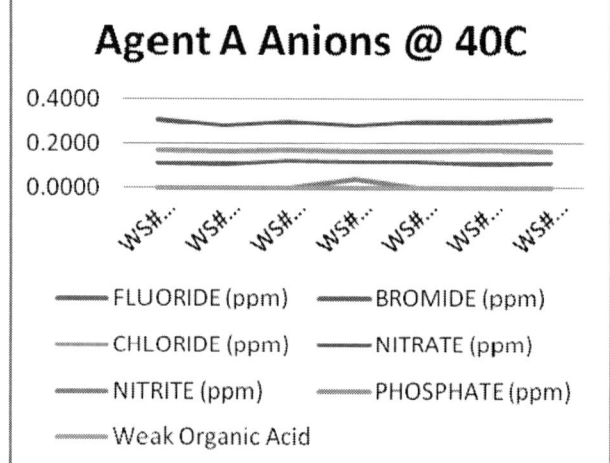

Figure 7. Agent A Anions @ 40C

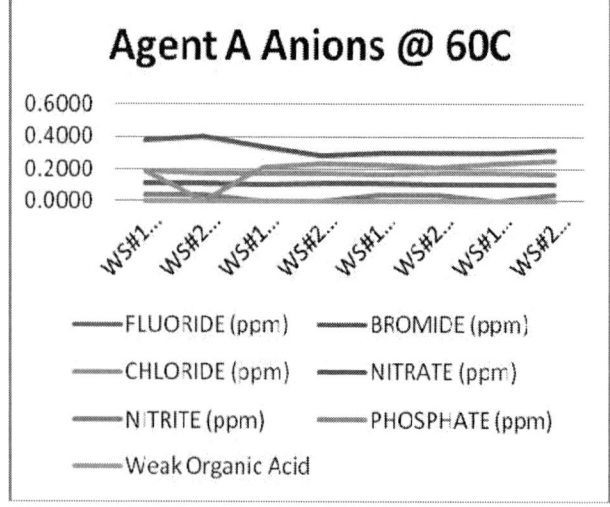

Figure 8. Agent A Anions @ 60C

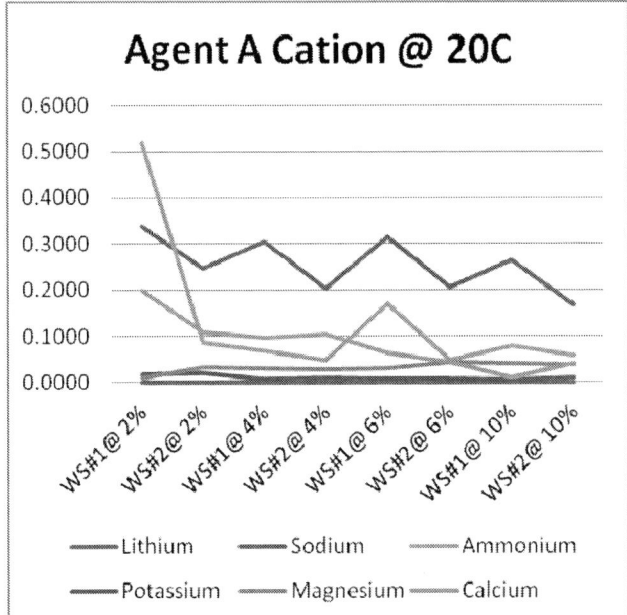

Figure 9. Agent A Cations @ 20C

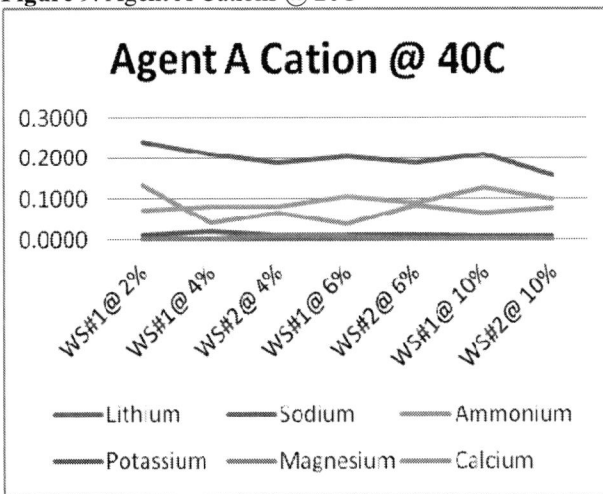

Figure 10. Agent A Cations @ 40C

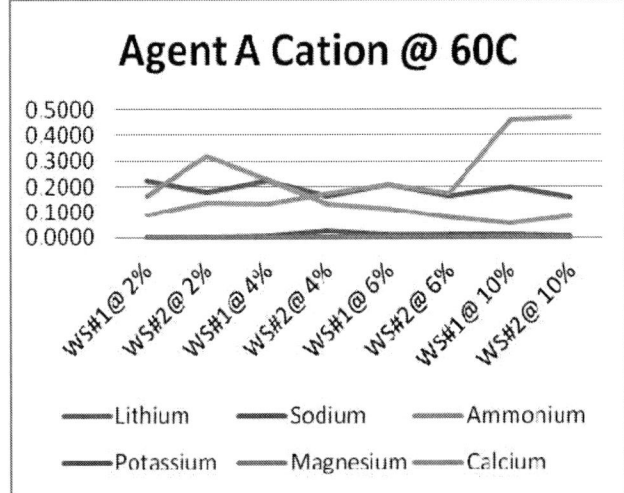

Figure 11. Agent A Cations @ 60C

Inferences that can be drawn from the Agent A data include less visual difference between the two fluxes them than when exposed to water alone and cleaning marginally improved with higher concentration.

Looking specifically at anions, the levels were very low overall. At the lowest temperature point of 20C, we do see the data spread for WOA. As temperature increases all the results trend together with WOA reduced to 0 ppm as temperature increased to 40 & 60C while nitrates and phosphates rose slightly at 40 & 60C.

Cation sodium and ammonium were slightly higher levels. Ammonium levels dropped at 4-8% concentration but slight rose at 10% concentration. Performance seems to improve at higher temperatures. WS#2 continued to have slightly lower levels than WS#1, while WS#1 had a little less overall visual residue.

Interaction and Main Effects Plots
Agents B & C showed slightly better performance, but rather than review those data points individual we will do so through the use of interaction and main effects plats to allow easy comparison.

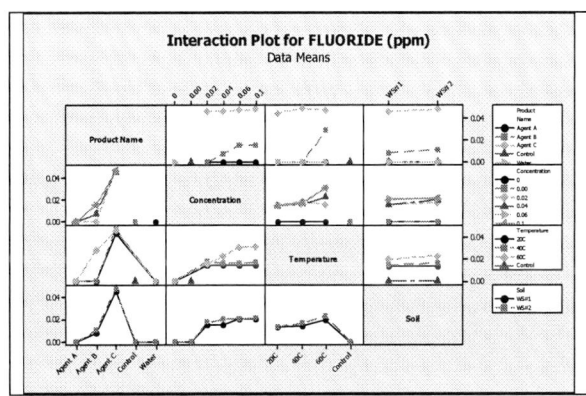

Figure 12. Interaction Plot for Fluoride

Here we see little response to Fluoride from water or Agent A & B. Agent C levels are a bit higher, though still very low across all temperatures and concentrations.

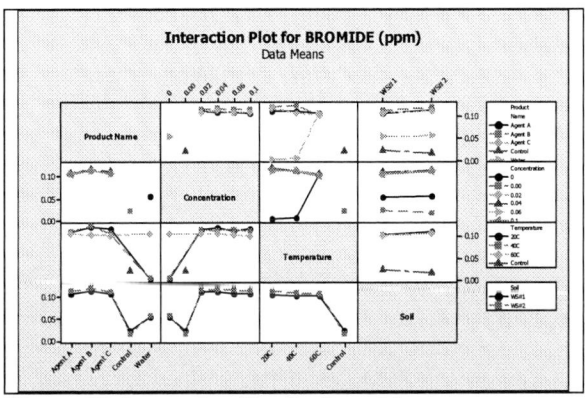

Figure 13. Interaction Plot for Bromide

Bromide seems to show identical affects for all cleaning

5

agents across concentrations and temperatures, with the water and control effects being very similar.

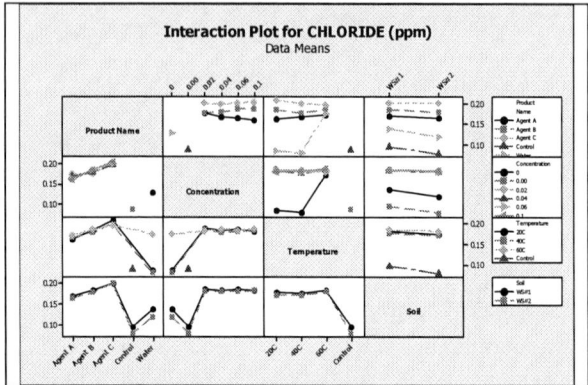

Figure 14. Interaction Plot for Chloride

With Chlorides we do see some spread in the data, though across a very small range. We see Agents A, B & C performing in that order consistently throughout the data though the gaps are very small.

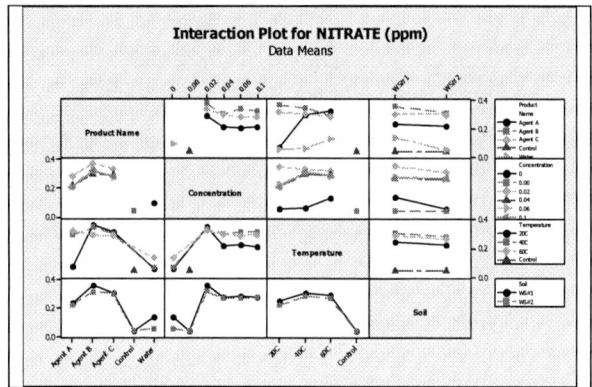

Figure 15. Interaction Plot for Nitrate

Similarly to bromides, nitrates seem to react to all Agents in a comparable fashion.

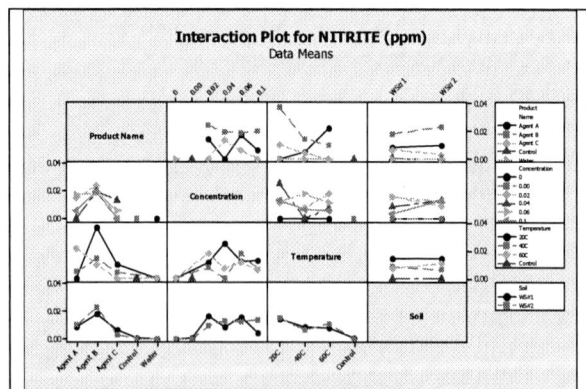

Figure 16. Interaction Plot for Nitrite

Agent B seems to respond to temperature when looking at the Nitrite data, but generally the materials perform comparably.

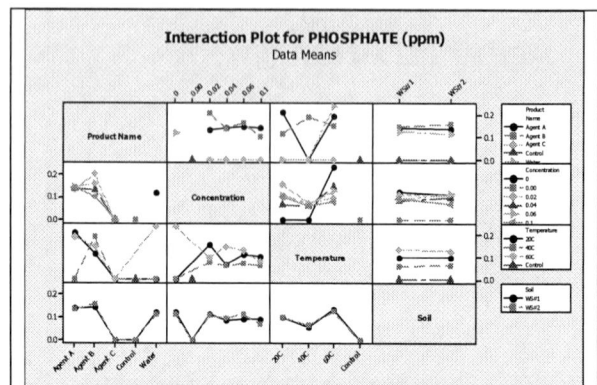

Figure 17. Interaction Plot for Phosphate

Agent C seems to have an advantage with phosphate ions, but the values are all quite low for agents.

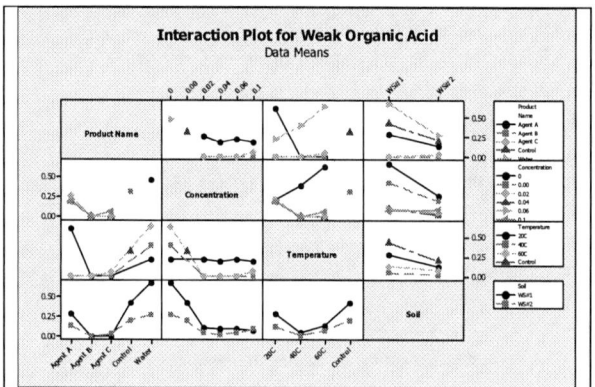

Figure 18. Interaction Plot for Weak Organic Acid

Water responds meaningfully raising the temperature from 20 C to 40C as one would expect. While WS#2 appears to have lower levels of WOA, the data is skewed by the control and water only results which are meaningfully poorer than all the Agent data. This is a meaningful observation.

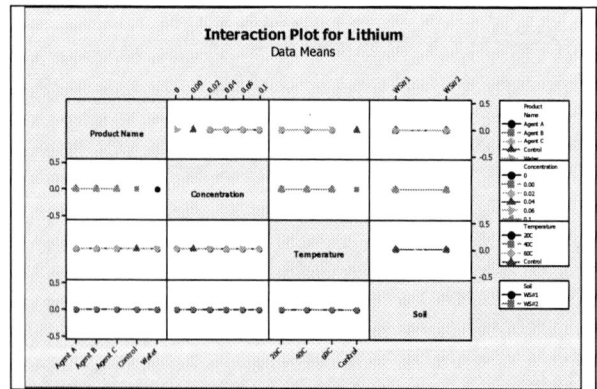

Figure 19. Interaction Plot for Lithium

Everything was 100% successful.

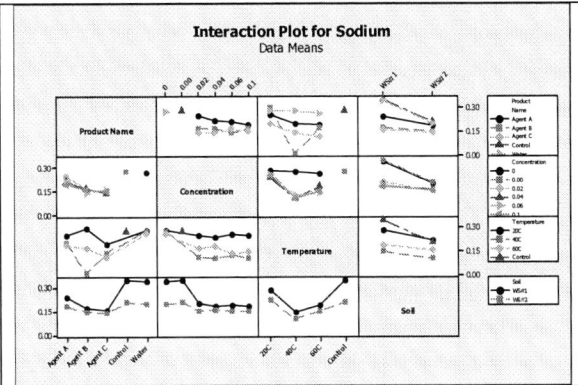

Figure 20. Interaction Plot for Sodium

While WS#2 was consistently better than WS#1, it was a slight difference. The major change in the plot was again driven by the control and water points. Sodium also trends better with increased temperature.

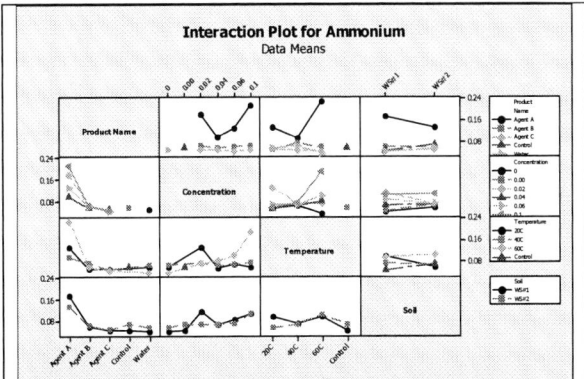

Figure 21. Interaction Plot for Ammonium

Agents B & C meaningful improved over Agent A. No other variables seem overly sensitive.

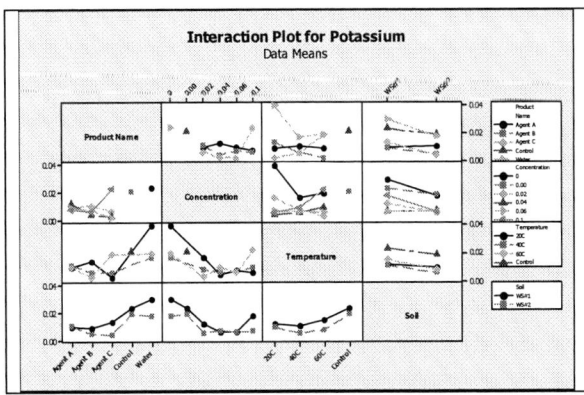

Figure 22. Interaction Plot for Potassium

Concentration seems to help, while temperature does not appear very responsive for the potassium ions.

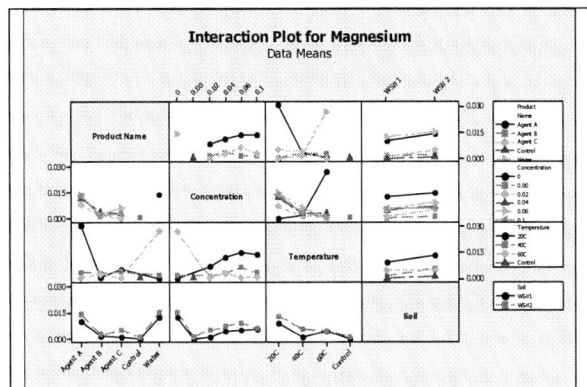

Figure 23. Interaction Plot for Magnesium

Concentration and temperature seem a bit responsive, but the levels seem quite small. Here is a case where WS#1 seems to do better than WS#2 in general.

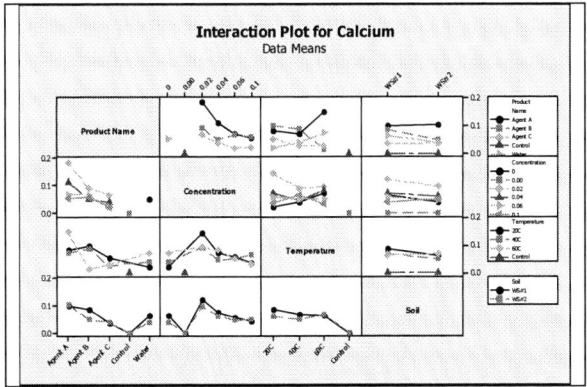

Figure 24. Interaction Plot for Calcium

Agent C once a gain a clear winner with little difference between WS#1 & WS#2.

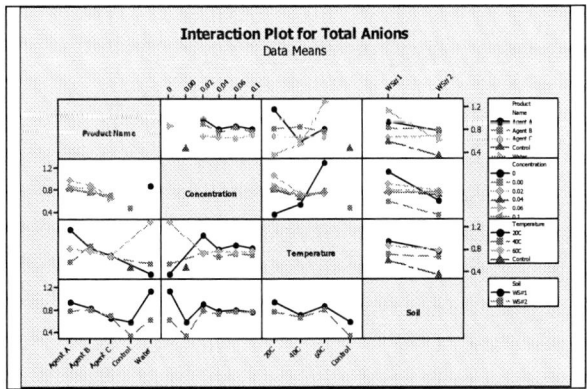

Figure 25. Interaction Plot for Total Anions

Looking at total Anions, Agent C seems to come out on top but it is shades of gray not a real breakout while increased temperature and concentration have modest overall contributions.

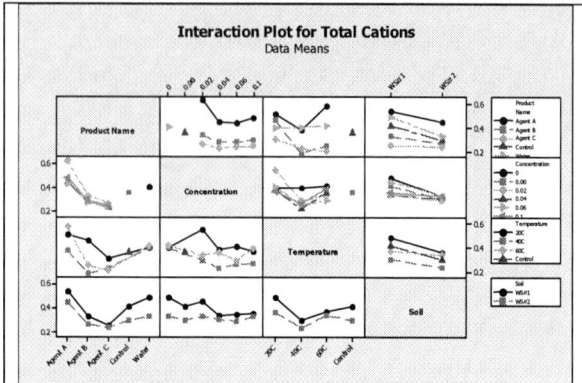

Figure 26. Interaction Plot for Total Cations

Once again, Agent C comes out on top for total Cations as well. Not by a wide margin, but a discernible margin.

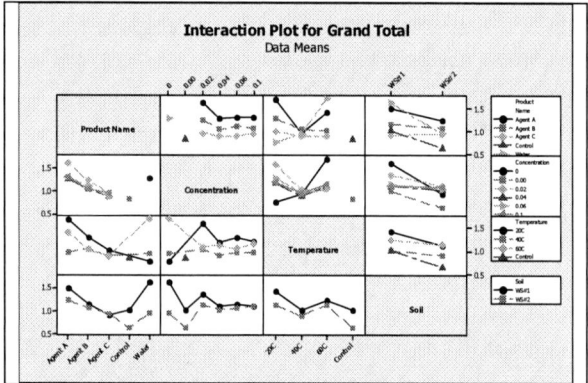

Figure 27. Interaction Plot for the Grand Total

Not surprisingly, Agent C again breaks out of the data by a small margin.

CONCLUSION

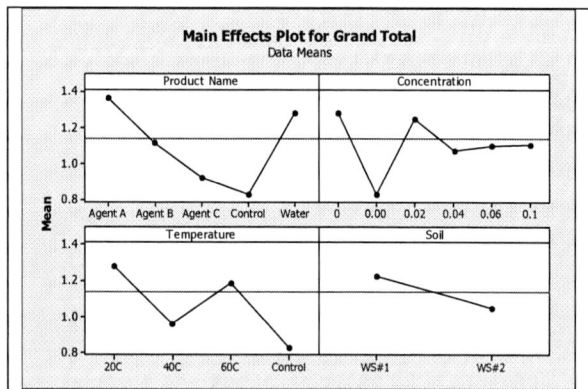

Figure 28. Main Effects for Grand Total

With the control data skewed, Agent C breaks out as the winner though temperature and concentration do not trend toward more is better.

The large data package in this DOE makes the analysis rather straight forward. As in most protocols, there are ambiguous results at times and not every dataset reaches the same conclusion. This point is key; any particular product

life cycle may have unique sensitivities important to its operating for everyday of its service life. Detailed data such as this, though expensive and time consuming to generate can be enormously instructive for such high value, long lived devices.

FUTURE RESEARCH

Work such as this has several potential paths forward. One is to include more soils into the current data matrix. Another is to keep the same dataset and move downstream into commercial grade cleaning equipment to evaluate the impact of meaningful mechanical energy. More importantly for the industry, as work such as this propagates industry standards will need to be developed and validated for these point source contamination levels.

ACKNOWLEDGMENTS

Data such as this does not come quickly, easily nor inexpensively. Much thanks goes to the leadership at Kyzen for allocating the substantial resources to complete this project. Those resources are in house IC equipment of course, but more importantly a talented team led by Dr. Mike Bixenman and David Lober that includes John Garvin and James Perigen.

REQUIREMENTS ON A CLASS "0" EPA – BASICS, STANDARDS, ESD EQUIPMENTS AND MEASUREMENTS

Dipl.-Ing. Hartmut Berndt
B.E.STAT European ESD competence centre
Kesselsdorf, Germany, Saxony
hberndt@bestat-esd.com

INTRODUCTION

Lately, more and more publications report about ESD requirements for an EPA „class 0". What does this mean? This point is not described in the ESD standards ANSI/ESD S20.20 and the IEC standard IEC 61340-5-1. The electronic industry follows these rules when it comes to the protection of electronic components and assemblies against electrostatic discharges. Only one standard divides electronic components into certain hazard classes, the HBM standard IEC 61340-3-1 (ANSI/ESDA/JEDEC JS-001-2012).

Table 1. Classification

Classification	Voltage range (volt)	Notes
0A	< 125	
0B	125 to < 250	
1A	250 to < 500	
1B	500 to < 1,000	
1C	1,000 to < 2,000	
2	2,000 to < 4,000	
3A	4,000 to < 8,000	
3B	≥ 8,000	

According to this classification, class "0A" means a maximum electrostatic voltage of 125 volt. Class "0B" means an electrostatic voltage between 125 and 250 volt. The typical requirement for an EPA according to the ESD standards (ANSI and IEC) is a maximum electrostatic voltage of 100 volt. At the moment, the requirements for the electronic industry are higher than for other industries. Most of the EPAs meet these requirements.

Some semiconductor manufacturers demand a voltage of 0 volt for their electronic components within the handling area of an EPA. Is it possible to implement such requirements? Most of the ESD equipment on the market only grants the requirements up to 100 volt. Thus, how does the material have to be developed? Currently, only special ionizers are suitable to meet the target. Typical ionizers only guarantee a minimum of 100 volt, high specialized ones 10 volt (residual charge or balance). An optimized ESD Control System for machines with focus on cost-effectiveness is presented later.

All electronic components and assemblies are exposed to risks of electrostatic discharges. Producers, suppliers, distributors and users have to perform the ESD control system during the whole manufacturing process, the measurements as well as during the applications. All active electronic components, beginning with simple diodes, transistors or complex inner circuits, require an extern ESD control system. In the next step, SMD resistors and condensers, and prospectively NEMS and MEMS are included in this danger category. Tests show that these passive components can be damaged through electrostatic discharges.

The structures of electronic components become smaller and smaller. 5 volt of an electrostatic charge are already enough to change the structures in small electronic components. The structures will achieve such small dimensions, so that electrostatic charges can cause permanent damages. In the year 2024 the sizes of the electronic components will be less than 10 nm. Then, electrostatic charges of 0.1 nC and electrostatic fields of 10 volt /cm (or 1000 volt /m) will be enough to damage ESDS permanently.

BASICS

The person is the greatest danger for electronic components. Electrostatic charges of a person are typically higher than 100 volt. The best way to reduce the electrostatic charge of a person is grounding. We have two ways for personal grounding: wrist straps and ESD shoes. The wrist strap contact with the skin of the person directly. So, the final charge of a person can be smaller than 100 volt. Many companies use ESD shoes for grounding. In figure 1, the grounding resistance is shown depending on the body voltage. Typical values for system resistance are higher than 10 MOhm. This means that the body voltage is higher than 100 volt.

The first and best way for grounding is the connection with ESD equipment of personnel grounding. Grounding only by table mats and floor materials is not enough. Staff is controllable and workstations are ESD conform. Nevertheless, how do machines, automated handling systems, packaging systems etc. work and which requirements they have to meet?

REQUIREMENTS ON ESD - MATERIAL AND DEVICES

Requirements for a Person

Persons have to be equipped with ESD required shoes and a conductive garment. The most important measure is the wristband.

The figure 1 shows the important relationship between the body voltage or electrostatic charge of a person and the resistance to ground. So, if we estimate a lower level for electrostatic charge, we need a lower limit for the maximum resistance between the person and the floor grounding connection. Figure 1 also shows that we need an alternate grounding path from the person to the floor grounding connection in the future. If the resistance between person and shoes becomes smaller ($< 1 \times 10^6$ Ω), the total resistance between person-shoes-floor material is going to be smaller at all ($< 1 \times 10^6$ Ω).

Table 2. Requirements for personnel equipment

Step	Requirements R_A		Notes
	Today	Future	
Wristband	$< 3.5 * 10^7$ Ω	$1 * 10^6$ Ω ?	
Shoes	$< 1.0 * 10^8$ Ω ($3.5 * 10^7$ Ω)	$1 * 10^6$ Ω ?	Higher requirements are necessary, when the grounding happens exclusively over the floor. Is the maximum resistance from $3.5 * 10^7$ Ω to high? Yes, the system resistance must be smaller than $1 * 10^7$ Ω.
Working clothes	$< 1.0 * 10^9$ Ω	?	The first value can only determinate the surface resistance. The second important value is the charge decay time or charge distribution time from the surface. The charge decay time must be smaller than 2 s (from 1000 v to 100 v)[1].
Note: A new measurement method has to be developed to measure the static decay time of working clothes. The existing methods are not adequate any more.			

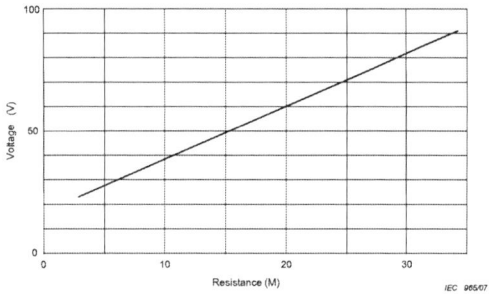

Figure 1. Body voltage relating to resistance to ground (IEC 61340-5-2 Ed. 1.0 TR, 2007)

Only a wristband can guarantee a permanent discharge of personal charge to a grounding point. Persons are connected by their wristband, so, no electrostatic charge can be generated. If a person has to walk around permanently because of his/her activity, the discharge can be performed through the shoes. In fact, the floor must be conductive. The shoes must have a defined resistance to ground (see table 2). Garments prevent the transfer of all electrostatic charges from normal daily clothes. If these measures are realized persons/employees are equipped against ESD.

Requirements for Working Places

Working places must be constructed like in the table 3 shown below. If it is constructed like this, no electrostatic charge can be developed. Furthermore, working place surfaces must guarantee that electrostatic charges can be eliminated safely. Additionally, working places must be equipped with a central grounding point like earth bonding points and earth bonding boxes. The resistance to ground of the working surface has a limited area. Table 3 shows different requirements for working places.

The resistance should not be too small. If it is too small a hard discharge (CDM method) can happen suddenly, which can damage ESDS. The limit of the upper resistance is defined in accordance to the fast and controlled, but still safe, discharge of electrostatic charge. At the same time the decay time is determined.

Table 3. Requirements for working place surfaces

Step	Requirements R_A		Notes
	Today	Future	
Working place surface	$> 7.5 * 10^5$ Ω and $< 1 * 10^9$ Ω	$< 1 * 10^6$ Ω ?	$1 * 10^9$ Ω is too high and produce more than 1000 volt of electrostatic voltage. Surfaces with a resistance to ground about $< 1 * 10^6$ Ω lead to electrostatic charges higher than 100 v. Additionally, the discharge behavior of the surfaces have to be determined.
Decay time	$< 2s$ (from 1000 volt to 100 volt)	$< 2s$ (from 100 volt to 10 volt)	at a resistance value higher than $1 * 10^9$ Ω. The measurement of the static decay time is necessary at $1 * 10^6$ Ω.

Requirements for Floors

The floor is an important part of an ESD area. It is necessary for persons who do not wear any conductive shoes and whose discharge mostly happens over the floor. The electrical characteristics are shown in table 4. There are many experiences with conductive floors. Basically, conductive coverings are suitable because hard coatings (epoxy) have additional problems with the contact resistance between person-shoes-floors is decisive. The reason for it is the basic principle of the discharge of charged persons. Tests with different floor materials showed that only a few of them are suitable. [2]

The additional measurements of the decay time are urgently necessary to qualify the material completely.

Table 4. Requirements for floors

Step	Requirements R_A		Notes
	Today	Future	
Floor	$< 1 * 10^9$ Ω	$1 * 10^6 \ldots 1 * 10^7$ Ω	The system resistance is the most important value in the future for personnel grounding.
Higher requirements	$< 3.5 * 10^7$ Ω	?	Today: at a maximum electrostatic charge of 100 volt from a person, 10 volt will be the maximum in the future.
Decay time	< 2 s (from 1000 volt to 100 volt)	< 2 s (from 100 volt to 10 volt)	The measurement of the static decay time is required above a resistance from $1 * 10^6$ Ω.

behavior. Some materials are not suitable. Previous tests made many questions like: Do the measurement probes and sample are suitable at all? Do the probes really establish the contact person-shoes-floor? Is the contact material of the probe maybe incorrect? Some interested parties have the opinion that contact material on probes do not agree with the reality. Other assumes that probes are not the reflection of the contact person-shoes-floor. A further question cause quite a stir: Can a person be standardized? Additional tests have been realized and will be realized in the future [1].

The basic requirements to conductive floors are not influenced through the tests. Electrostatic charges should be discharged over a conductive floor.

Extensive attempts show, that higher requirements have to be fulfilled at working places, where people work by standing. A higher resistance ($> 3.5 * 10^7$ Ω) would develop electrostatic charge higher than 100 volt. There are different types of floors – floor coverings and floor coatings, which can be thin or thick. But the contact

Requirements for an EPA

For having an optimized protection of ESDS, ESD working places and working areas are necessary. The basic equipment: an ESD working place, which contains a conductive surface covering, a wristband and a grounding system. All equipment must be connected with a grounding point. That grounding point guarantees the same potential at all points of the working place.

The installation of ESD areas (EPA) is wiser. Because of the design of all materials and equipment, electrostatic potential above 100 volt cannot be developed. Nevertheless, if some should be developed caused by unsuitable packaging materials, one can discharge them without any danger.

After having equipped everything according to the ESD requirements - all persons, working places and so on - new sources of electrostatic charge will be seen. Persons and working places must be handled like the ESD requirements. The charges can be controlled.

Requirements for Packaging Material
The packaging requirements correlate with the IEC 61340-5-1 [4] (enumeration according to the standard) and will be introduced in the new packaging standard IEC 61340-5-3 (see table 5 and table 6).

Table 5. Packaging material inside and outside of an EPA

6.1 Inside an EPA	6.2 Outside an EPA
Packaging used within an EPA (that satisfies the minimum requirements of ANSI/ESD S20.20) shall be: 1. Low charge generation. 2. Dissipative or conductive materials for intimate contact. Items sensitive to < 100 volt Human Body Model may need additional protection depending on application and program plan requirements.	Transportation of sensitive products outside of an EPA shall require packaging that provides: 1. Low charge generation. 2. Dissipative or conductive materials for intimate contact. 3. A structure that provides electrostatic discharge shielding.
Notes: If electric field shielding materials are used to provide discharge shielding, a material that provides a barrier to current flow (insulator) must be used in combination with the electric field shielding material. Where this standard does not provide a test method, the user must determine the electrostatic discharge shielding properties of the packaging. See Appendix G for guidance about determining discharge shielding properties.	

Table 6. Requirements for packaging material

Material property	Resistance limits	Charge decay time	Requirements for class zero	Notes
Electrostatic conductive	$1 \times 10^2 \, \Omega \leq R_S < 1 \times 10^5 \, \Omega$	-	$(1 \times 10^2 \, \Omega) \leq R_S < 1 \times 10^4 \, \Omega$	Lower resistance range, add charge decay time
Electrostatic dissipative	$1 \times 10^5 \, \Omega \leq R_S < 1 \times 10^{11} \, \Omega$	$(> 1 \times 10^9 \, \Omega)$ < 2 s	$1 \times 10^4 \, \Omega \leq R_S < (1 \times 10^{11} \, \Omega)$	Lower resistance range, add charge decay time
Electrostatic shielding	< 50 nJ	-		Shielding or field shielding properties
Note: R_S ... surface resistance (IEC 61340-2-3)				

A third requirement is the question of tailoring. In the future, companies will have more responsibility for the art of the packaging material.

ESD Control Program
The introduction and the control of these 5 steps were already described last year in the concept "5 Steps Plan of an ESD Control System" [1]. The result is the following ESD control system:
1. Analysis of ESDS, their damage limits and the existing manufacturing process.
2. Creation of a program and the introduction steps of the ESD control system.
3. Personnel training
4. Introduction of the ESD control systems
5. Control and certification of the introduced ESD control systems.

The introduction of this ESD control system is more complex than the single system requirements of the IEC 61340-5-1 and the control program of the ANSI/ESD S20.20. Only both standards and the additional existing concept guarantee a safe ESD control system as well as the protection of ESDS against electrostatic charges. We cannot find enough information and requirements for the machines in the existing standards.

Requirements for Machines
The first and only requirements are demands for a grounding of all metal parts as well as for the avoidance of plastic usage, which could generate electrostatic charges and fields. Experiences show that this is not enough for the protection of ESDS in automated handling machines and systems. ESDS will not be damaged by the operator, but by machines. The transport operation of an ESDS in a machine can happen as following:

1. Removal of the ESDS out of packaging is the first sub-process. The ESDS has an isolating case, so it will be electrostatically charged during the removal out of the reel or the tray.
2. The electrostatically charged ESDS will be transported to the PCB. Thereby a further electrostatic charge can develop. The movement at high speed Pick-and-Place System should be enough of the generation of electrostatic charges.
3. By placing it on the PCB, different potential between the ESDS and the PCB exist. So, the potential

difference leads to a discharge, which will damage the ESDS.

These examples show that electrostatic charges always develop when ESDS are parted or transported. Electrostatic charges will always generate because of the reason that components as well as PCBs are made of an isolating material. Other acts and production steps show, that this is not the only possibility for the generation of electrostatic charges in a production process. Further critical steps are for example: the printing of PCBs, the labeling of PCBs and assemblies as well as test constructions.

Manual handling of individual components is not common anymore. PCB assemblies are handled mainly by equipment and the final phases of mechanical assembly are done by both humans and robots. In consequence of this the Human Body Model (HBM) is not valid ESD simulation model as much as previous. The main electrostatic risk during automated manufacturing is with Charged Device Model (CDM) type of electrostatic discharges. The additional model, but not standardized yet, is the Charged Board Model (CBM).

In the CBM type of ESD the assembled Printed Wiring Board (PCB) or some of the mechanics parts can be charged during handling. The discharge to ground or between the objects can happen at the same time. CBM type of discharge is typically more seriously than other models for components due to high capacitance and high stored charge of PCB assemblies or mechanics.

There are some main ESD control principles which are important in ESD Protected Area (EPA) as well as in automated process equipment:
1. All conductive and dissipative items are grounded.
2. Materials or parts which are in contact with ESDS must be made of electrostatic dissipative material.
3. Non-essential insulating materials are excluded.
4. Where insulating materials or parts are needed, the possible charges must be minimized by special measures, like ionization, shielding or coating.

Enclosures of machines are normally made of conductive material. The conductive enclosure should have a straight and reliable connection to ground and the distance of the insulating parts should be long enough in order not to create high electrostatic fields close to ESDS. Special attention should be paid on grounding of parts which are separated from the enclosure or are movable, like adjustable conveyor.

There are a lot of materials which can be in contact with ESDS items. Components to be placed are stored in reels with plastic tapes covered and nozzle picks the component from reel. Components are placed on the PCB and PCB is

contacted with conveyor belts and possible support pins, gripper, clamps etc. All these materials should be made of electrostatic dissipative material at least in contact area and a resistance to ground value should be between $1 * 10^6$ Ω and $1 * 10^9 \Omega$.

Components and PCB material have plastic, insulating material and they can become charged by tribocharging, e.g. by rubbing against conveyor belt, touching on other product parts or in routing process. The charged ESDS item can subject to CDM or CBM risk. All rotating and sliding elements form an ESD risk. The tribocharging during automated manufacturing should be minimized and metal contact to ESDS should be prevented. Normally, these preventions are not enough. Thus, an ionizer should be installed in the area of rotating material.
Ionizers are applied sometimes to remove electrostatic charges from machines. Electronic components and PCBs cannot be grounded. Thus, ionization is the only method minimizing electrostatic charges at the moment. Intelligent ionizers are able to detect electrostatic charges in machines and to generate equivalent charges for their decrease either.

Requirements/Questions for an ESD control program for machines:
1. Are all parts grounded?
2. Is there no plastic material charged? Is only ESD plastic material used?
3. How does delivery of ESDS and PCBs happen?
4. Is the packaging material ESD conform?
5. How are the requirements for the packaging material for non ESDS defined?
6. How is the transport der ESDS inside of the machine defined?
7. If non ESD material is used (i.e. for high voltage wires), do the transportation ways of the ESDS have enough distance from this?
8. Are the PCB and the ESDS of the same potential when they get in contact? (i.e. in a pick-and-place-machine)
9. Are the PCB and the ESDS discharged enough?
10. Are the transport conveyors, belts and systems between the machines on the same electric potential?

Based on these questions, ESD requirements will be created for machines.

MEASUREMENTS
The most instruments on the market for measure electrostatic voltage, electrostatic charge and for resistance are not qualified for these correct measurements. Only special contact volt meters (CVM) or electrostatic static volt meters (EVM) are able to measure this very small electrostatic charge.

Figure 2. Measuring PC conductor with high an impedance contact volt meter [13]

The first step is the measurement with a contact volt meter. Furthermore, high sensitive electrostatic volt meter can be used. They do not damage ESDS during the measurement.

CONCLUSIONS
The requirements "0 volt" can be achieved, when die maximum value will be required. Unfortunately, today we do not have any ESD material, which requires these limits.

The biggest problems are machines and automated handling equipment (AHE), because very small charges are generated in these machines, independent from persons. These small and fast discharge procedures are energy-intensive. They cause damages of electrostatic sensitive devices and assemblies. The grounding of all metal parts does not suffice. New processes, which either discharge very small electrostatic charges fast or prevent theses discharges, have to be developed.

The only way to meet such requirements is precision-ionization. All other ESD equipment have more than 0 volt. Even with a limit value of 125 volt (Level 0A) it is hard to find suitable ESD material. A further attempt is to classify ESD control areas in different zones.

1. Basic control requirements
2. Advanced control requirements
3. Extended control requirements

Until now, there are not any different requirements for these areas, defined in an ESD control program.

REFERENCES
[1] H. Berndt, Five Steps for the Introduction of an ESD Control System, Proceedings APEX 2004, Anaheim, CA, U.S.A.
[2] H. Berndt, A study of the Variables of Electrodes used in the Measurement of Table and Floor Materials and How They Affect the Test Results, 23. EOS/ESD-Symposium 2001, Portland, OR

[3] H. Berndt; VDE-Schriftenreihe - Normen verständlich - Band 71 Elektrostatik - Ursachen, Wirkungen, Schutzmaßnahmen, Messungen, Prüfungen, Normung, VDE-Verlag, 3. Auflage, 2009
[4] IEC 61340-5-1 Electrostatics - 08.2007: Part 5: Specification for the protection of electronic devices from electrostatic phenomena, Section 1: General requirements
[5] IEC 61340-5-2 Electrostatics – 08.2007: Part 5: Specification for the protection of electronic devices from electrostatic phenomena, Section 2: User guide
[6] ANSI/ESD S20.20-2007 ESD Association Standards for the Development of an Electrostatic Discharge Control Program for – Protection of Electrical and Electronic Parts, Assemblies and Equipment's
[7] ANSI/ESD SP10.1-2007 ESD Association Standard practice for Protection of Electrostatic Discharge Susceptible Items – Automated Handling Equipment (AHE)
[8] A. Olney, B. Gifford, J. Guravage, A. Righter, Real-World Charged Board Model (CBM) Failures, 25. EOS/ESD Symposium 2003, Las Vegas, NV
[9] JEDEC Standard JESD22-C101-E, Field Induced Charged Device Model Test Method for Electrostatic-Discharge-Withstand; Thresholds of Microelectronic Components, December, 2009
[10] D.L. Lin, FCBM – A Field-Induced Charged-Board Model for Electrostatic Discharges," IEEE Transactions on Industry Applications, Vol. 29, No. 6, pp. 1047-1052, November/December 1993.
[11] International Technology Roadmap for Semiconductors (ITRS), Factory Integration, Update 2011
[12] H. Berndt, Electrostatic Discharge (ESD) and the Technology Roadmap to 2020, Pan Pacific Microelectronic Symposium, January 2008
[13] White Paper 2: A Case for Lowering Component Level CDM ESD Specifications and Requirements, April 2010

DESIGN AND FABRICATION OF ULTRA-THIN FLEXIBLE SUBSTRATE

Ming-Kun Chen[2], Yu-Jung Huang[1], Yi-Lung Lin[2], and Shen-Li Fu[1]
[1]Department of Electrical Engineering, I-Shou University
Kaohsiung, Taiwan, R.O.C.
yjhuang@isu.edu.tw
[2]Advanced Semiconductor Engineering Test RD
Kaohsiung, Taiwan 811, R.O.C

ABSTRACT

We present the development of a flexible polyimide substrate that act as the conduit for power and signal transmission with the additional features of precise control of the metal trace geometries. Using microfabrication techniques, a metal trace is fabricated that possesses a flexible polyimide-based interconnection. The performance of the design was measured with multimeter. The HFSS simulation result of copper-plated interconnection with a core thickness of 12 um has an impedance value of 50 ohm at 1 GHz. Investigated of microfabrication were performed using an energy-dispersive spectrometry (EDS) and Zygo Optical Profilometer in order to verify the visual aspects of the Cu interconnection, such as identification the elemental composition of materials, roughness and thickness. The experiment is conducted to study the effect of the process parameters on the Cu film surface properties. The results obtained in this work can be applied to the fabrication of flexible microelectronic devices.

Key words: 3D Integration, Flexible Polyimide Interposer (FPI), fine line wiring, energy-dispersive spectrometry (EDS).

INTRODUCTION

The industry market is now faced with the increasing importance of a new trend, "More than Moore" (MtM), where value-added applications are provided by incorporating functionalities that do not necessarily scale according to "Moore's Law". Historically, Moore's law has been predicted that the number of components per chip would double every 12 months. Performance and productivity of microelectronics has been increased continuously over more than four decades due to the enormous advances in photolithography, wafer size, process technology, and device. This has led to the development of technologies that can lead to the ultra-miniaturization of electronic systems. However, the performance improvement gained in transistor scaling is insignificant compared to the negative effects of interconnect scaling. The delay of global interconnects increases with technology scaling. The ITRS roadmap predicts Three-dimensional (3D) integration as a key technique to overcome this so-called "wiring crisis" [1]. 3D technology, an alternative solution to the scaling problems, is a well-accepted approach for so-called "More than Moore" applications [2].

3D integration is an emerging technology that vertically stacks and interconnects multiple device layers. In the 3D integration, silicon interposer and Through-Silicon-Via (TSV) are primary enablers that can take full advantages of 3D ICs. 3D ICs with TSVs offer improved electrical performance due to the short interconnect between stacked ICs. 3D interposers based on silicon or glass substrate aim at replacing traditional printed circuit board (PCB) laminate or ceramic technologies for the sake of extreme miniaturization and performance. However, a silicon substrate cannot be utilized as a good medium for signal transmission since interconnects on the silicon substrate suffer from substrate losses caused by the penetration of the electric and magnetic field. Glass has many advantages as an interposer material over silicon; namely ultra-high resistivity and availability in thin and large sizes. Glass has been studied as an interposer material, mainly focusing on metalized through-glass-via (TGV) in thick glass substrates using laser ablation [3-6].

To implement 3D stacking technology as a low-cost and easy-to-use mounting method, a new interposer is necessary. The thin flexible PI films have desirable properties for use in the electrical and electronics industry because they are a group of good thermal stability, high flexibility, low dielectric constants, excellent mechanical strength, low loss tangent and electrical insulating properties [7-9]. Developments have lately been made with various embedding technologies, such as Chip-In-Polymer [10, 11], Chip-In-Substrate [12], or flexible bumped tape interposer (FBTI)[13]. The development of a flexible polyimide interposer (FPI) is reported in this study.

This study focuses on an approach based on the use of polyimide substrate for 2.5D/3D ultra-thin packaging applications. We present our investigations of the factors associated with fabrication of high-density Cu interconnect structures and Cu via using polyimide film and electroplated Cu conductor lines. We use a 3D-profilometer microscope for characterizing roughness and slope errors to determine the influence of each technological step on the surface quality.

DESIGN

3-D TSVs are incorporated into IC packages as a mean to interconnect two or more stacked die. In addition, a vias can go through the bulk silicon of the lower die to connect to the package substrate. A variation on this idea is the notion of 2.5-D, where devices are sitting side by side on a common interposer. This interposer can be used to fan out or reroute the electrical traces of a device while routing the traces to the package substrate below, connected by means of microbumps. For example, a silicon interposer can be placed between a die and an organic substrate. A silicon interposer has fine pitch interconnects on its top surface to connect with and redistribute signals from the die above. As shown in Figure 1, the interposer's interconnects redistribute signals to TSVs running downward through the interposer and connecting with the substrate. TSV interposers provide flexibility for the integration of die from different semiconductor technology nodes and deliver advantages in miniaturization, thermal performance and fine line/width spacing in a semiconductor package [14].

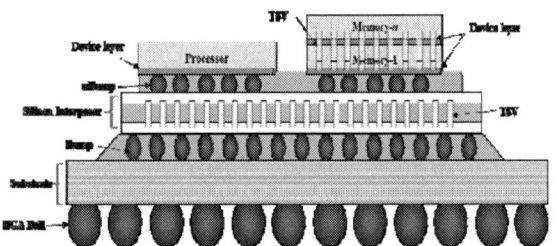

Figure 1. Schematic Drawing of 3D Stacking of the Silicon Interposer with TSV

A construction of Cu fine pitch pattern on the FPI includes a PI base film, a seeder layer copper, and a layer of electrodeposited copper. The FPI was made using a copper-coated PI film composed of 12.5 μm of polymerized PI layer with a size 20 × 30 mm². The manufacture of these fine pitch patterns makes use of photolithographic techniques for fine pattern generation, continuous vacuum techniques for seeder layer, and special electrodeposition methods for copper build-up. The main experimental fabrication process of interposers is shown in Figure 2 (a) and (b).

A test pattern for further electroplating process was fabricated using standard photolithographic techniques. For the formation of the test pattern used in the electroplating of Cu, a seed layer was coated on the surface of the PI substrate by an evaporator. Before the experiment, the flexible substrates were baked in a hot plate for 10 minutes at 110 □ to remove any surface absorbed moisture. To start the fabrication process the PI substrate was ultrasonically cleaned by ethanol and then rinsed with DI water allowed by drying using nitrogen gas. The seed layer of 200nm Cu is deposited on a flexible substrate using a thermal-evaporation process, without using a catalyst or pre-deposited buffer layers. The substrates were spin-coated with a patterned layer of thick AZ4620 photoresist (AZ Electronic Materials). The PR of 6 μm was spun onto the substrate at 3000 rev. per min for 30s using a spinner. The substrate was then soft baked on

an open hotplate at 90 °C for 3 minutes. The substrate was exposed with a dose of UV illumination of 350-450 nm wavelengths at 100mJ/cm2 intensity, using an M&R Nano Technology Co. AG350 Mask Aligner and Exposure System and a plastic mask. The mixture of AZ 400K and DI water (1:4) was used to develop the exposed pattern for about 5 minutes.

The samples were immersed in an activator for 1 min and cleaned with distilled water, nitric acid (3% in volume) and acetone. In the electroplating process, Cu was deposited by the direct current on the Cu/PI substrate in an electrolyte at the 25 °C and 50 °C temperatures. We used a commercial Cu sulfate solution from A Gold Jet Tech Inc. in Taiwan. The Cu electroplating bath includes electrolyte consisting of PTH-502A, 10% with H_2O and PTH-501B, 10% with H_2O. To obtain Cu layers with a 2~15 μm thickness, the current was varied in the range of 20~30 mA and the plating time was controlled in the range of 200~800 seconds.

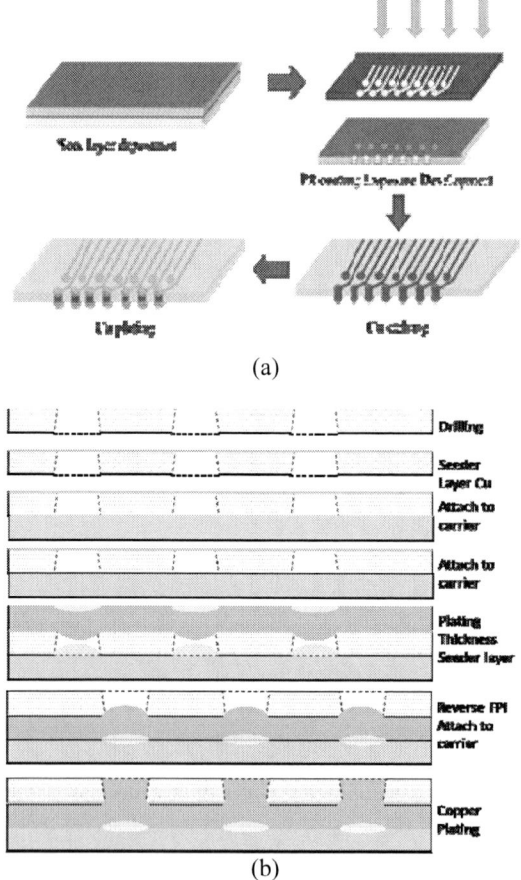

Figure 2. Schematic Illustrations for the Fabrication Processes of a FPI (a) and Bottom-up Cu Electroplating Via Process (b)

Bottom-up Cu electroplating of via is proposed in this paper for high aspect ratio via filling [15]. The bottom electrode could be made by sputtering metals on polymer and then remove the polymer to form the film type bottom electrode. Many studies report bottom-up fill of Cu in electroplating baths using additives, but few reports about bottom-up fill of Cu in electroless plating solutions have been published [16]. It lost ground to electrolytic plating because of concerns due to hydrogen evolution, complexity in process control, low deposition rates, and potentially adverse environmental impact. They can be hampered by poor contacts to the surrounding layers and an incomplete fill. We used Cu electroforming method after the seeder layer formation by the heat evaporation. Figure 2 (b) illustrates the process flow chart of DC Cu electroforming vias after the FPI holes made by laser ablation process. Excellent selectivity is as expected within the whole FPI and all of vias are covered by electroforming Cu with good uniformity. Figure 3 shows the design of the interposer mask.

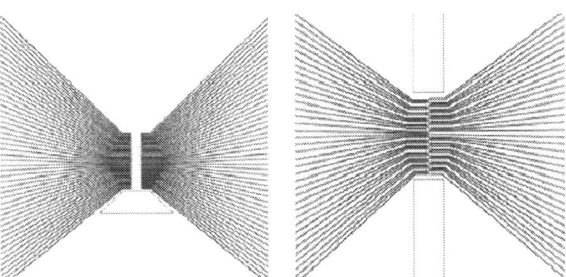

Figure 3. Illustrative Drawing of Two-layer Cu Interposer

To verify the FPI structures in Figure 2, we simulated the transmission coefficient (S21) of microstrip lines structures. The simulations are performed using an Ansoft's High Frequency Structure Simulator (HFSS). The FPI substrate used in the simulation has the same parameters as the Taiflex PI with a core thickness of 12.5 um, a dielectric constant of $\varepsilon r=3.4$, and a loss tangent of 0.003. In general, two variables determine the characteristic impedance: line width and gap width. In the simulations, the width of the center metal trace in all the structures is 12 um. The reflection coefficients are reduced by -3 dB at 4.65 GHz under the 2.5 cm length of FPI interconnection as depicted in Figure 4.

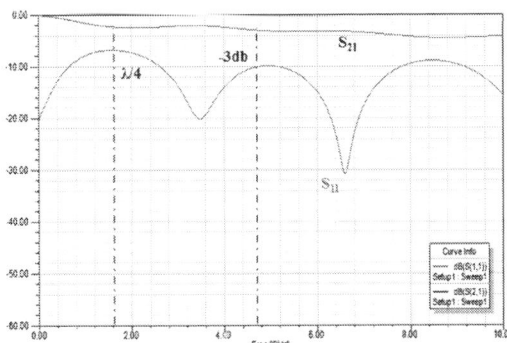

Figure 4. Simulation Results of the Frequency Dependent Scattering Parameters for the Structures deplicted in Figure 3

FABRICATION RESULTS
Electroplating Cu
Optical images of the Cu evaporation-coated PI surface after electroplating Cu deposited for different times were fabricated in Figure 5. The surface microstructure was observed using a Nikon Eclipse ME600 optical microscope (with a digital camera, Sony DFW-SW910), 2400× magnifications with attached image analysis software (Sony IIDC), after suitable calibration as shown in Figure 6(a). It can be seen that there were no voids at the wetting interface. The incorporation of Cu in the electroplated Cu surface was verified by the presence of the Cu peak in the EDS spectrum as indicated in Figure 6(b).

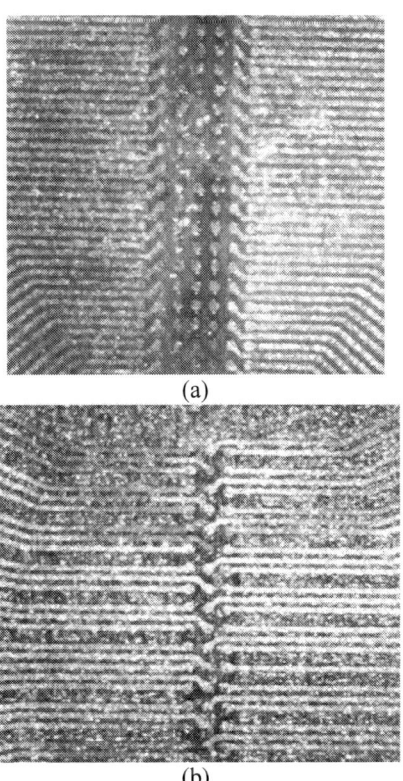

(a)

(b)

Figure 5. Optical Image of Fabrication Results for Top side(a) and Bottom Side(b)

There are two parameters that must be adjusted in order to improve the uniformity of Cu: the current density and the temperature of the bath. The first test concerns the study of the variation in applied current value. The thickness of the Cu coatings deposited was determined by ZYGO NewView 7000 series 3D optical surface profilometer (Zygo Corporation). A 10* objective and 2*zoom were used for a lateral resolution of 1.12 nm. Various samples were measured to ensure that the images were representative of the trace surface. The profile result of FPI interconnection obtained by the proposed method is shown in Figure 7, where Peak to Valley Distance (PV) was found to be 5.140 um with a Ra Value as 1.804 um. The Cu thickness proved to be linear as a function of the deposition time with zero incubation time at 50 °C temperatures and the agitation speed of 50 rpm. Figure 8 shows the thickness and roughness of the plated FPI interconnection at room temperatures under 30mA of plating current. The Ra value of trace surface increased as the neutralization time increased and the Ra of the 800 seconds under 300mA current was 1.5 um.

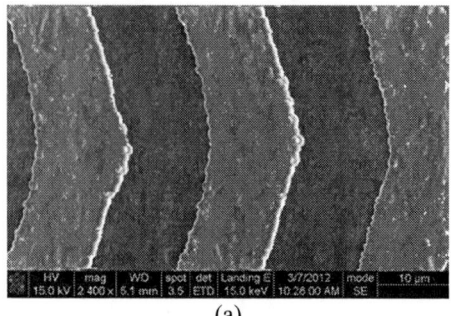

(a)

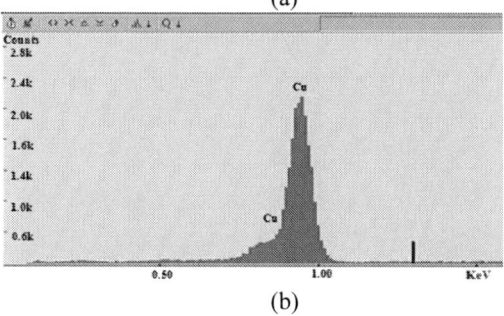

(b)

Figure 6. SEM Image of the Surface of Interconnection-coated Copper at 2400× magnifications (a) and an EDS Spectrum on the Metal Trace (b)

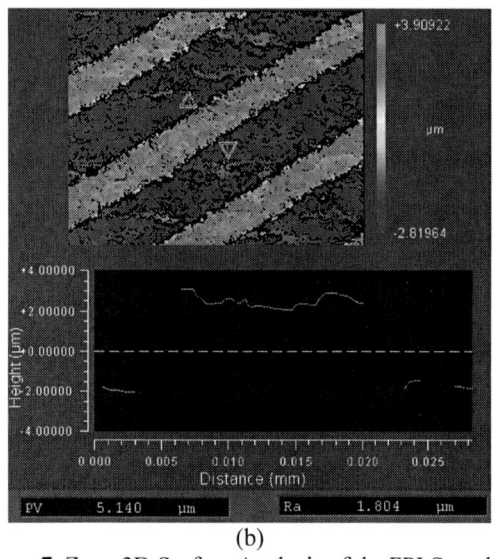

(b)

Figure 7. Zygo 3D Surface Analysis of the FPI Sample. Peak to Valley Distance (PV) was Found to be 5.140 um and Ra Value was 1.804 um.

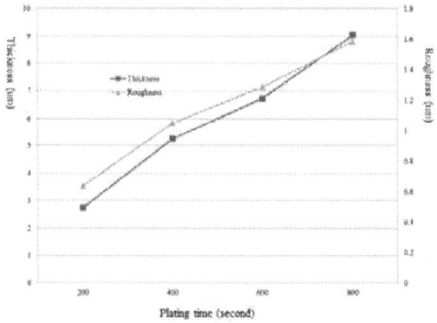

Figure 8. Variation of the Copper Thickness and Surface Roughness under 30mA of Plating Current

CONCLUSION

A fine pitch patterned processing sequence for the preparation of interposer used for 3D integration applications was developed and experimentally evaluated. The main features of this technology are the preparation of conductor lines on the FPI, which enable the chip connect to the interposer.

ACKNOWLEDGEMENT

The authors wish to acknowledge the assistance and support of National Science Council, R.O.C., under Grant NSC 101 - 2221 - E - 214 – 077.

REFERENCES

[1] International Technology Roadmap for Semiconductors (ITRS) (2007) http://www.public.itrs.net.

[2] P. Garrow, C. Bower, P. Ramm, "Handbook of 3D Integration", Wiley-VCH, 2008 (ISBN: 978-3-527-32034-9).

[3] L. Brusberg, H. Schröder, M.Töpper,and H. Reichl, "Photonic System-in-Package technologies using thin glass substrates," in Electronics Packaging Technology Conference, 2009. EPTC '09. 11th, 2009, pp. 930-935.

[4] H. Schröder, L. Brusberg, R. Erxleben, I. Ndip, M. Töpper, NF Nissen, H. Reichl, "glassPack; A 3D glass based interposer concept for SiP with integrated optical interconnects," in Electronic Components and Technology Conference (ECTC), 2010, pp. 1647-1652.

[5] L. Brusberg, H. Schröder, N. Arndt-Staufenbiel, M. Wiemer,, "3-D Thin film interposer based on TGV (Through Glass Vias): An alternative to Siinterposer," in Electronic Components and Technology Conference (ECTC), 2010 Proceedings 60th, 2010, pp. 66-73.

[6] L. Brusberg, H. Schröder, M. Töpper, N. Arndt-Staufenbiel, J. Röder, M. Lutz, H. Reichl,, "Thin glass based packaging technologies for optoelectronic modules," in Electronic Components and Technology Conference (ECTC) 59th, 2009, pp. 207-212.

[7] M.S. Doyle, W. Martin, D. Pease, T. Timpane,, "Low-loss flex circuit interconnect: Development of reduced insertion-loss flexible packaging," in Proc. 57th ECTC, 2007, pp. 1870-1876.

[8] H. Braunisch, J.E. Jaussi, J.A. Mix, M.B. Trobough, B.D. Horine, V. Prokofiev, L. Daoqiang, R. Baskaran, P.C.H. Meier, D.H. Han, K.E. Mallory, M.W. Leddige,, "High-speed flex-circuit chip-to-chip interconnects," IEEE Transactions on Advanced Packaging, Vol. 31, No. 1, pp. 82-90, Feb. 2008.

[9] E. McGibney, J. Barton, L. Floyd, P. Tassie, J. Barrett,, "The High Frequency Electrical Properties of Interconnects on a Flexible Polyimide Substrate Including the Effects of Humidity," IEEE Transactions on Components, Packaging, and Manufacturing Technology, Vol. 1, No 1, pp. 4-, Jan. 2011.

[10] R. Aschenbrenner, A. Ostmann, A. Neumann, H. Reichl,"Process Flow and Manufacturing Concept for Embedded Active Devices," Proc. Of the IEEE Electronics Packaging Technology Conference, 2004, pp. 605- 609.

[11] A. Ostmann, A. Neumann, S. Weser, E. Jung, L. Böttcher, H. Reichl, "Realization of a Stackable Package Using Chip in Polymer technology," Polytronic Conference, June 23. – 26. 2002.

[12] Y. H. Chen, J.R. Lin, S. Chen, C.T. Ko, T.Y. Kuo, C.W. Chien, S.P. Yu, "Chipin-Substrate Package," CSP Technology.

[13] Kazuhito Hikasa, Toshiaki Amano, Toshiya Hikami, Ken'ichi Sugahara and Naoyuki Toyoda, "Development of Flexible Bumped Tape Interposer," http://www.furukawa.co.jp.

[14] J. U. Knickerbocker, et al, "Development of next generation system-on-package (SOP) technology based on silicon carriers with fine-pitch interconnection," IBM Journal of Research and Development, Vol. 49, No. 4/5, pp. 725-754, 2005.

[15] H.H. Chang, Y.C. Shih, C.K. Hsu, Z.C. Hsiao, C.W. Chiang,Y.H. Chen,and K.N. Chiang "TSV Process Using Bottom-up Cu Electroplating and its Reliability Test," Proc. 2nd Electronics System-Integration Technology Conference, Greenwich, UK, 2008 pp 645-650.

[16] Z. Wang, O. Yaegashi, H. Sakaue, T. Takahagi, and S. Shingubara, "Bottom-Up Fill for Submicrometer Copper Via Holes of ULSIs by Electroless Plating," Journal of The Electrochemical Society, Vol. 151, No. 12, pp. C781-C785, 2004.

SIDE WALL WETTING INDUCED VOID FORMATION DUE TO SMALL SOLDER VOLUME IN MICROBUMPS OF Ni/SnAg/Ni UPON REFLOW

Y. C. Liang[1], C. Chen[1,*], and K. N. Tu[2]

[1] Department of Materials Science and Engineering, National Chiao Tung University,
Hsinchu, Taiwan, R.O.C.
[2] Department of Materials Science and Engineering, University of California at Los Angeles
Los Angeles, CA, USA
*chih@mail.nctu.edu.tw

ABSTRACT

A processing failure of void formation has been observed in 3D IC microbumps due to small solder volume. We prepared the sandwiched Ni/Sn2.3Ag/Ni microbumps with 4 μm and 11 μm thick solders and reflowed them at 260 °C to study the mechanism of void formation in the processing. Due to the thin solder, intermetallic compound formation of Ni_3Sn_4 from the two interfaces of the solder joint can physically bridge each other. When that happens, the degree of freedom of motion in the direction normal to the interfaces is removed. Consequently, when the remaining molten solder is drained by side wall reaction, large voids form in the joint. This is a unique mode of processing failure because of the smaller and smaller volume of solder joints in the trend of miniaturization.

Key words: Intermetallic compounds; Soldering; Nickel

INTRODUCTION

For high-density packaging in microelectronic industry, the three dimensional integrated circuits (3D IC) by vertically stacked silicon chips is expected to achieve higher performance than the conventional flip-chip technology. This is because in order to accomplish the multi-functional requirements for future generation electronics, interconnections with high input/output counts and fine pitch are needed. Therefore, fine pitch interconnections with through-silicon-vias (TSVs) of Cu vias for 3D IC of Si chips has been developed recently.[1] Microbump technology is required to join the vias between Si chips. Lead-free solder bumps about 10 μm in height and in diameter were adopted in microbumps.[2] In contrast with conventional flip-chip solder bumps, which have a height and diameter of 100 μm, a microbump has a much smaller solder volume, about 1000 times smaller. In the transition from flip chip technology to microbump technology, the solder volume change has caused new processing issues as well as reliability issue. For example, under the same reflow condition of time and temperature, a much larger volume fraction of intermetallic compounds (IMCs) is formed in microbumps. Because of the need to reduce the fraction of IMCs, a few microns thick Ni layer under-bump-metallization (UBM) has been coated at the

surface of the Cu vias as a diffusion barrier.[3,4] Thus, the interfacial reactions between Pb-free solder and Ni UBM have attracted a great deal of attention. In this paper, we report a processing failure in the Ni UBM microbump.

In the literature, several reports have addressed the metallurgical reactions between Pb-free solders and Ni UBM in flip chip technology.[5-20] IMC of Ni_3Sn_4 forms, and the interfacial morphology of Ni_3Sn_4 depends on the reaction conditions and plays a crucial role in affecting the mechanical reliability of the solder joints.[3,17,18] How does the small solder volume in microbumps affect the morphology and growth kinetics of Ni_3Sn_4 requires a systematic study. Indeed very few literatures covered the topic.[21,22] It is expected that after a longer reaction time, the entire Pb-free solder can be transformed into IMC completely. In other words, the IMC from both sides of the solder joint joined to each other, and the joint became an IMC joint. In this study, we investigated the interfacial morphology of Ni_3Sn_4 IMC in a sandwiched structure of Ni/Sn2.3Ag/Ni microbump during various reflowing times at 260 °C. The thickness of the solder between two Ni UBM layers is 4 μm and 11 μm. We observed very large void formation in the thinner solder joint during reflow.

EXPERIMENTAL

Sandwiched Ni/Sn2.3Ag/Ni microbumps were fabricated by joining two Ni/Sn2.3Ag samples as depicted in Figure 1. A layer of 0.1 μm thick Ti was deposited onto an oxidized Si wafer first to serve as an adhesion layer and followed by sputtering a Cu seed layer of 0.2 μm thick. After that, photolithography was employed to pattern cylinders of 100 μm in diameter for the electroplating of 3 μm Ni UBM and the Sn2.3Ag solder of three sets of thickness of 1 μm, 2 μm, and 10 μm. After electroplating of the Ni and the solder, the wafers were reflowed at 260 °C for 1 minute (the first reflow) to ensure solder cap formation on the Ni. Then they were diced to be 2.3 × 2.3 mm Si chips. After that, one chip with 2 μm thick solder were flipped over and aligned with another chip. The schematic experimental setup is shown in Figure 1(a). Then they were joined together at 260 °C for 3 minutes (the second reflow) to form microbump solder joints with 4 μm thick solder. For comparison, microbump solder joints with 11 μm thick solder were also prepared by joining samples of 10 μm thick and 1 μm thick solder using the same method, as depicted in Figure 1(b). The solder thickness was measured to be 4.2 ± 0.1 and 11.1 ± 0.2 μm for the two sets of samples, respectively. The pitch between adjacent microbumps was 200 μm. Metallurgical reactions were investigated for additional reflow at 260 °C for 5, 10, 30, 60 and 120 minutes (the third reflow) on a hotplate, and then air-cooled at a cooling rate of 5 °C/sec. We note here that the reflow time in this paper represents the total reflow time, which includes the 1 minute reflow after the electroplating, the 3 minutes reflow during joining the microbump samples, and the additional reflow to investigate the metallurgical reactions. Therefore, the reflow time under investigation is actually 4, 9, 14, 34, 64, and 124 minutes.

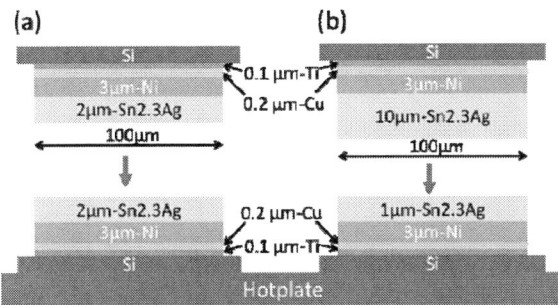

Figure 1. Schematic diagram for the microbump structures and the jointing setup used in this study. (a) 4-μm-thick solder sample; (b) 11-μm-thick solder sample.

Cross-sections of the samples after the reflow were mechanically polished, and the microstructure of Ni_3Sn_4 IMC on the cross-sections was observed by scanning electron microscopy (SEM) with a back-scattered electron image (BEI) detector. The composition of the IMC was examined by energy dispersive spectroscopy (EDS). Growth of the IMC was quantified by calculation on the basis of image analysis software, which measured the IMC area on the cross-section and then divided by the interfacial length between the IMC and the Ni UBM.

RESULTS AND DISCUSSION
For the 4 μm thick solder sample, the solder transformed completely to Ni_3Sn_4 IMC after reflow for 34 min, and many voids were observed in the middle of the joint. Figures 2(a) to 2(c) illustrate the cross-sectional SEM images of the sample after 4, 14, and 34 min respectively at 260 °C. The as-joined microbump, shown in Figure 2(a), has Ni_3Sn_4 IMC formed at both the top and the bottom interfaces. When the reflow time increased to 14 min, an obvious loss of the solder is seen in Figure 2(b) due to necking formation in the periphery of the joint. The loss

of solder in the periphery of the microbump can be attributed to the out-flowing of the solder because of side wall wetting. The side wall of the cylindrical Ni UBM was a free surface and it can be wetted by the solder to form Ni_3Sn_4 IMC. It is a driving force for solder to flow out during reflow, which has been reported in previous literatures.[23,24] Actually, the sputtered Cu seed layers beneath the Ni UBM were consumed by the out-flowing solder too. After a 34 min reflow as shown in Figure 2(c), many large voids were observed in the middle of the joint. This microbump was cross-sectioned by focused ion beam (FIB) to prevent artificial damage from mechanical polishing.

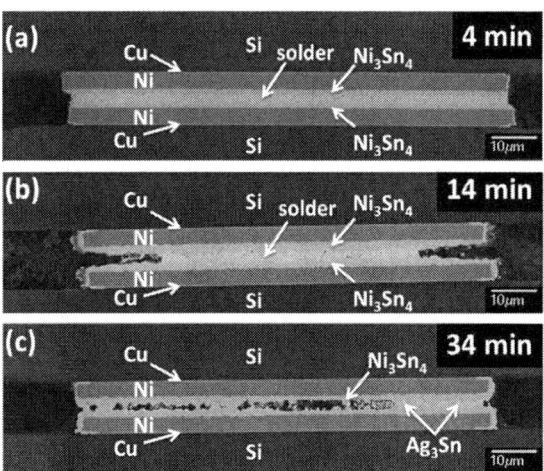

Figure 2. Cross-sectional SEM images showing the microbumps with 4-μm-thick solder subjected to a (a) 4 min (as-jointed); (b) 14 min; (c) 34 min reflow at 260 °C.

When the solder joint has a large volume as in flip chip technology, the side wall wetting will not lead to necking formation and void formation. But when the solder volume is small, void formation can occur. Three reasons may cause the serious void formation. First, Sn atoms may diffuse out to react with the Cu seed layer beneath the Ni UBM. Figures 3(a) to 3(c) represent the enlarged SEM images on the periphery of the joints for the samples shown in Figs. (2). It is clear that the Sn atoms diffuse along the lateral Ni_3Sn_4 IMC and migrate to the Cu seed layer to form $(Cu, Ni)_6Sn_5$ IMCs. As demonstrated in Fig. 3(b), the IMCs at point B and point C were Ni_3Sn_4 confirmed by EDS; whereas the IMCs at point A and point D were detected to be Cu_6Sn_5 and $(Cu, Ni)_6Sn_5$, respectively. Furthermore, as shown in Fig. 3(c), the IMCs at point F and point G were Ni_3Sn_4, and the IMCs at point E and point H were $(Cu, Ni)_6Sn_5$. Second, molar volume shrinkage takes place when Sn reacts with Ni to form Ni_3Sn_4 IMC.[22] But we propose below a third reason which is unique due to the small solder volume.

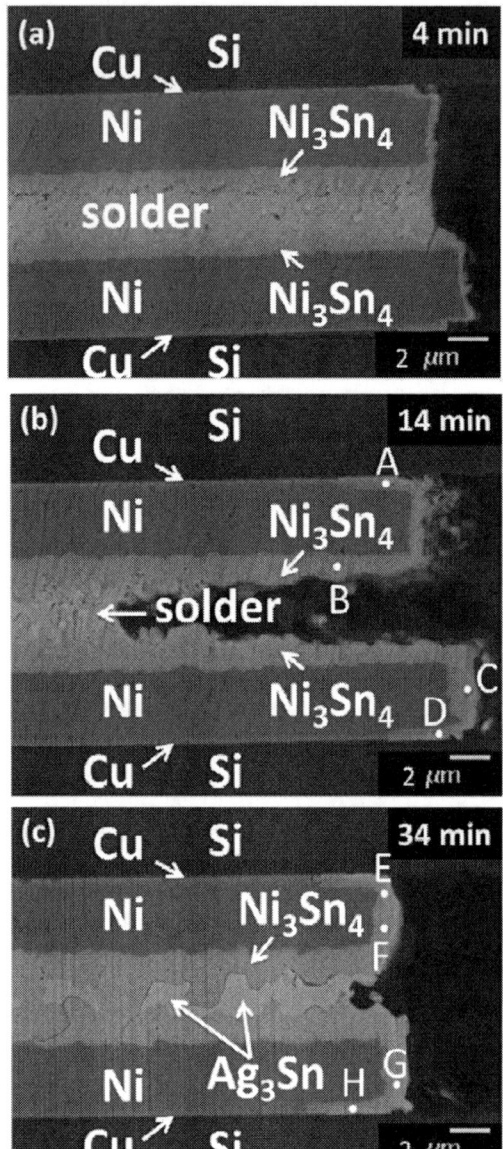

Figure 3. Enlarged cross-sectional SEM images showing the periphery of microbumps with 4-μm-thick solder after reflow for (a) 4 min (as-jointed); (b) 14 min; (c) 34 min at 260 °C.

In Fig. 2(b), while there is a serious necking formation in the periphery of the solder joint, there is no void in the middle of the joint. We found that in Fig. 2(b), the Ni_3Sn_4 IMCs on the top side and on the bottom side have not bridged together, thus the molten solder in the joint has the freedom to move or shrink in the direction normal to the interface of the solder joint. However, after the 34 min reflow, when the Ni_3Sn_4 IMCs on the both sides bridged together, large voids appeared in middle of the joint, as illustrated in Fig. 2(c). This is because while the flow of molten solder is unlimited, the freedom of solder joint to shrink normal to its interface is limited. Thus, when the molten solder is drained by the side wall reaction, void must be formed in the middle of the solder joint. We will discuss the point later.

The locations for the Ag_3Sn precipitates appear differently at different stages. As shown in Figure 3(a), finely-dispersed Ag_3Sn IMC can be observed in the solder matrix in the as-joined sample. After an additional 10 min reflow, as the solder gradually converted into Ni_3Sn_4 IMC, the Ni_3Sn_4 IMCs at the two interfaces thickened at about the same rate as shown in Figure 3(b). The Ag_3Sn precipitates grew larger and they still distributed randomly in the solder matrix. However, when the reflowing time reached 34 min, the solder layer transformed completely into Ni_3Sn_4 IMC, as shown in Figure 3(c), and some large Ag_3Sn IMC can be observed to appear near the edges of the microbump. This microbump was polished by FIB. Since the Ag atoms in the solder behaved as an inert element during the Sn/Ni interfacial reaction, they were constantly rejected and dissolved into the remaining molten solder during the Sn/Ni reaction. Eventually, all the Ag atoms precipitated out as the large Ag_3Sn IMC grains and they tend to adhere to Ni_3Sn_4 IMC. Therefore, the mechanical strength of the heterogeneous phase boundaries between Ag_3Sn and Ni_3Sn_4 is a key factor affecting the reliability of the microbumps in 3D IC applications.

As for the 11 μm thick solder microbumps, the side wall wetting and the necking formation at the periphery of the solder layer occurred at a much longer reflow time. As demonstrated in Figures 4(a) to 4(d), the solder did not shrink obviously until the reflow time reached 64 min. When the reflow time reached 124 min, the morphology of the Ni_3Sn_4 grains became faceted as shown in Figure 4(d).

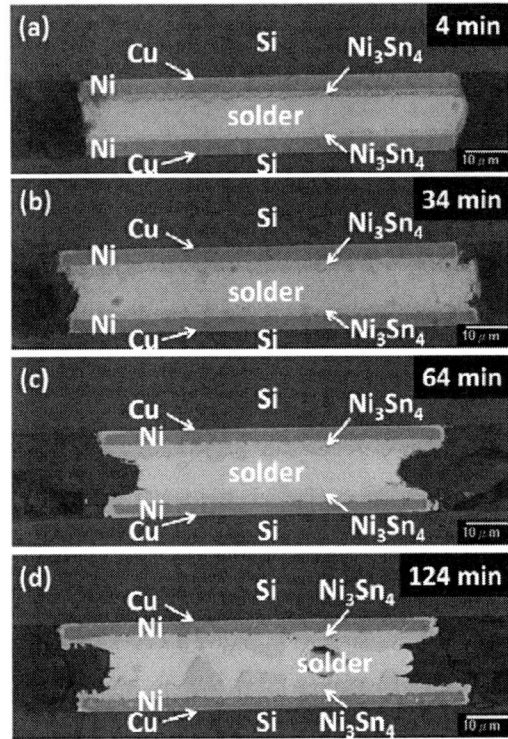

Figure 4. Cross-sectional SEM images showing the microbumps with 11-μm-thick solder after reflow for (a) 4 min (as-jointed); (b) 34 min; (c) 64 min; (d) 124 min at 260 °C.

As shown in Figure 2, it is intriguing that the necking located in the periphery of the microbumps for the 4 μm thick solder sample reflowed after 14 min. Yet, a large amount of voids scattered in the microbumps after reflowed of 34 min. This may be attributed to the following mechanism. Once the Ni_3Sn_4 IMCs bridged the joint; the physical height of the microbump is fixed. The joint cannot shrink anymore in the vertical direction. As the interfacial reaction continues, the molten solder becomes thinner. The molten solder may be drained much easier due to capillary force. As the molten solder is drained by side wall reaction, voids must form near the center region of the joint. For a larger and thicker solder joint, the IMC on the two sides of the joint will not be able to bridge each other, hence no void will form by this mechanism. Obviously, if we can prevent side wall reaction, no such kind of void formation will occur.

CONCLUSIONS

In summary, the interfacial reactions at 260 °C in the Ni/Sn2.3Ag/Ni microbumps with 4 μm and 11 μm thick Sn2.3Ag solder have been studied. The effect of small solder thickness on void formation is significant. For the 4 μm thick solder, when the reflow time reached 14 min,

serious necking or shrinking of the solder from the periphery was observed due to side wall wetting. After a 34 min reflow, voids formed in the center of the microbump. This is because the Ni_3Sn_4 IMCs from the upper and lower interface of the microbump have bridged together, it removed the degree of freedom of the microbump to move or to shrink in the normal direction. Hence a drain of the molten solder by the side wall wetting will leave voids in the center of the joint. This is a processing failure of the microbump in 3D IC applications. This is a unique mode of processing failure because of the smaller and smaller volume of solder joints in the trend of miniaturization. In the thicker 11 μm solder, it took much longer reflow time to allow the bridging of IMC.

ACKNOWLEDGEMENT

The financial support from the National Science Council, Taiwan, under the contract NSC 98-2221-E-009-036-MY3, is acknowledged.

REFERENCES

1. J. C. Lin, W. C. Chiou, K. F. Yang, H. B. Chang, Y. C. Lin, E. B. Liao, J. P. Hung, Y. L. Lin, P. H. Tsai, Y. C. Shih, T. J. Wu, W. J. Wu, F. W. Tsai, Y. H. Huang, T. Y. Wang, C. L. Yu, C. H. Chang, M. F. Chen, S. Y. Hou, C. H. Tung, S. O. Jeng, and D. C. H. Yu, in Proceedings of International Electron Devices Meeting, IEEE, 2.1.1 (2010).
2. K. N. Tu, Microelectron. Reliab., 51, 517 (2011).
3. K. N. Tu and K. Zeng, Mater. Sci. Eng. R., 34, 1 (2001).
4. P. G. Kim, J. W. Jang, T. Y. Lee, and K. N. Tu, J. Appl. Phys., 86, 6746 (1999).
5. M. He, A. Kumar, P. T. Yeo, G. J. Qi, and Z. Chen, Thin Solid Films, 462-463, 387 (2004).
6. G. Ghosh, J. Appl. Phys., 88, 6887 (2000).
7. P. L. Tu, Y. C. Chan, K. C. Hung, and J. K. L. Lai, Scr. Mater., 44, 317 (2001).
8. G. Ghosh, Acta Mater., 49, 2609 (2011).
9. J. F. Li, S. H. Mannan, M. P. Clode, K. Chen, D. C. Whalley, C. Liu, and D. A. Hutt, Acta Mater., 55, 737 (2007).
10. G. Ghosh, Acta Mater., 48, 3719 (2000).
11. W. M. Tang, A. Q. He, Q. Liu, and D. G. Ivey, Int. J. Miner. Metall. Mater., 17, 459 (2010).
12. M. L. Huang, T. Loeher, D. Manessis, L. Boettcher, A. Ostmann, and H. Reichl, J. Electron. Mater., 35, 181 (2006).
13. S. J. Wang, H. J. Kao, and C. Y. Liu, J. Electron. Mater., 33, 1130 (2004).
14. R. Labie, W. Ruythooren, and J. V. Humbeeck, Intermetallics, 15, 396 (2007).
15. W. J. Tomlinson and H. G. Rhodes, J. Mater. Sci., 22, 1769 (1987).
16. C. M. Chen and S. W. Chen, Acta Mater., 50, 2461 (2002).

17. M. O. Alam, Y. C. Chan, and K. C. Hung, J. Electron. Mater., 31, 1117 (2002).

18. J. W. Yoon, S. W. Kim, and S. B. Jung, J. Alloys Compd., 385, 192 (2004).

19. C. Y. Liu, H. W. Tseng, and J. M. Song, Electrochem. Solid-State Lett., 13, H298 (2010).

20. R. S. Cheng, H. J. Chang, T. C. Chang, and J. H. Chou, Electrochem. Solid-State Lett., 15, H75 (2012).

21. J. F. Li, P. A. Agyakwa, and C. M. Johnson, Acta Mater., 59, 1198 (2011).

22. H. Y. Chuang, J. J. Yu, M. S. Kuo, H. M. Tong, and C. R. Kao, Scr. Mater., 66, 171 (2012).

23. D. W. Zheng, W. Wen, and K. N. Tu, Phys. Rev. E, 57, 3719 (1998).

24. C. Y. Liu and K. N. Tu, Phys. Rev. E, 58, 6308 (1998).

A MECHANISTICALLY JUSTIFIED MODEL FOR LIFE OF SnAgCu SOLDER JOINTS IN THERMAL CYCLING

P. Borgesen[1], L. Yang[1], A. Qasaimeh[1], L. Yin[2], and M. Anselm[3]
[1]Department of Systems Science & Industrial Engineering
[2]Department of Mechanical Engineering
Binghamton University
Binghamton, NY, USA
[3]Universal Instruments Corporation
Conklin, NY, USA
pborgese@binghamton.edu

ABSTRACT

We have shown the life of a SnAgCu solder joint in a typical BGA or CSP assembly in thermal cycling to scale with the time to completion of a network of high angle grain boundaries across the high strain region of the joint. This provides for a credible materials science based model. In-depth studies did however show this to require significant temperature variations. Isothermal cycling may also lead to recrystallization, albeit at a much lower level depending on alloy, processes, and cycling parameters, but a quantitative model would need to be completely different. The question therefore arises as to how large a cycling temperature range is required for our model to apply. We present results indicating that repeated cycling between 20°C and 60°C should be sufficient, i.e. the model should allow for extrapolation of accelerated test results to realistic service conditions.

Many practical applications involve a combination of thermal excursions and mechanical cycling, and there is little doubt that thermal cycling induced recrystallization will tend to lead to much faster crack growth through the solder in subsequent vibration, etc. We discuss how this greatly complicates the definition of a conservative but still practical accelerated test protocol.

Key words: Reliability, thermal cycling, recrystallization, fatigue, service conditions, Pb-free solder.

INTRODUCTION

Almost all reliability assessment is, at least implicitly, focused on the anticipated life in service. In the microelectronics industry the most accurate predictions of long-term life are usually based on accelerated testing together with some way of extrapolating the test results to the different combinations of stress, temperature, time, atmosphere, humidity, etc. characterizing service conditions. This of course requires a quantitative model.

The overwhelming majority of reliability tests conducted by the industry are so-called 'engineering tests' in which the focus is usually not on the quantitative prediction of life in service. However, even relative comparisons between alternatives or against generic specifications rely on an assumed correlation between the accelerated test results and performance under realistic service conditions. In the absence of a very large empirical data base for the specific materials and designs in question, which is not yet available for lead free solder joints in long term service, confidence in such a correlation requires a mechanistically justified model.

When it comes to the long-term life of microelectronics assemblies a major concern, aside from defects which are addressed by mitigation rather than life assessment, is wear out of the solder joints. Thermal mismatch induced stresses usually lead to failure of such joints by solder fatigue. Significant knowledge and understanding has by now been gathered with respect to damage in, and failure of, SnAgCu solder joints. A quantitative model is emerging, but for now the most useful result is that the mechanistic insight allows for better confidence in comparisons. Thus, we have used this to argue that Ball Grid Array (BGA), Chip Size Package (CSP), and Thin Small Outline Package (TSOP) assemblies may be compared in conventional accelerated thermal cycling [1], whereas Quad Flat No-Lead (QFN) and surface mount passive assemblies are likely to have different rate controlling mechanisms and thus acceleration factors [1, 2].

A more general concern, and one shared by all reliability engineers, is that the damage mechanism controlling failure in an accelerated test is ***different*** from the one dominating in service. It is in fact not enough that the same damage mechanisms are involved, a safe correlation between test results and performance in service requires the same mechanisms to be rate controlling. We shall illustrate this concern in our discussion of damage due to combinations of thermal excursions and mechanical cycling below.

First, however, the following section briefly outlines our current understanding of thermal cycling induced damage and failure. After that we describe and discuss experiments intended to ascertain that the same mechanisms are likely to be rate controlling in accelerated thermal cycling and in a relatively benign 'office environment' where on-off cycles may only lead to variations from room temperature to an operating temperature of 60°C.

THERMAL CYCLING MODEL

Realistic SnAgCu solder joints are almost always the result of a single solidification event leading to a so-called 'cyclic twinning' structure of the Sn grains. The up to three unique Sn grain orientations differ from each other by roughly 60° and are separated by low-angle twin boundaries. In relatively large, BGA and CSP scale, solder joints we end up with either a single Sn grain or a so-called 'beach-ball' structure (Fig. 1). Depending on the combination of solder alloy, pad finishes, impurities, and solder volume smaller joints may exhibit an interlaced twinning structure with a lot more twin boundaries (Fig. 2). In neither case, however, do the boundaries appear to facilitate crack growth in cycling.

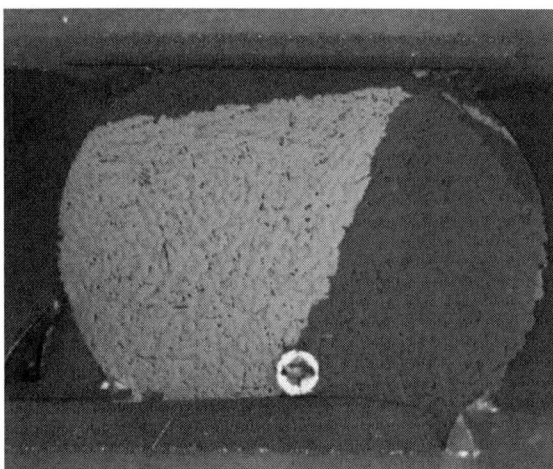

Figure 1: Cross polarizer image of SnAgCu BGA joint cross section showing boundaries ('beach ball') between three large Sn grains [1].

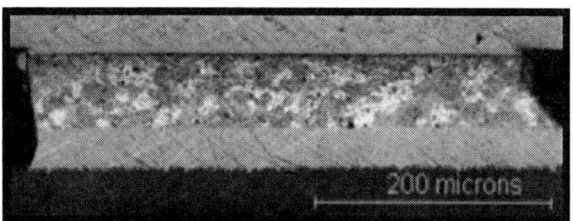

Figure 2: Cross polarizer image of LGA solder joint showing interlacing of the three Sn grain orientations [1].

Using an optimized dye-and-pry approach we showed [3] crack initiation to occur very early in thermal cycling of BGA assemblies, after less than 5% of the characteristic life, $N_{63.2}$. However, transgranular crack growth remained relatively slow for quite a while after that. We believe this to be the reason for reports of a significant crack initiation time. An eventual strong acceleration of the crack growth was found to coincide with the completion of a network of high angle grain boundaries across the high strain region of the joint due to recrystallization of the Sn grains [4], and cracks are indeed commonly seen to have propagated along the new grain boundaries (Fig. 3). The same behavior was observed for TSOP joints (Fig. 4), and dependencies on thermal cycling parameters suggested similar acceleration factors as well [1].

A similar scenario appears to apply to interlaced twinning structures in small solder joints, such as in LGA and flip assemblies, except that here recrystallization seems to be delayed by cycling induced migration of the twin boundaries first driving the structure towards more of a 'beach ball' configuration [2, 5]. Not surprisingly, acceleration factors therefore appear to be slightly different [5].

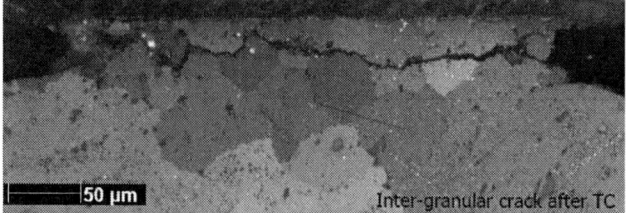

Figure 3: Cross polarizer image of SnAgCu BGA joint with crack along grain boundaries in recrystallized region after accelerated thermal cycling [1].

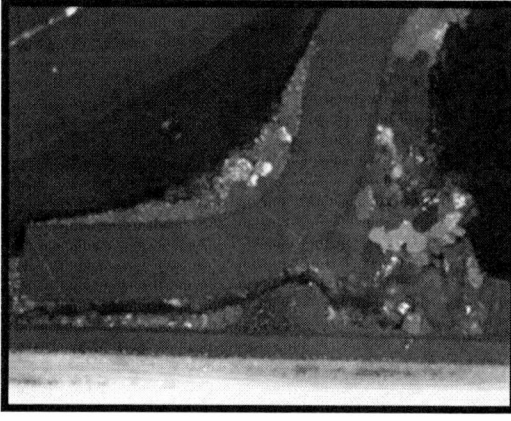

Figure 4: Cross polarizer image of SnAgCu TSOP joint with crack along grain boundaries in recrystallized region after accelerated thermal cycling [1].

While the observation of fatal cracks through a recrystallized region is not anything really new [6-12], the discovery that the completion of the network of grain boundaries in BGA joints would invariably occur within 25-50% of $N_{63.2}$, independently of the cyclic strain range, temperature range, and dwell time, is [4]. This offers intriguing possibilities. Recrystallization appears to be a rate limiting mechanism as far as failure is concerned, so if we can predict the time to completion of recrystallization across the joint, we can predict life. In fact, even if the fraction of life is not always the same under conditions outside of the parameter regime considered by Yin et al. [4], recrystallization may still be rate limiting. For small solder joints indications would be that twin boundary motion together with recrystallization might be rate limiting.

A general, ***quantitative*** model of the recrystallization still offers significant challenges. However, progress towards a practical model continues. Meanwhile, the purpose of the present effort is to ascertain whether recrystallization is still likely to be rate controlling as far as the long-term life under 'office environment' type service conditions is concerned.

EXPERIMENTAL

The present experiments employed specially balanced model test vehicles designed, among other, to minimize warpage in assembly and test. The components were fabricated by sandwiching 0.5mm thick bare Si die between 0.4mm thick FR-4 laminate substrates with a rigid flip chip underfill material (Fig. 5). 20mil (0.5mm) diameter Sn-3.0Ag-0.5Cu wt.% (SAC305) solder balls were reflowed onto electroless nickel immersion gold (ENIG) coated solder mask defined pads on the components. The 256 pads were arranged in 4 perimeter rows with an 0.8mm pitch (Fig. 6). Flux was printed onto the pads before placement and reflow was done in a nitrogen ambient with less than 50 ppm O_2 using a Vitronics - Soltec 10-zone full convection oven and a profile with a 245°C peak temperature and 45 seconds above liquidus.

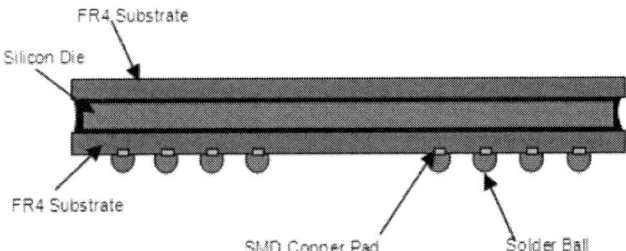

Figure 5: Side view sketch of fully balanced model BGA component

Figure 6: Bottom view of model BGA component

SAC305 solder paste was then printed onto 16mil (0.4mm) diameter OSP coated Cu pads on a 4-layer 62mil (1.55mm) thick printed circuit board, and the components were placed and attached in the same kind of reflow process (above).

Assemblies were subjected to thermal cycling with 10°C/min ramp rates and 10 minute dwells at the maximum and minimum temperatures, and samples were removed after different numbers of cycles for cross sectioning and cross polarizer microscopy. Three different profiles were considered: 0/100°C, -40/60°C, and 25/100°C.

RESULTS AND DISCUSSION

The long-term fatigue life of SnAgCu solder joints in BGA, CSP, TSOP, LGA and flip chip assemblies in thermal cycling appears to be governed by cycling induced recrystallization of the Sn grains, which ends up forming a network of high angle grain boundaries across the high strain region of the joint, followed by rapid crack growth along these boundaries [2, 3, 4, 13, 14]. This failure mechanism competes with transgranular crack growth. Cracks initiate almost immediately, long before any significant recrystallization, but in conventional accelerated thermal cycling they propagate only slowly until the completion of the above mentioned grain boundary network [3]. The question remains, however, whether the same is also true for the much milder thermal cycles characteristic of long-term service conditions.

Effective recrystallization does require significant temperature variations. Isothermal cycling of SAC305 solder joints at any temperature between room temperature and 125°C did not lead to significant recrystallization before failure [15, 16]. Room temperature cycling of SAC205 did cause more recrystallization, but not enough to form a network of grain boundaries before failure by transgranular crack growth [17].

Systematic studies of combinations of pre-aging, room temperature shear cycling and short anneals at elevated temperatures showed that as-reflowed SAC105 joints recrystallized more readily than SAC305, and that recrystallization of the latter is significantly enhanced by aging induced coarsening of the secondary precipitates [14, 18]. It is well established that strain enhanced diffusion in thermal cycling provides for rapid precipitate coarsening [19-22]. Importantly, significant recrystallization required repeated alternations between the pile-up of dislocations at low temperature and the coalescence and rotation of dislocation cell structures at high temperatures.

This explains why completion of a recrystallized region across the joint can be viewed as a rate controlling mechanism for failure in thermal cycling, while that is not so for failure in isothermal cycling.

Effect of High Temperature: For a given combination of pre-aging, shear fatigue cycling, and annealing times the most effective recrystallization was achieved with an annealing temperature of 100°C [14]. Annealing at 70°C or 125°C for the same amount of time gave less recrystallization. The question thus arises whether an operating temperature of, for example, 60°C in service would be sufficient to cause effective recrystallization before failure by transgranular crack growth. Not only is the 'annealing' temperature itself reduced, but so is the cycling induced precipitate coarsening.

Figure 7: Cross polarizer images of SAC305 corner joints after 400 cycles of 0/100°C (left) or -40/60°C (right).

Figure 7 shows representative cross sections of corner joints in our BGA assemblies after completion of 400 cycles of 0/100°C and -40/60°C. Neither of the joints is close to failing yet, but the 0/100°C cycling has led to significant recrystallization in the high strain region at the top. Empirical experience suggests that lower temperatures should lead to slower failure, for the same ΔT, and indeed the -40/60°C cycling led to little or no recrystallization at this stage.

Figure 8: Cross polarizer images of SAC305 corner joints after 900 cycles of 0/100°C (left) or -40/60°C (right).

After 900 cycles there seems to be evidence of extended recrystallization for both temperature ranges (Fig. 8).

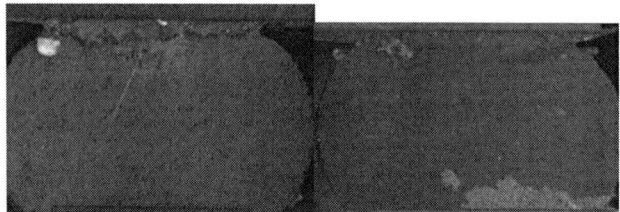

Figure 9: Cross polarizer images of SAC305 corner joints after 1300 cycles of 0/100°C (left) or -40/60°C (right).

Figure 10: Cross polarizer image of SAC305 corner joint after 1300 cycles of -40/60°C.

After 1300 cycles of 0/100°C recrystallization has led to a continuous network of grain boundaries across the joint, and crack growth along it is almost complete (Fig. 9 left). The joints subjected to 1300 cycles of -40/60°C, on the other hand, still showed only limited recrystallization (Fig. 9 right).

Only after 1730 cycles of -40/60°C is the network of new grain boundaries across the joint complete and significant crack growth is evident (Fig. 10). The joint is still quite far from failing, but it is clear that a 10 minute dwell at 60°C after each excursion to low temperatures is sufficient to cause effective recrystallization and enable rapid crack growth.

Effect of Low Temperature: Of course, annealing at a sufficiently high temperature is not enough to cause extensive recrystallization. Such annealing has to continuously alternate with excursions to a sufficiently low temperature to ensure the effective pile up of dislocations. The question thus arises whether turning off our computer and let it cool to room temperature is sufficient for this.

Figure 11 shows representative cross sections of SAC305 solder joints after exposure to 971 cycles of 25/100°C. In this case recrystallization has led to a continuous network of grain boundaries across the high strain region of each joint, while crack growth is still quite limited. It thus appears that room temperature is sufficiently low to ensure effective pile up of dislocations.

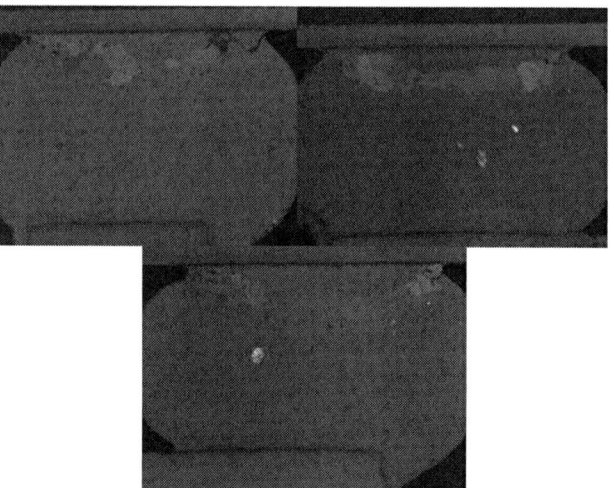

Figure 11: Cross polarizer images of SAC305 joints after 971 cycles of 25/100°C.

Combinations of Thermal Excursion and Mechanical Cycling: As long as vibration, bending, shock, etc. remain completely negligible, such as might be expected for servers and desk top computers, it thus appears that our emerging thermal cycling model may apply under even rather mild service conditions. Of course, different combinations of strain ranges, dwell times, heating/cooling ramps, alloys and pad finishes may affect this.

A more general concern is that many practical applications involve a combination of thermal excursions and mechanical cycling [13, 23]. As documented in detail in a forthcoming publication life in isothermal cycling is determined by transgranular crack growth and ends after the accumulation of a given amount of inelastic energy (work).

There is an obvious need for a practical approach to the assessment of reliability under a realistic combination of repeated mechanical loading and thermal excursions. Experimental assessments may, however, be dangerously misleading unless it can somehow be ensured that the consecutive or concurrent combination closely mirrors the combination of concern in service [13, 23].

A first requirement is that the failure mode is the same [23]. An initial shock loading may for example not lead to damage visible in any cross section but still cause solder joints to fail prematurely by pad cratering in subsequent thermal cycling [24]. On the other hand thermal cycling induced recrystallization may significantly weaken *and* soften the solder itself, so that the failure mode in subsequent bending or vibration changes from cratering or intermetallic bond failure to solder fatigue.

Even a combination of, say, thermal cycling with a level of vibration that would have been sufficiently mild to ensure solder failure by itself may easily lead to a change in failure mode. Recrystallization is almost certain to allow for much faster crack growth, and no longer transgranularly, in subsequent vibration.

Finally, even before recrystallization thermal cycling already leads to significant coarsening of the secondary precipitates, and thus to strong changes in the solder properties. The primary effect is here the softening of the solder, leading to much greater inelastic energy deposition (work) per cycle in vibration. Unlike in thermal aging strain enhanced precipitate coarsening is concentrated in the high strain region of the joint, leading to much greater increases in the inelastic energy deposition there.

Overall, while we have established the generic understanding to rationalize test results and discuss potential effects of combined mechanical and thermal loading [13], much more systematic studies are required before we can propose relevant accelerated test protocols. For a start, we suggest that systematic studies of the effects of temperature and strain ranges (S-N curves for different temperatures) after thermal cycling induced recrystallization may help define a meaningful conservative test. However, such a test would only apply to solder failures and not account for competing failures through the intermetallic bond or by pad cratering. Also, it would not account for potential order of magnitude effects of amplitude variations on solder fatigue life [25-29].

CONCLUSIONS
The rate of failure of SAC305 solder joints in BGA, CSP, TSOP, flip chip, or LGA assemblies in thermal cycling is controlled by the recrystallization induced formation of a network of grain boundaries across the high strain region. This mechanism does require numerous repeated alternations between pile up of dislocations at low temperature and the coalescence and rotation of dislocation cell structures at high temperature. However, a maximum dwell temperature as low as $60^{\circ}C$ and a minimum dwell temperature as high as $25^{\circ}C$ appear to be sufficient for this to be the dominant mechanism, i.e. conventional accelerated thermal cycling should be relevant for thermal mismatch induced failure under even relatively benign service conditions.

Many realistic service conditions involve a combination of thermal excursions and repeated mechanical loading, e.g. vibration. Depending on the details this may lead to failure much faster than predicted based on Miner's rule of linear damage accumulation. Research to establish an accelerated test approach aimed at conservative assessments for those cases where solder fatigue is the dominant failure mode is proposed.

ACKNOWLEDGMENTS
This research was supported by the AREA Consortium and by the U.S. Department of Defense through the Strategic Environmental Research and Development Program (SERDP). The help of Michael Meilunas, Universal Instruments, with component fabrication and assembly is gratefully acknowledged.

REFERENCES
[1] L. Wentlent, L. Yin, M. Meilunas, B. Arfaei, and P. Borgesen, "Damage Mechanisms and Acceleration Factors for No-Pb LGA, TSOP, and QFN Type Assemblies in Thermal Cycling", Proc. SMTA Int. (2011), pp. 101-110

[2] L. Yin, M. Meilunas, B. Babak, L. Wentlent, and P. Borgesen, "Effect of Microstructure Evolution on Pb-free Solder Joint Reliability in Thermomechanical Fatigue", Proc. 62nd ECTC, 2012, pp. 493-9

[3] A. Qasaimeh, S. Lu, and P. Borgesen, "Crack Evolution and Rapid Life Assessment for Lead Free Solder Joints", Proc. 61st ECTC, 2011, pp. 1283-90

[4] L. Yin, L. Wentlent, L. Yang, B. Arfaei, A. Qasaimeh, and P. Borgesen, "Recrystallization and Precipitate Coarsening in Pb-free Solder Joints during Thermo-mechanical Fatigue", J. Electronic Materials 41, Issue 2 (Feb. 2012) pp. 241-252

[5] S. Joshi, B. Arfaei, M. Obaidat, A. Alazzam, M. Meilunas, L. Yin, M. Anselm, and **P. Borgesen**, "LGAs vs. BGAs – Lower Profile and Better Reliability", Proc. SMTAI 2012

[6] P. T. Vianco, J. A. Rejent, and A. C. Kilgo, "Time-Independent Mechanical and Physical Properties of the Ternary 95.5Sn-3.9Ag-0.6Cu Solder", J. Electr. Mater. 32 (2003) pp. 142-151

[7] P. Lauro, S. K. Kang, W. K. Choi, and D.-Y. Shih, "Effect of Mechanical Deformation and Annealing on the Microstructure and Hardness of Pb-Free Solders", J. Electr. Mater. 32 (2003) pp. 1432-1440

[8] S. Terashima and M. Tanaka, "Thermal Fatigue Properties of Sn-1.2Ag-0.5Cu-xNi Flip Chip Interconnects", Mater. Trans. 45 (2004) pp. 681-688

[9] D. W. Henderson, J. J. Woods, T. A. Gosselin, and J. Bartelo, et al., J. Mater. Res. 19 (2004) 1608

[10] J. Karppinen, T. Laurila, and J. K. Kivilahti, "A Comparative Study of Power Cycling and Thermal Shock Tests", in Proc. 1st Electron. Systemintegr. Technol. Conf. (2006) pp. 187-194

[11] J. Sundelin, S. Nurmib, and T. Lepisto, "Recrystallization Behavior of SnAgCu Solder Joints", Mater. Sci. Eng. A 474 (2008) pp. 201-207

[12] T. T. Mattila and J. K. Kivilahti, "The Role of Recrystallization in the Failure of SnAgCu Solder Interconnections Under Thermomechanical Loading", IEEE Trans. Comp. & Packag. Techn. 33 (2010) pp. 629-635

[13] P. Borgesen, L. Yang, A. Qasaimeh, and B. Arfaei, "Damage Accumulation in Pb-free Solder Joints for Complex Loading Histories", Proc. Pan Pacific Microelectronics Symposium, Hawaii, Jan. 2011 (CD)

[14] A. Qasaimeh, Y. Jaradat, L. Wentlent, L. Yang, L. Yin, B. Arfaei, and P. Borgesen, "Recrystallization Behavior of Lead Free and Lead Containing Solder in Cycling", Proc. 61st ECTC, 2011, pp. 1775-81

[15] T. K. Korhonen, L. Lehman, M. A. Korhonen, and D. W. Henderson, "Isothermal fatigue behaviour of the near-eutectic Sn-Ag-Cu alloy between -25–125°C", J. Electron. Mater. 36 (2007) pp. 173–178.

[16] A. Mayyas, L. Yin, and P. Borgesen, "Recrystallization of Lead Free Solder Joints – Confounding the Interpretation of Accelerated Thermal Cycling Results", Proc. ASME Int. 2009, IMECE2009-12749

[17] B. Arfaei, Y. Xing, J. Woods, J. Wolcott, P. Tumne, P. Borgesen, and E. Cotts, "The Effect of Sn Grain Number and Orientation on the Shear Fatigue Life of SnAgCu Solder Joints", Proc. 58th ECTC, 2008, 459-465

[18] A. Qasaimeh, 'Study of the Damage Evolution Function for SnAgCu in Cycling', Ph. D. dissertation, Binghamton University, May 2012

[19] P. Lauro, S. K. Kang, W. K. Choi, and D. Y. Shih, J. Electr. Mater. 35 (2006) 250

[20] I. Dutta, J. Electr. Mater. 32 (2003) 201

[21] I. Dutta, D. Pan, R. A. Marks, and S. G. Jahav, Mater. Sci. Eng. A 410-411 (2005) 48

[22] S. Allen, M. Notis, R. Chromik, and R. Vinci, J. Mater. Res. 19 (2004) 1425

[23] V. A. Raghavan, B. Roggeman, M. Meilunas and P. Borgesen, "Effects of 'Latent Damage' on Pad Cratering: Reduction in Life and a Potential Change in Failure Mode" (accepted for publication in Microelectronics Reliability)

[24] T. T. Mattila and M. Paulasto-Krockel, "Toward Comprehensive Reliability Assessment of Electronics by a Combined Loading Approach", Microelectronics Reliability 51 (2011) pp. 1077-1091

[25] L. Yang, V. Raghavan, P. Borgesen, B. Roggeman and L. Yin, "On the Complete Breakdown of Miner's Rule for Lead Free BGA Joints", Proc. SMTA Int. 2009

[26] L. Yang, L. Yin, B. Roggeman, and P. Borgesen, "Effects of Microstructure Evolution on Damage Accumulation in Lead Free Solder Joints", Proc. 60th ECTC, 2010

[27] Y. Jaradat, J. Chen, J.E. Owens, L. Yin, A. Qasaimeh, L. Wentlent, B. Arfaei, and P. Borgesen, "Effects of Variable Amplitude Loading on Lead Free Solder Joint Properties and Damage Accumulation", Proc. ITHERM (2012) pp. 740-4

[28] Y. Jaradat, J. E. Owens, A. Qasaimeh, B. Arfaei, L. Yin, M. Anselm, and P. Borgesen, "On the Fatigue Life of Microelectronic Interconnects in Cycling With Varying Amplitudes", Proc. SMTAI 2012

[29] L. Yang, L. Yin, B. Arafei, B. Roggeman and P. Borgesen, "On the Assessment of the Life of SnAgCu Solder Joints in Cycling with Varying Amplitudes" (accepted for publication in IEEE Transactions on Components and Packaging Technologies)

A NEW MANUFACTURING MODEL FOR SUCCESSFULLY COMPETING IN HIGH LABOR RATE MARKETS

HOW TO MINIMIZE LABOR AND MATERIAL, THE CONTROLLABLE CONTRIBUTIONS TO A HIGH-TECH ELECTRONIC PRODUCT'S COST, AND ASSESS A MANUFACTURING REGION'S BUSINESS CLIMATE

Tom Borkes
The Jefferson Project
Orlando, FL, USA
jeffer2@earthlink.net

ABSTRACT

This paper presents the results of a study that challenges a widely accepted tenet of industry "wisdom." Specifically, the axiomatic claim that to be most competitive in the global high-tech electronic product marketplace, the assembly of the product must be done in the lowest available labor rate environment. This "given" is thought to be especially certain for high volume applications.

A new assembly high labor rate model is developed on the basis of the key cost variables, **LMNOP: (L)**abor, **(M)**aterial, and **NOP (N)**ational **(O)**ut-bordering **(P)**redisposition).

The paper demonstrates that for **L**:

1. The ratio of the labor cost to the total product cost can be made relatively small through the development of high yield, automated assembly processes, coupled with severely reduced non-value added costs (e.g., indirect labor, overhead, G & A, ICT, rework, et al.).

2. When labor content is minimized, the effect on total cost by even a large manufacturing labor rate disparity approaches zero. Labor rate difference in many applications is shown to be a distraction, a red herring of sorts, cloaking the true root causes of non-competitiveness and masking an understanding of the total competitive landscape.

3. To achieve this level of labor cost reduction, certain conditions must be present and several long-standing paradigms must be challenged, replacing them with principles arrived at through common sense and out-of-the-box thinking. These are:

    - The counterintuitive recognition that for automated, high-tech electronic product assembly it is more costly to hire low wage equipment operators than it is to hire multi-functional engineer/operators.

    - Achieving true yield rates of 99.6% through the development of statistically capable processes, and the control of these processes by employing proactive techniques, rather than using traditional reactive strategies.

    - Utilizing continuous flow manufacturing (CFM) for ALL production applications. This means work-in-process (WIP) is minimized by pulling, not pushing, the product through the factory, and balancing product flow by knowing takt times for all process steps.

    - Designing products using the Principles of **DF MATERS (D)**esign **(F)**or **(M)**anufacturing, **(A)**utomation, **(T)**est, **(E)**nvironment, **(R)**eliability and **(S)**erviceability)

    - Having a multi-skilled workforce that has been taught manufacturing skills in the real world must be available. The long-term vehicle for developing this leading edge, world class workforce is a system of education that can fertilize, incubate and hatch this talent, allowing graduate to hit the ground running.

4. The traditional hierarchical organizational model with its pyramid of people into groups, groups into sections, sections into departments, and collections of departments under a director, must be totally dismantled. Just two groups replace it: a. self-managed product teams, and b. a leadership group

This paper demonstrates that for **M**:

1. Material cost includes raw and material management.

2. Since material cost is by far the greatest contributor to total product cost, minimizing labor cost without addressing material cost is like buying a car at a low price without an engine – you've saved money, but you will not be going very far.

3. Without publicizing it, material manufacturers and their distributors have been fairly and unfairly pegging component pricing to the regions where the components are being assembled. This is not an issue for multi-national corporations with global buying power, but becomes a heavy competitive burden for Tier 3, 4 and 5 companies that only purchase and assemble components in high labor rate regions.

4. A low maintenance, highly reliable material planning and management system must be part of the enterprise resource planning infrastructure.

Finally, the paper documents the results of an **NOP** analysis: the tendency of the business climate created by a country to cause a company to either manufacture or seek manufacturing outside of its borders. The conclusion is that the largely uncontrollable business climate where the manufacturing is conducted has a significant influence on cost and competition. Tax rates, cost of money, exchange rates, currency stability, regulations and other factors are addressed. The paper concludes with a discussion on the effect these factors have in typical risk/reward, cost/benefit studies by those whose capital is put at risk.

Key words: U.S. competitiveness, offshore manufacturing, Concurrent Education, electronic components

INTRODUCTION

The TOTAL cost of producing a high-tech electronic product is what makes an assembly operation competitive or not. Chasing low labor rates around the world has been the sport of manufacturing mavens from high labor rate markets since Henry Ford developed his division of labor assembly model and production line workers' wages slowly increased. [1] "Labor, Labor, Labor, how can we compete with $--- per hour labor." (note: fill in the $ yourself). That drum beat has continued to be heard in the United States from New England starting in the 18th century, to the South, to the deep South, to Japan, to the Caribbean, to Mexico, to China, and on to Vietnam in the 21st century. For both Original Product Developers (OPD) who build the products they design themselves, and Electronic Manufacturing Service (EMS) providers who build OPD products on a contract basis, comparative models trying to characterize the relative cost of electronic product assembly should include ALL costs.

There are only two reasons to out-border production:
1. Your desire to sell your product into that country or nearby markets – Brazil is a good example, but not for shipping and other logistics reasons, but for national trade policy reasons – more on this in the NOP discussion.
2. You cannot compete on cost with a low labor rate country. Why are you unable to compete – is it L, M, NOP or some combination of all of these?

Today's manufacturing experts say that high volume product assembly applications will never return to high labor rate regions because of the labor rate issue. They say product assembly in these areas will be limited to low volume and prototype quantities. But what production requirement results in higher labor costs on a per unit basis – building one hundred or one million of a product? The economies of scale, amortizing set-up, non-recurring engineering and tooling labor costs should make building large numbers cost less. High volume production in high labor rate markets should be more competitive than low volume production. [2]

The Whole is Equal to the Sum of the Parts

All the contributors to the cost of producing an electronic product can be classified by putting them into three buckets: LMNOP (Labor, Material, and National Out-Bordering Predisposition). This paper identifies and analyzes each of these cost contributors in their respective buckets.

ANATOMY OF THE COST OF A HIGH TECH ELECTRONIC PRODUCT

Manufacturing companies producing electronic products are usually ranked in tiers according to sales volume. One such ranking is as follows (in USD):
Tier 1: greater than $2.0 billion
Tier 2: between $500 million and $2.0 billion
Tier 3: greater than $100 million and less than $500 million
Tier 4: greater than $30 million and less than $100 million
Tier 5: less than $30 million [3]

Table 1 presents the cost structure for a very successful consumer electronic product. The ODP responsible for the product design out-borders all product assembly labor to a Tier 1 EMS, as opposed to doing the assembly internally. Some of the source material data presented in Table 1 have been reassigned into the categories used in this paper.

Table 1. Electronic Tablet Cost Breakdown,
Source: Kenneth Kraemer, University of California, Irvine

Retail Price: (100%)	$499
Costs to the OPD	
Bucket 1 (L)	
EMS Labor EMS):	$33 (6.6%)
OPD General, Selling & Administration	$75 (15.0%)
Bucket 2 (M)	
Material: Raw $154 (30.9%) + Material Markup $87 (17.4%) =	$241(48.4%)
Total Costs to the OPD:	$349 (70%)
ODP Net Profit before Taxes (EBIT)	$150 (30%)

What about cost bucket 3: NOP (National Out-Bordering Predisposition)? The costs associated with this variable are baked into the EMS labor rate and OPD overhead. An OPD in-border outsourcing often results in a lower overhead to absorb but requires buying EMS labor at a much higher rate. In 2008, using the most current data available, the spread in the fully burdened (including profit) electronic assembly EMS hourly labor sell rates between the high ($50.43) and low labor rate ($6.00) markets was a striking $43.57 per hour. [4]

One appropriate question is how much of this spread results from a raw hourly labor rate difference versus differences between fixed and variable overhead burdening in these low and high manufacturing labor rate regions? It is just these fixed overhead costs that in part form a nation's out-bordering predisposition (NOP).

Another important question in the in-border versus out-border decision-making process is whether there are other controllable ways to close the fully burdened labor cost disparity between high and low cost labor regions. Finally, what is the best value proposition for an OPD: building their own products or employing an EMS?

30% NET MARGIN: AN OPD VERSUS AN EMS AND IN-BORDERING VERSUS OUT-BORDERING

The previous section documents the cost structure for a real-world electronic product. In this case, the product's sell price minus its cost results in a net profit of 30%. This incredible return plays a significant role in making the OPD the most valuable company in the world.

An OPD sells a product, and an EMS sells a service. For both an OPD and an EMS, competition drives price, not production cost. Part of the OPDs value is its intellectual property (IP) estate. In the EMS world there is little IP. An OPD can use its IP to separate itself from its competition. It can typically demand higher margins based on a more valuable IP position. This IP can translate into features that the consumer is willing to pay more for. An EMS does not have this leverage. Consequently, EMS companies generally operate at much lower margins than OPD companies. Like the low margins associated with food pricing in the retail supermarket business, they need high sales volume to compensate and make a reasonable return. Actual EMS margins, therefore, are established almost exclusively by cost. Costs incurred by an EMS are determined by:

1. How well they choose their OPD customers. EMS companies are constrained by the automated equipment capacity they own. If they fill that capacity with marginally profitable business they have nothing else to sell. Examples of *high cost* customers are those whose products are poorly designed, or are difficult and expensive to deal with.
2. Assembly yields, efficiencies, i.e., performance to standard cost, and equipment utilizations.
3. How well they manage material.
4. Payroll and direct labor overhead.
5. Whether they are "board stuffers" or full service providers

Because of fierce competition an EMS company is generally working on thin, booked business margins. There is not much operational room between a positive financial outcome and metaphorically wrapping a $5 bill around each unit you ship – the more product you ship the more money you lose – perhaps it is better to have the equipment lay idle. That is, unless things are so bad that shipping product at a loss at least contributes to some payroll and overhead absorption. This situation typically does not last long as this *death spiral* generally leads to a predictable bad terminal outcome.

There are exceptions to the customary EMS paper-thin margins. For example, an EMS company that specializes in providing leading edge assembly technologies and regulated product assembly can achieve higher margins (e.g., medical, military). Another area of margin expansion opportunity is being able to excel in a challenging scheduling and product mix environment. Dealing with low volume / high mix products and customers who change their ship schedule frequently are examples.

The OPD for the product in Table 1 is realizing a net profit of 30%. The EMS assembling the product may net 2-3%. All things being equal, there is no inherent process advantage for an OPD assembling a product they have designed, or contracting the assembly to an EMS. The process and the equipment are the same. The capital equipment depreciation and many of the other product assembly costs that are deferred by an OPD when out-sourcing to an EMS are paid indirectly since the EMS must imbed these in the burdened labor sell rate. In other words, the 30% net is not because the OPD is using an EMS for its product assembly. However,

there is one category of overhead cost that may not be a dollar for dollar transfer for EMS to OPD. This is the NOP (National Out-bordering Predisposition). If any raw high and low labor compensation cost disparity can be reconciled (see "An Alternative High Labor Rate Cost Model" below), the cost of doing manufacturing in the OPD home country may be adversely burdened by government policy (e.g., taxes, regulation, ability to borrow money, etc.), this will dissuade production and encourage OPD out-bordering (and, ironically EMS out-bordering). Finally, there are several additional factors, that should be considered in the OPD outsourcing value proposition.

Pro OPD outsource:
1. An ODP may use an EMS to mitigate some risk:
   - Equipment is not at risk of sitting idle.
   - Material and labor cost adders based on poor product design and forecast weakness are reduced to the extent the liability for these items are not stipulated in the contract with the EMS.
   - An ODP controls its production schedule.

Anti-OPD outsourcing:
1. An EMS adds an additional layer of cost markup to the price of an OPD product.
2. Additional OPD costs associated with managing the EMS.
3. An ODP using an EMS stretches their supply chain, adding to the risk of supply and product fulfillment interruption.

THE COMPETITIVE LANDSCAPE

In the U.S. (a high labor rate region), the 1975 contribution from all its value-added manufacturing as a percentage of its total GDP was ranked 16th in the world. In 2004, the U.S. ranking for this relative manufacturing *health* metric dropped to 73rd. In 1975, the contribution from all value-added manufacturing done in China (a low labor rate region) as a percentage of its total GDP was ranked 30th in the world. In 2004, as a percentage of a country's total GDP, manufacturing in China ranked 2nd in the world (Table 2).

Table 2. World Change from 1975 to 2004 between a High and Low Labor Rate Country as a Function of Value-added Manufacturing as a Percentage of Total GDP [5]

	1975	2004
U.S.	16	73
China	30	2

In this table, *Manufacturing* refers to industries belonging to ISIC Section C and made up of manufacturing divisions 15-37. [6] *Value-added* is defined as the net output for the ISIC Section after adding up all outputs and subtracting intermediate inputs. It is calculated without making deductions for depreciation of fabricated assets or depletion and degradation of natural resources. These statistics include all manufacturing – whether it can be easily automated, such as circuit board assembly, or labor intensive, such as installing shingles on the roof of a new home or fabricating and assembling a weather satellite (any manufacturing application where either the product design does not lend itself to automation or there are only low quantities required). The specific ISIC division that addresses electronic product manufacturing is division 26, *Manufacture of computer, electronic and optical products*, with Group 261 and Class 2160 addressing the *Manufacture of electronic components and boards*. The International Standard Industrial Classification (ISIC), revision 4, defines these categories.

The example of the ascendency of manufacturing activity in low labor rate environments, accompanied by the exodus of manufacturing jobs out of high labor rate markets as represented in Table 2 by China and the United States, is well known. The consumer's constant reminder of this shift has been in the relentless trend of what is by the now the ubiquitous product marking: *Made in (pick your low labor rate country of choice)*.

With progressively more and better assembly automation available, the other shift in high labor rate regions has been in a consistently decreasing labor contribution to total product cost – at least on paper. This has been a coping tactic used to combat the out-boarding low labor rate mania. Is the increased level of automation being fully exploited by high labor rate countries? This question is answered in the section of this paper entitled, *An Alternate High Labor Rate Model.*

If everything else is assumed equal between two companies, how does the material content of a product affect today's competitive landscape between high and low labor rate regions? The answer seems obvious. The cost of purchasing the material to build equal quantities of the product should be the same wherever the product is assembled – same material, same quantity, same cost. The shipping cost differential for large material quantities is negligible. However, the cost of managing the material, an overhead cost, is weakly influenced by the cost of the labor hours needed for this task. The material question is an important one because the material cost dominates the total product cost. But is the cost of material the same notwithstanding the location of product assembly? This question is answered in the section of this paper entitled *Procuring the Material.*

With the opening of the global marketplace, *competition* has come to mean primarily competition between nations. However, the individual states in the United States also compete with each other. Each state has the incentive to create the most attractive manufacturing business climate to lure and persuade OPD and EMS assembly operations to put down stakes within their boarders. This was a key component to what the founders called federalism: The concept that states that were free to compete with each other and were effectively independent laboratories to try out new ideas and policies. The states that create a less expensive business environment promoted manufacturing and were more successful. The techniques that work in one state could be copied by another state – the ones that do not work could be ignored. This concept has been replaced to a large extent by National government policy that equally burdens all states by imposing additional layers of regulation. State and local policies are still differentiators between the individual states,

but not when competing with another country that may have much lower national regulation costs. National policy trumps state policy and cannot be defied by an individual state. This is part of the NOP (National Out-boarding Predisposition) and will be discussed in detail in the final section of the paper.

ELECTRONIC PRODUCT COST: A GENERAL VIEW
The cost of producing an electronic product is simply the sum of labor cost and the material cost. In most electronic product assembly applications the material cost is generally somewhere between 50% and 80% of the total product cost. For a specific electronic product, the actual material/labor cost split depends primarily on a combination of the following factors:

1. The functionally of the product as it dictates the requirement for high tech and custom (more expensive) material, both for the circuit board and higher level assemblies.
2. The amount of post-circuit board (higher level or box build assembly) that is required and the corresponding labor rates that are applied.
3. To the extent justified by production volume, the degree to which the labor intensive box build processes can be automated.
4. The design of the product as it affects the ability to successfully automate (with low yield loss) the assembly, test, etc. processes (DF MATERS). [7]
5. The manufacturer's overhead and other indirect costs that are absorbed in the labor rate and cause it to be inflated – the higher these costs, the higher the labor rate, and hence, the lower the material cost as a percentage of the total cost.

The portion of total product cost that the circuit board(s) contribute is normally characterized with even higher material-to-labor cost ratios – generally, with material comprising 75-95% of the total circuit board cost because of the inherent level of standardized automation readily available (e.g., the automated placing and soldering of most electronic components). Factors 1, 4 and 5 apply in establishing where material cost as a percentage of total cost falls for a specific circuit board assembly within the 75-95% range.

As stated earlier, although the material portion of the cost dominates, it would seem reasonable to assume that the most controllable portion of the total cost is the smaller labor portion. This part of the cost can be affected by good process development and control, competitive labor rates, and the extent to which labor content can be squeezed out through automation. This assumption would be true if the larger material portion of the cost for a given product assembly application were independent of where the product was assembled. In other words, the same bill of material (BOM) costs the same whether it is purchased in a high or low labor rate assembly environment – but it is not. The question then becomes: Why? In the section of this paper entitled *Procuring the Material*, an attempt is made to determine the material suppliers' cost differences between locations in low and high labor rate regions. These differences are analyzed to see if they justify a difference in the material pricing quoted to the product assemblers in the different labor rate areas – and, if so, how much. Models are developed to determine the cost of doing business by polling the two primary sources of electronic components, the ODM (Original Device Manufacturers) and their franchised distributors. The users, OPD (Original Product Developers), also known as OEM (Original Equipment Manufacturers), and the EMS (Electronic Manufacturing Service) providers were also polled to help comprehend the historic price difference between purchasing the same components in low cost and high cost labor markets, as well as the current state of purchasing material. The material section concludes that while traditionally there may have been a significant price advantage to buying components in low cost labor environments, this material disparity has largely disappeared. In addition, the *indirect* material costs incurred by the Product assembler (inspection, non-conformance, kitting inventory, attrition, PPV, rework, scrap, etc.) are analyzed, and a material management strategy is developed to minimize these elements of cost. Figure 1 shows how from an accounting perspective, value is added to the purchased material along the assembly process, the corresponding costs are credited and debited into and out of accounting centers, and used to absorb labor and overhead costs based on performance to standard costs (the basis of the selling price of the product to the customer). At the end of the process when the product is either shipped or put in finished goods inventory, the result is either a P (Profit) or L (Loss).

Electronic Product Assembly: Establishing Labor Cost
There are different methods of estimating and accounting for a product's labor cost. The labor cost is generally the number of direct labor hours needed to build the product multiplied by the loaded labor rate. The direct labor hours normally consist of the assembly and test *touch* labor (Raw "L"). The loaded labor rate is the average direct labor rate factored up

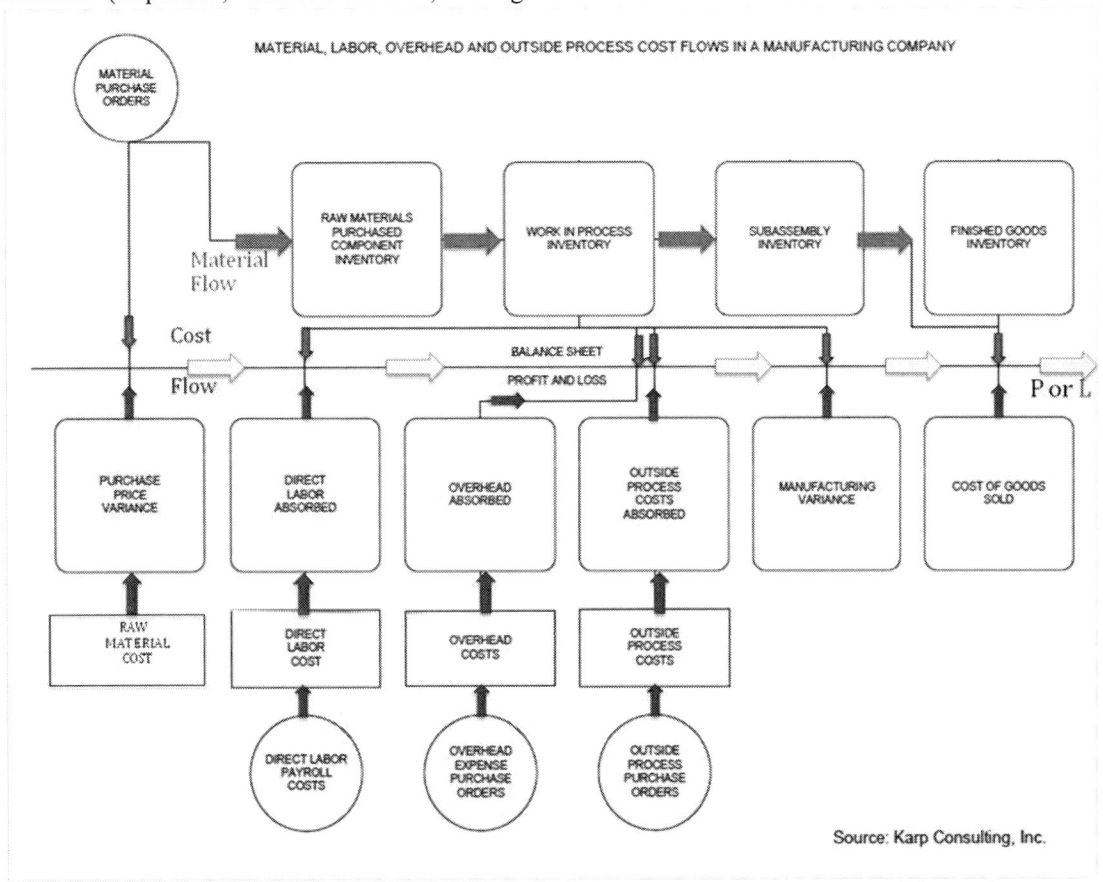

Figure 1. An Accounting Cost Flow From the Purchase of Material to the Sale of the Product

to absorb the overhead (including consumable material such as solder), indirect and other non-product related labor costs, i.e., non-direct costs. Also loaded as a percentage of labor cost are the SG&A (Selling, General & Admin) costs. This labor loading is needed to pay for non-project related management, marketing, sales and other non-product related costs that are required to operate the business. Finally, the loaded labor cost is marked up with profit to establish the labor sell price. Over a year, *selling* enough loaded labor hours, by virtue of selling the assembled products they are part of, will pay for all the direct, indirect, overhead and SG&A costs incurred and provide the company with net earnings – a profit. Other cost accounting systems, such as activity-based costing, convert more indirect and overhead costs into direct costs. For example, to estimate the cost to assemble a board with SMT components, all the direct labor (machine operators, hand solder personnel, etc.), indirect labor (process development engineering, etc.), overhead (SMT equipment depreciation, etc.) and other activity allocated costs are estimated. Then, a theoretical de-rated average SMTA placement rate and an equipment utilization rate are established. The total activity cost for SMT board assembly is divided by the number of SMT components for all products that the pick and place machine is expected to place. This, then, becomes the cost per component. For a specific board, the number of components is multiplied by the cost per component. For example, the labor cost estimate for a board with 500 components processed through the SMT activity cost center of $0.0025 per component would be $1.25. An accounting system is set up to collect the actual costs as they are incurred.

Whatever labor system is utilized, all labor costs must be accounted for. The success of paying for all the direct and absorbed labor costs and making the projected profit is pegged to the operation's ability to:

1. Assemble the product within the labor hours that are embedded in the product sales price.
2. Build and sell the volume of products on which the loaded labor rate was based.
3. Not exceed the estimated costs (both direct and non-direct) that determined the estimated labor rate, and hence, the labor cost.

The labor estimating and accounting process can seem complex and convoluted because of its mathematically indeterminate nature. A particular product's estimated labor cost is dependent on the overall operation's loaded labor rate. The loaded labor rate, in turn, is a function of the operation's estimated total labor cost and an estimate of the total labor hours that will be *sold*, in the products that are assembled – one of which is the product whose labor cost is being estimated. Mathematically expressed:

C_P = Cost of the Product ($)
C_M = Cost of the Material (in $)
C_{PL} = Total Labor Cost of the Product ($)
C_{DL} = Direct Labor Cost of the Product ($)
C_{PLT} = Total Annual Labor Cost ($)
C_{DLT} = Total Annual Direct Labor Cost ($)
R_L = Labor Rate ($/hr)
C_{IL} = The Portion of the Total Non-Product Specific Labor and Facility Overhead Cost that will be loaded or absorbed in the Total Labor Cost ($)
C_{ILT} = Total Annual Non-Product Specific Labor and Facility Overhead Cost that will be loaded or absorbed ($)
H_{RL} = Direct Labor Needed to Assemble the Product (hr)
H_{RLT} = Direct Labor Needed for all the Products that will be assembled over the year (hr)
C_P = $C_{PL} + C_M$

C_{PL} = C_{DL}	+	C_{IL}	Eq. 1
C_{PL} = H_{RL}	x	R_L	Eq. 2
C_{PLT} = C_{DLT}	+	C_{ILT}	Eq. 3
C_{DLT} = H_{RLT}	x	R_L	Eq. 4
C_{DL} = C_{DLT}	x	(H_{RL} / H_{RLT})	Eq. 5
C_{IL} = C_{ILT}	x	(H_{RL} / H_{RLT})	Eq. 6

R_L = $(C_{DLT} + C_{ILT}) / H_{RLT}$ Eq. 7

- To determine the total labor cost of a product (C_{PL}), seven equations need to be solved. There are more unknowns than equations.
- For a given product, the number of direct labor hours it will take to assemble the product (H_{RL}) is determined during the quoting process, i.e., it is a known independent variable.
- The total amount of non-product-specific labor and facility overhead cost needed to run the operation over a year (C_{ILT}) is budgeted, i.e., it is an estimated independent variable. It is the material management variable that is loaded into the labor rate.

How is the total labor cost, C_{PL}, determined?
First, the labor rate is calculated (Eq. 7):
R_L = $(C_{DLT} + C_{ILT}) / H_{RLT}$

To solve for R_L, the total direct labor hours that will be expended (i.e., *sold*) for *all* products over the year, H_{RLT}, must be estimated. Also estimated is the total annual non-product-specific labor and facility overhead cost that will be absorbed or loaded, C_{ILT}, in the labor rate.

For example: It is estimated that a product assembly facility's non-direct cost for the year (C_{ILT}) will total $2.5 million. The facility employs 50 direct labor people who make an average of $20 / hr. Therefore, the yearly direct labor cost is $2M (Raw "L"). The direct assembly time estimated for the product being quoted (H_{RL}) is 0.50 hr. per unit. The business forecast for the year estimates all the products sold for the company will require 100,000 hours of direct labor (H_{RLT}). What should be charged for labor (C_{PL}) to assemble the product?

To calculate the labor rate:
R_L = $(C_{DLT} + C_{ILT}) / H_{RLT}$ (Eq. 7)
C_{DLT} = 50 x 2000 hr x $20 per hr = $2.0M
C_{ILT} = $2.5M
H_{RLT} = 100,000 hrs
R_L = $(2M + 2.5M) / 100,000$ = **$45 / hr**

For a product that will require 0.5 hr per unit,
The labor cost for the product is:
C_{PL} = H_{RL} x R_L (Eq. 2)
C_{PL} = 0.5 hr x $45 per hr.   = **$22.50**

The non-product-specific cost absorbed in the total product cost:
C_{IL} = C_{ILT} x (H_{RL} / H_{RLT}) (Eq. 6)
C_{IL} = 2.5M x (0.5 / 100,000)   = **$12.50**

The direct labor part of the total product cost is:
C_{DL} = C_{DLT} x (H_{RL} / H_{RLT}) (Eq. 5)
C_{DL} = $1M x (0.5 / 50,000)   = **$10.00**

Checking the result:
C_{PL} = C_{DL} + C_{IL} (Eq.1)
C_{PL} = $10.00 + $12.50   = **$22.50**

To pay for the annual non-product-specific company costs (C_{ILT} = $2.5M), enough products must be sold over the year that contain at least the number of direct labor hours that the labor rate (R_L) was based on (H_{RLT} = 100,000). For every product sold in the above example, $12.50 of non-product-specific company cost is paid for. To pay for the total $2.5M in annual non-product-specific company costs through the sales of only the example product, 200,000 units would have to be sold over the year.

Using an activity-based system, a total labor cost estimate for a product is done on an activity basis. For example, the SMT activity uses a cost per pick-and-placed component to absorb the SMT activity costs (including the indirect, overhead and other non-direct cost allocations). It relies on paying for those costs by *selling* the actual placements made through selling the products of which they are a part.

Product labor costs incurred for assembly in high labor rate regions of the world can compete with their low labor rate counterparts. The impact of the labor *rate* disparity on labor cost can be negated if the direct labor hour *content* is minimized. This can be done if the available automation is exploited and the controllable non-direct costs that inflate the labor rate are significantly reduced. [8] In fact, contrary to *industry expert* opinion, product assembly in high labor rate environments should be more competitive for high volume applications than for low volume applications. [9] That is, if the other larger part of the product cost is not dependent on where the assembly takes place.

Electronic Product Assembly: Establishing Material Cost
The material used for electronic products is some combination of electronic components, circuit boards, I/O devices (keypads, displays, speakers, microphones, etc.), molded plastic or metal housings, cables and other materials that create the product's functionality. The manufacture of these materials has largely tracked that of the assembly industry continuously migrating to low labor rate localities.

The material cost a product assembler incurs can be divided into two parts: raw material (raw "M") and material management. Material management costs (costs other than that of the "raw M") are analogous to indirect labor costs. These generally include:

1. Cost of Carrying the Inventory –
   The cost of the money needed to finance the material until the finished product is paid for.
   Usually this is expressed as a percent of the raw "M" cost, representing how much interest the dollar value of the material would have earned if it had been in an interest-bearing bank account. The most commonly used metric for this variable is the number of *inventory turns*. Inventory turns = the annual cost of the material / average value of the inventory over the year.
2. Material purchasing –
   This is the cost of the procurement personnel (buyers) who order the material.
3. Material planning –

The cost of the MRP (Material Resource Planning) activity that is charged with scheduling material for manufacturing to meet product shipping requirements and minimizing inventory. This activity is generally a subset of the ERP (Enterprise resource planning) activity.

4. Incoming inspection – The cost of the labor used to in-ship and inspect the material after it is delivered.
5. Attrition – raw "M" lost because of:
   a. electrical, mechanical or cosmetic non-conformance due to damage or malfunction,
   b. loss of material in the assembly process (e.g., dropped by the placement machine),
   c. inventory not reconciled,
   d. pilferage/theft.
6. Scrap –
   a. material lost in post-solder assembly related to issues such as manufacturing defects resulting in the need to discard components or entire assemblies,
   b. excess material that becomes obsolete (more material purchased than was required and/or cancelled customer orders).
7. Shipping – Since material is almost always FOB the manufacturer, the product assembler pays the freight. Also, any additional costs incurred to expedite deliveries versus normal freight charges must be included.
8. Distributor loaded cost – This is the cost that is added by the distributor to the component price to absorb the distributor's indirect costs and other operating expenses.
9. Distributor markup – This is the difference in the distributor's selling price and distributor's loaded cost. It represents the EBIT (earnings before interest and taxes). The resulting net earnings after interest and taxes are the distributor's profit.
10. Purchase price variance (PPV) – The difference in the raw "m" between the standard cost the product assembler's quote to the customer is based on, and the actual purchased price of the material.
11. Currency exchange rate – The cost of purchasing material with a currency other than the currency of the country where the material was produced.
12. Trade policy / Corporate tax rate – The policy between importing and exporting countries and the taxes the government imposes on earnings.

Establishing a price point for a product
There are three basic factors that influence the price point an OPD will set for a product:

1. What the OPD's targeted customer base is willing to pay (the higher the price, the lower the demand).
2. The cost of the product (based on a desired margin: the higher the cost, the higher the price).
3. Price of competing products

Item 2 is determined by the contents of cost bucket 1: (L), the labor, and cost bucket 2: (M), the material.

ELECTRONIC PRODUCT COST: THE MODELS
Generalized Labor Cost Model

It is difficult to develop a series of models that everyone can agree with. At best, we have to blend industry and government data. These data are themselves averages of different regions, industry sectors and assembly work that is done: either by an OPD - a company that designs and manufactures their product, or, an EMS provider - a contract assembler who builds products for OPDs. For example, the model for the automotive electronic assembly sector may be more heavily burdened with employee benefits if the assembly is being done by the OPD rather than at an EMS. Also, relative currency fluctuations and other NOP-related policies can play a significant role in assembly cost.

Table 3 provides some historical labor rate and trend data for China. Technically, the labor cost of a product does not include profit or fee. Adding the overhead costs that need to be absorbed, plus profit, to the raw material and labor of a product results in the selling price. For the purposes of this analysis, however, profit will be considered part of the overhead and loaded accordingly. Another way of saying this is that the models will result in the labor sell price. In addition, in China, the cost of labor is strongly influenced by whether the assembly is being done in urban areas or by TVEs (town and village enterprises).

Table 3. **China Manufacturing Hourly Rates, 2002-04 [10]**

Year	Basis (Yuan)	Basis (U.S. $)	Index (U.S. = 100)
2002	4.73	0.57	3
2003	5.17	0.62	3
2004	5.50	0.66	3

In 2012, data provided by the Chinese government updated the year-to-year change in electronic assembler compensation in China through 2008. These data are included in Table 4. In 2004 the hourly compensation (wages plus benefits - note: not the fully burdened labor sell rate) to an electronic product assembler in China was the equivalent of $0.66 USD. In 2008 this compensation rate almost double to $1.36 (about a 25% per year increase). The prediction is that this trend of global catch-up will continue. China is projected to experience compensation cost increases averaging 13.6% per year for the foreseeable future. [11]

Table 5 presents similar data for India, another low labor rate region. In India, the compensation rate rose from the equivalent of $0.85 USD in 2004 to $1.17 in 2007 (about a 13% per year increase). [12]

In the United States, a high labor rate region, the average hourly compensation (not the fully burdened labor sell rate)

Table 4. Hourly compensation costs in manufacturing for China, in U.S. dollars, 2003-2008

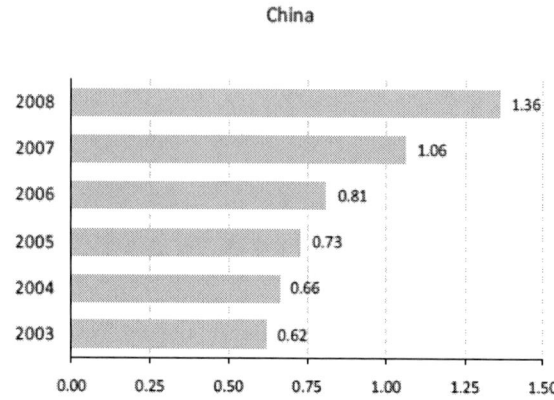

Table 5. Hourly compensation costs in manufacturing for India, in U.S. dollars, 2003-2008

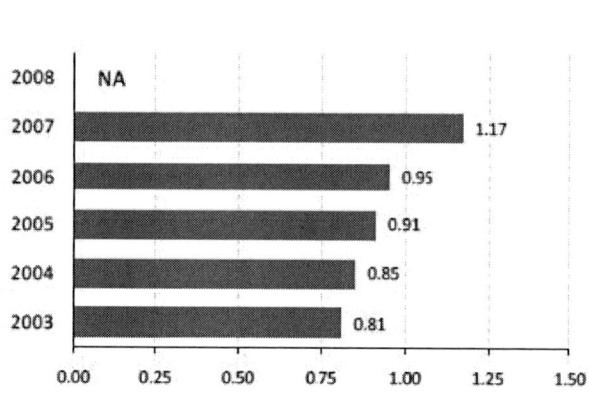

for electronic production and nonsupervisory employees went from $23.05 in 1997 to $34.74 in 2010 (about a 9% per year increase) [Table 6]. In 2007, the average US hourly compensation was $26.63 and the fully burdened labor sell rate was $50.43 [13]. In 2012, the average compensation rate for electronic production and nonsupervisory employees with a 32% benefit load was $23.58 x 1.32 = $31.13/hr., a drop of 1.8% from 2010. [14]

A better employment category to compare a high labor rate region with a low labor region is the region's mean earnings (wages) for electrical and electronic equipment assemblers. This is the category for most of the direct labor. In 2011 these earnings were $15.17/hr [15] in the United States. Applying a 32% benefit rate to these earnings results in a compensation rate of $20.02/hr. This is a good estimate of the direct labor before overhead burdening and will be employed in the high labor rate model that follows.

Table 6. Hourly compensation costs in manufacturing, U.S. dollars, and as a percent of costs in the United States [16]

	Hourly Compensation Costs			
	in U.S. dollars		U.S.=100	
	1997[1]	2010	1997[1]	2010
Norway	26.38	57.53	114	166
Switzerland	30.00	53.20	130	153
Belgium	29.12	50.70	126	146
Denmark	24.09	45.48	105	131
Sweden	24.97	43.81	108	126
Germany	29.15	43.76	126	126
Finland	22.35	42.30	97	122
Austria	25.52	41.07	111	118
Netherlands	23.40	40.92	102	118
Australia	19.10	40.60	83	117
France	24.88	40.55	108	117
Ireland	17.03	36.30	74	104
Canada	18.84	35.67	82	103
United States	**23.05**	**34.74**	**100**	**100**
Italy	19.67	33.41	85	96
Japan	22.28	31.99	97	92
United Kingdom	18.50	29.44	80	85
Spain	13.92	26.60	60	77
Greece	11.56	22.19	50	64
New Zealand	12.37	20.57	54	59
Israel	12.32	20.12	53	58
Singapore	12.15	19.10	53	55
Korea, Republic of	9.36	16.62	41	48
Argentina	7.43	12.66	32	36
Portugal	6.38	11.72	28	34
Czech Republic	3.24	11.50	14	33
Slovakia	2.86	10.72	12	31
Brazil	7.07	10.08	31	29
Estonia	NA	9.47	NA	27
Hungary	3.05	8.40	13	24
Taiwan	7.04	8.36	31	24
Poland	3.13	8.01	14	23
Mexico	3.47	6.23	15	18
Philippines	1.28	1.90	6	5

NA=data not available.
[1] With the exception of Estonia, 1997 is the first year data for all countries are available to BLS.

TOTAL L

The "labor rate" or "compensation cost" for a region of the world is the manufacturing statistic that is primarily publicized by the media. It is usually given as the reason a region either attracts or repels manufacturing activity. It is clearly part of the total L, but by no means all of it. All other non-product specific material related indirect and overhead costs must be absorbed into the compensation cost to result in the fully burdened labor rate. If profit or fee is included in the loading, the result is a fully burdened labor sell price in $ / labor hour. Total labor cost is driven by four factors:

1. Direct assembly labor rate: compensation cost ($/hr.)
2. Labor content (hr)
3. Overhead Absorption: including costs imposed by the government where the production is taking place or NOP costs (% of direct labor $/hr)

4. Labor management: efficiency and yield (%)

Direct Labor

Product assembly including functional test is a value-added enterprise. The direct labor used to assemble and test a product clearly adds value to the OPD or the EMS customer. The cost of this part of the labor is a function of the direct assembly labor rate and is driven by the region of the world where the assembly is being done. However, Factors 2, 3 and 4 are largely controllable. Not only is direct labor the value-added part of the total labor cost, it is necessary because it is the basis for absorbing indirect and overhead labor costs (Factor 3). These costs are often of questionable value. Without direct labor, who pays for the management of an assembly operation? Certainly not the customer.

COST BUCKET 1: L (LABOR)

There are several ways to account for cost, e.g., activity based costing and labor-loaded costing. The point is that regardless of the system that is used, all the costs must be included, resulting in Total L. For this exercise, a labor loading system will be used. The same hypothetical circuit board is costed in both the high and low labor rate companies. Two generalized labor cost models that represent the existing low and high assembly labor environments will establish the gap between the models, define the elements that contribute to the gap, and provide visibility to which elements, if any, are controllable.

The Existing High Direct Labor Rate Model

Assumptions (in USD):

1. Wage (earnings): $15.17/hr [17]
2. Benefits are 32% of wages (earnings or raw labor rate: $4.85/hr)
3. Employee Compensation (Earnings + Benefits) for U.S. manufacturing: $20.02/hr
4. Overhead rate for full labor burdening is 250% of wages (earnings or raw labor rate)
5. Benefit cost is included in overhead rate
6. All indirect labor is included in overhead rate
7. Overhead includes SG&A (sales, general and administrative costs – generally, a percentage of labor and material)
8. Overhead includes material handling, inspection and attrition (usually loaded as a percentage of the raw material cost)
9. Profit or fee is included in overhead rate to result in a fully burdened labor selling price
10. Assumptions 3 through 8 produce a fully burdened labor selling rate of $37.93/hr
11. Fixed overhead (facility/equipment) is 4% of fully burdened labor
12. Variable overhead (increases with increased product volumes) is 56% of fully burdened labor and includes all indirect labor
13. Touchup labor cost is $5.00/solder joint
14. In-Circuit Test (ICT) yield loss labor costs are $25.00/ board to troubleshoot, rework and retest
15. Functional test yield loss labor costs to troubleshoot, rework and retest are $50.00/board

Indirect Labor and Overhead Absorption

Some of the other non-direct labor that is typically included in the indirect labor costs and need to be absorbed into the fully burdened labor sell price include:

- Personnel to load bills of material into ERP/MRP
- A procurement department to get quotes and order material
- Industrial engineers who quote labor
- A master scheduler and planners who plan and release work orders to production
- Material handlers (in-shipping, material inspectors, pack and ship)
- Inventory and stock room personnel
- Production planners who release work orders

- Process engineers who develop assembly process and write methods sheets
- Kitting people who pull and kit material for released work orders
- People who deliver the kits to the appropriate equipment and work stations
- People who set up the stencil printers
- Set-up people who load material on component placement equipment
- In-process inspectors
- Technicians who troubleshoot the automated equipment process when it is producing defects
- People who perform maintenance on the production equipment
- Supervisors and managers for procurement, production, process engineering, test engineering, and quality assurance
- Human resources
- Factory safety officer
- Office and manufacturing cleaning personnel
- IT people to maintain and upgrade computer equipment

For each of these indirect labor employees described above, besides salaries and hourly wages, the following costs and benefits for each employee must be absorbed in the fully loaded labor selling rate:

- Medical insurance
- Unemployment compensation tax
- Worker compensation insurance
- Social Security tax
- Medicare taxes
- Holiday pay
- Vacation pay and O.T. premium
- Sick pay
- Pension or retirement plan contributions
- Training costs

Fixed overhead includes:

- Building costs
- Utilities: Power, natural gas, water, and sewer for the operation
- Computer and communication systems for the facility
- Spare parts for the operations and facilities
- Depreciation on the assembly equipment and facilities
- Insurance and property taxes on the assembly equipment and facilities
- Safety and environmental costs

Applying these high labor rate assumptions to a business model for a hypothetical Tier 2 circuit board assembly operation (in USD):

- Sales/year = $1B
- The circuit boards have a 75% to 25% raw material-to-fully burdened labor ratio cost mix
- Of the 25% fully burdened labor cost, 50% is machine-based labor, 50% is hand-based labor

- Raw material cost/year = $750M

- Total burdened labor cost/year = $250M
- Total unburdened (raw) labor cost/year = $100M
- Total absorbed overhead cost/year = $150M

- Average board price **$100**
- Number of boards/year 10M

- Material $/board = $75
- Labor $/board = $10
- Overhead $/board = $15
- Fully burdened labor rate = $37.93/hr
- Raw direct labor (compensation) rate = $15.17/hr
-
- Total absorbed overhead (includes material related labor and attrition costs, SG&A and profit) = $22.76/hr
- Standard labor hr/board = $10 per board/$15.17 hr = 0.6592 labor hr/board

- Fully Burdened Labor Price/board = $37.93/hr x 0.6592 hr = $25/board

The Existing Low Labor Rate Model

2012 Employee Compensation Rates for Manufacturing in China: Average in USD: $1.67/hr [18]

Compensation is defined as whatever is paid to or for the workers in money or in kind according to relevant regulations, including:
- Wages
- Bonuses
- Free Medical Services
- Medicine
- Transport subsidies
- Social insurance
- Housing fund

With the weakening of the dollar, the willingness of the Chinese government to float their currency, and the continued upward labor cost pressure in urban centers, we will use an employee compensation rate of $4.00/hr. USD.

Assumptions: (in USD)
1. Employee Compensation (Earnings + Benefits) for China manufacturing: $1.67. As stated above, this analysis will use an employee compensation rate of $4.00.
2. Overhead rate of 300% will be used for full labor burdening and "reality" factor considerations.
3. Since it is difficult to know what part of the $4.00 compensation rate is raw labor and what part benefits, the entire compensation rate will be burdened with the 300% overhead rate
4. Overhead rate includes SG&A (sales, general and administrative costs – generally, a percentage of labor and material)
5. Overhead rate includes material handling, inspection and attrition (usually loaded as a percentage of the raw material cost)
6. Profit or fee is included in overhead rate to result in a fully burdened labor sell price

7. Assumptions 1 through 5 produce a fully burdened labor sell rate of US $12.00 / hr.
8. Average touchup labor costs $0.10 per solder joint

These assumptions are now applied to the same general business model used for the high labor rate circuit board assembly operation:
- The standard labor hour usage per board is 2 times greater than in the high labor rate cost model because of the availability of inexpensive labor versus the cost needed to develop, control and maintain an automated process capability = 2 x 0.6592 = 1.3184 hr/board

- Unburdened labor rate = $4.00/hr
- Fully burdened labor rate = $12.00/hr
- Unburdened labor cost/bd = 1.3184 hr x $4.00/hr = $5.27/board
- Fully burdened labor price/board = 1.3184 hr x 12.00/hr = $15.82/board

- Overhead absorbed/board = $10.55/board

- Average board price = $75 + $15.82 = **$90.82**/board
- Number of boards assembled/year = 11.0M
- Sales / year = $1B
- Total raw material/year = $826M
- Total burdened labor cost/yr = $174M

OBSERVATIONS ON THE EXISTING MODELS

The result of this modeling is that a $100 circuit board assembled and tested in a high labor rate environment sells for $90.82 if it is made in a low labor rate environment. To demonstrate the viability of a high labor rate region successfully competing with a low labor rate region it is necessary to show that the resulting $9.18/board gap in price can be overcome.

Examining the cost elements of the two current models, the following observations are made:
1. The labor selling rate gap between the two models permits the low labor rate companies to *throw* a lot more labor at the assembly. In this low labor rate model, higher skilled (and cost) labor that can develop capable and controllable assembly processes may or may not be available. Whether they are or not, an alternate strategy is to merely address in-process quality issues with labor-intensive, post-automation rework touch labor.
2. When faced with low yields in a high labor rate environment, the high cost of troubleshooting, reworking and retesting assembly defects (and scrapping material) can be a significant factor in the inability to compete. The low labor rate competition may simply mask this root cause of the failure to compete.
3. Of the elements that contribute to the labor cost in the high labor cost model, the largest controllable elements are direct raw labor cost and indirect labor absorption cost.

4. Soft considerations such as the logistics challenges of assembling products in remote locations, the cost of doing business (increased travel costs and time), measuring and analyzing performance in real time and the cost of changes to products are difficult to quantify, but are real.

AN ALTERNATE HIGH LABOR RATE COST MODEL

Can we develop an alternate high labor rate model that closes the cost gap between the current high and low labor cost models (about a 37% labor cost reduction using the low labor rate model for our hypothetical board)? The industry today answers emphatically: No! But, before accepting this conclusion an attempt to drill down into the two primary cost differentiators is in order: the difference in the raw labor and the difference in overhead rates.

Those who have been working in a high labor cost electronic product assembly environment have been working at cross-purposes. On one hand, they have embraced an assembly technology that continues to get more and more complex. This complexity is a function of three factors that have emerged primarily as a result of the evolution in electronic component design and packaging (Figure 2). These are:

1. Advances in robotics and other forms of machine automation.
2. An increased complexity in the assembly process.
3. The requirement to understand the physics involved in the process. When we were exclusively hand soldering, the terms *thixotropic*, *rheology* and even *hydroscopic* were terms rarely heard around the workbench.

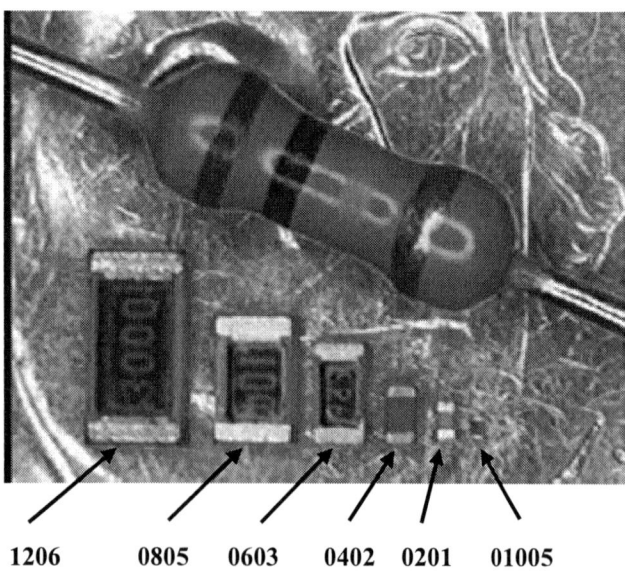

| 1206 | 0805 | 0603 | 0402 | 0201 | 01005 |

(English Designations)

Figure 2. Passive SMT Component Family & Their Patriarch: The Axial-leaded Resistor (All on a Jefferson Nickel)

With these requirements comes the need to develop a statistically capable process for the automation. A process window must be developed that is wide enough to contain the natural variation that will occur over time in the process variables. To do this successfully for automation, the physics that underlies the process and the equipment that conducts the process must be well understood. Hand inserting and hand soldering the two leads of an axial leaded resistor to a circuit board might take 30 seconds of labor. Having a high speed robot place an SMT resistor in solder paste that was printed by machine, and then melt the solder in a reflow oven, might take 0.1 seconds when processed with the rest of the components on the circuit board – and the raw labor cost for this operation, if the board is handled by machine and travels on automated conveyors, is $0.00! This is only true if the resistor was soldered correctly. If not, and the solder joints need to touched up or the wrong value resistor was soldered to the board, the labor rework cost, especially in a high labor rate location, will add up rapidly. These defects have a much lower cost impact if the rework is done in a low labor rate operation. Therefore, in high labor rate regions process capability and control is paramount. A C_{pk} of 1.33 minimum is required (4 sigma, 63 ppm defect rate).

Ironically, this added complexity has been addressed by a relentless management quest to find less expensive, low skilled labor to deal with the low labor rate competition! Some of that inexpensive labor is needed to accommodate a circuit board design that can't be fully automated. But, unfortunately, even for the part of the design that can be automated, the advanced skill sets required to create capable and controllable assembly processes are either not available, or management is unwilling to pay for them. The result in many cases is increased cost for touch-up, rework and material scrap. "Keep looking for that low cost direct labor. We have no choice! Build! Build! Build! Rework! Rework! Rework! Ship! Ship! Ship! Ah, we met our monthly sales goals, maybe by wrapping a $5 bill around each board that was shipped!" In this case, the goal should be to ship LESS next month since the more boards shipped, the more money is lost. A company in this mode of operation becomes trapped in a death spiral. They will either go under, be sold or go offshore. Do you think the consolidation that we have seen over the last 10 years, companies gobbling up other companies – is a result of good fiscal performance? In most cases, it's a quick way to affect the bottom line. Add someone else's puny net profit to our puny net profit and survive another year.

Increase profit by reducing cost? How about having the direct labor pay the company to work here? Both have about the same chance with the existing management team hard at work steering this ship. The way they see it: "It's that cheap labor we have to compete against. Let's move offshore or get sold." What if we take a deep breath, take a step back and consider a new labor strategy.

The new strategy has three basic elements:

1. *Transformation of the direct labor workforce.*

U.S. industry has a history of assembling products in geographic areas whose labor costs can successfully meet competitive pressures. It is interesting to note that in most cases this happens only when a particular industry is threatened by the competition's lower prices. For example, shoemaking and textile industries thrived in New England in the 19th and early 20th centuries. Prices rose. Organized labor put further pressure on cost. Manufacturing continued. Southern U.S. companies and foreign factories began to produce products for lower costs because of the availability of cheap hourly labor. Automation reduced labor content, volume production increased and costs decreased, but the industries were still basically labor intensive. It was then that textile and shoe manufacturing left New England and moved to the South – then, they moved to the Deep South, then the Caribbean, then Mexico, then South America, then the Pacific Rim, China and Vietnam.

Cars and electronics followed. Again, improvements in automation slowed the transition, but still the drumbeat continued, *reacting* to the low labor rates that the competition acquired access to – not *anticipating* them. "It's time to move manufacturing again." This process was, and continues to be, repeated over and over even though the capability and quality of the automation in some industries provides the opportunity to reduce the theoretical labor content to a very small percentage of the total product cost. When this happens, the labor rate plays a relatively small competitive role. This is certainly true of most electronic circuit boards – but we don't exploit the automation fully. Why? The answer is: it's harder than looking for cheap sources of labor, we don't have the time, we don't have skills and we certainly don't have the vision and courage. No one ever says this, of course.

Notice also that the companies that are most successful in high labor cost regions seem to be small operations with flat organizations (*lean*, we like to say these days). This creates the perception that high labor cost operations are good at competing on low volume and prototype work, but the industry experts maintain, "Sorry, with those labor rates we need to go offshore for the high volume stuff."

The Low Volume / High Volume Paradox

In *Paper or Plastic? Choosing to Move Offshore* [19] a challenge is made to the logic of those who have relegated all future high volume manufacturing to low labor rate geographic regions. "…Finally, think about this – the 'experts' say, 'Future volume manufacturing will all be done 'over there' - we just can't compete in high volume manufacturing.' Oh, really? I thought high volume manufacturing requires LESS labor per assembly, not more, since NRE, fixture cost, set-up time etc. is spread over a large number. Since we pay more for domestic labor (for example, in a high labor rate region like the U.S.), we should be able to compete more effectively when building products domestically with less labor dollars per assembly. If we automate and just let the line run, doesn't the offshore *low cost* labor advantage asymptotically go to zero?"

The unspoken little secret is that for high labor rate cost environments, the higher the volume, the greater the impact poor yields have on their ability to compete. The answer to the paradox is that the higher the volume, the more defects there are that need touch up and rework – and it costs high labor rate operations a lot more for rework because of its labor intensity.

Figure 3 illustrates the two basic paths to reduce raw (unloaded) labor costs:

1. Reduce the hourly labor rate applied to the unit labor hours. ($/hr.)
2. Reduce the unit labor hour content. (hr.)

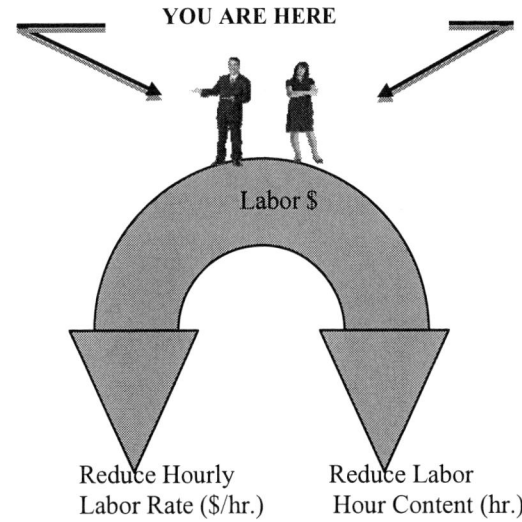

Figure 3. Two Paths to Reduce Labor Cost

Traditionally, we have tried to compete (reduce labor cost), primarily by taking the left path.

Do We Sell In-Circuit Test or Products?

In-circuit test (ICT) adds no value to the customer. The customer wants a product or circuit board that does what the product or board performance spec. says it should do. This is usually determined by a functional test. Why, then, do we do ICT? Unfortunately, it is usually used as a way to separate the good boards from the bad boards we build. In other words, we use ICT as a coping strategy to deal with an assembly process that is not capable or in control, or both.

If ICT yield rates above 99% can be achieved, the cost to do the test does not pay back; i.e., finding one defective board for every 100 that are built. Without the need to do ICT, the need to do post ICT troubleshooting, rework and retest is eliminated.

2. *A corporate model that focuses on the assembly of the customers' products and not on technical disciplines or departments within the organization.*

We have evolved into an industry of indirect labor specialists. We have a corporate structure that puts each of us

43

into our own silo. Our particular silo (department) tells us what specific role we will play in the company's operation. This is consistent with the division of labor and assembly line product flow of the early electronic assembly factory floor. Operator Number One inserted components R6, C12, U4, U16, and Operator Number Two inserts R1, R5, C6, U2 and so on, as downstream operators continued the process until the board was complete. In a similar way, a marketing person generates a product specification. From the specification, the electrical engineer designs the circuit, creating a schematic, and passes it on to the CAD person who lays out the board. The CAD person passes the design package to an industrial engineer who methodizes the design for production. The bill of material goes to someone in the procurement department to order the bare board and components. The design package goes to the electrical test department to have in-circuit and functional test developed, etc. Each department is like an island or a community with its own identity – success being measured by how well they do their specific jobs. The customers do not buy specific jobs – they buy products.

This organizational fiefdom promotes department focus and competition, many times at the expense of the customers' products. This structure is very expensive with much of the indirect labor adding questionable value.

Instead of trying to cope with these issues as we have a history of doing, the new model dismantles the traditional hierarchical structure. Just two groups replace all departments. The new multi-skilled, engineering-based direct labor is organized into self-managed customer product teams. A small leadership group serves as an enabling function, providing the product teams with the skill sets and tools they need for success. This permits a dramatic reduction (20%) in overhead cost because of a combination of the reduction of indirect labor, and the aforementioned significant increase in yield.

3. *Creating an educational environment that serves the needs of the new model.*

This type of fundamental change described above does not come easily. It is a daunting task to reduce labor cost sufficiently to compete with labor rates in the order of a few dollars per hour by taking the right path in Figure 3 (reducing labor hour content). A prerequisite is having a labor force that meets the demands of the new model. The current academic community is incapable of providing this workforce. Educating in one community (academia) and sending the educated to work in another community (the real world) has created an ever-increasing gap between academic preparation and industry need. High tech electronic product assembly simply changes too quickly to have its needs provided in an environment where it can take 2 – 3 years to get a curriculum changed. We need to create a *teaching hospital* of sorts for the high tech electronic assembly industry. A learning community should be established that wraps a school around a for-profit contract manufacturing facility, where students can be taught in a real-world

environment for the full tenure of their post-secondary education [20].

The High Labor Rate Cost Model Revisited

Assumptions (in USD):

1. Median Raw labor wage (Project Engineer with BS in Engineering: 10-19 years experience): $39.05/hr [21]
2. Benefits are 32% of raw labor: $12.50/hr
3. Employee Compensation (Earnings + Benefits) for U.S. Project Engineer: $51.55/hr
4. Overhead rate for full labor burdening is 200% of raw labor
5. All indirect labor is included in overhead rate
6. Overhead rate includes SG&A (sales, general and administrative costs – generally, a percentage of labor and material)
7. Overhead rate includes material handling, inspection and attrition (usually loaded as a percentage of the raw material cost)
8. Profit or fee is included in overhead rate to result in a fully burdened labor selling price
9. Assumptions 3 through 8 produce a labor selling rate of $78.10/hr
10. Fixed overhead (facility/equipment) is 4% of fully burdened labor
11. Variable (controllable) overhead is 26.1% of fully burdened labor and includes all indirect labor
12. Touchup labor costs $5.00 per solder joint
13. In-Circuit Test (ICT) yield loss labor costs $25.00 per board to troubleshoot, rework and retest
14. Functional Test yield loss labor costs $75.00 per board to troubleshoot, rework and retest

Applying these assumptions to the same Tier 2 high labor cost circuit board assembly operation business model (in USD):

- Sales/year = $1B

- The circuit boards have a 75% reduction in direct labor hours from the original high labor model because of:
  1. Full exploitation of automation (boards designed for automation)
  2. The near elimination of touchup and rework: Assembly yields of 99.5% - only 1 board in 200 requires touchup or rework – statistically capable processes are developed and kept in control by proactively monitoring process parameters in real time
  3. high functional test yields eliminating the value of In-Circuit Test

- Overhead rate is reduced 20% by organizational restructuring
- Fully burdened labor rate = $78.10/hr
- Raw direct labor rate = $39.05/hr

- Total Absorbed Overhead (includes material related labor and attrition costs, SG&A and profit) = $39.05/hr
- Average Labor hr/board = 0.6592 x 0.25 = 0.1648 labor hr/board

- Material $/board = $75
- Unburdened Labor $/board = $6.44
- Overhead $/board = $6.44

- Average board price = $75 + $6.44 + $6.44 = **$87.88**

The revised high labor rate model replaces the large, low rate labor force with a small, high rate group of mostly engineers who are multi-skilled and self-managed. This results in an increase in average labor compensation rate (from $20.02/hr to $51.55/hr), but reduces overall labor cost by reducing labor content, including the elimination of in-circuit test as part of the assembly process. The cost components for the three labor rate models are summarized in Table 7.

Table 7. Labor Rate Model Comparison (USD)

Model	Material Cost per Board	Direct Labor Cost/Bd	Absorbed Overhead Cost/Bd	Circuit Board Price
High Labor	75.00	10.00	15.00	100.00
Low Labor	75.00	5.27	10.55	90.82
Revised High Labor	75.00	6.44	6.44	87.88

L – SUMMARY OF RESULTS

This exercise has demonstrated that even a large disparity in labor compensation rates between high and low labor rate regions can be neutralized. This is possible because of the ability to reduce labor content by exploiting the available automation and achieving a significant reduction in yield loss with the corresponding labor costs reductions involved in troubleshooting, rework and retest. The yield improvements are a result of a small, multi–skilled workforce that has been educated in a real world production environment. The product team that is comprised of this world-class workforce understands process capability and has the skill sets to develop robust assembly processes that contain process variation well within the upper and lower process spec limits. They have a solid foundation in the physics of soldering, are fluent in material science and understand circuit theory. This permits rapid root cause failure analysis. They understand statistics and the theory of variation, as well as team dynamics. Their collective output is greater than the sum of their parts. An operational infrastructure is in place that among other things permits real time measurement and proactive process control. This labor content reduction is accompanied by a significant reduction in overhead costs. It is accomplished by breaking free of the traditional manufacturing corporate model. Quantitatively, in many applications these improvements create the ability to reduce the labor content by 75% and reduce the non-direct labor overhead burden by 20%. However, in some particular applications it may not be possible to reduce the labor content by 75%. For example, a circuit board may have many large, analog devices that require hand insertion, or the design requires discrete wiring (e.g., coax) be hand soldered, or the higher level, box build assembly requires significant hand labor. In these cases, reduced potential for reducing labor content in the product must be offset by further reductions in overhead. The 37% reduction in labor cost using the low labor rate model can be neutralized in a high labor rate environment for many product applications. If labor were the only cost factor involved in the *high* versus *low* labor rate competition the analysis would be complete. Unfortunately, it is not.

COST BUCKET 2: M (MATERIAL)
INTRODUCTION

A revised high labor rate model has been established to demonstrate that, for many electronic product assembly applications, low labor rate competition can be neutralized through automation, organizational rethinking and changing how we educate our workforce. This new, high labor rate model is competitive because, instead of fixating on the lowest labor hour *rate*, it significantly reduces labor cost through focusing on minimizing labor hour *content* by exploiting the available automation. In addition, factory *indirect costs* and *overhead* are minimized by dismantling the traditional hierarchical organizational structure and replacing it with one that best serves building products in an automated environment.

Attention is now given to analyze the second assembly cost consideration in competitive electronic product manufacturing – material cost. The conclusion is that any significant price differential between the cost of *purchasing* the same electronic components in high and low labor rate product assembly global regions is artificial. In other words, any relative pricing differences are not justified by the actual cost to manufacture and deliver the components to the assembly factories in different labor rate areas. In a similar sense to the unjustified difference in the price of purchasing the same piece of automation equipment in high and low labor rate markets (an historic anomaly that has never been adequately justified in terms of cost), this paper concludes that relative electronic component pricing is determined primarily by *what the local market will bear (read: is willing to pay)* at best, and by political reasons at worst. This has contributed to the exodus of electronic product manufacturing to low labor rate regions as much as, or more than, the well-publicized labor rate disparity. Inflated profit margins under the guise of higher prices caused by higher costs such as shipping and overhead have, in some cases, resulted in material manufacturers and distributors charging a 20-50% premium in high labor rate regions. These material cost differences can exacerbate labor cost differences, contributing to the allure of low labor rate electronic product manufacturing. Government protectionist policies, currency exchange rates and other non-labor or material-related cost considerations will be addressed in detail in the final section of this paper on NOP: National Out-bordering Predisposition. Finally, it is recommended that material procurers in high labor rate environments insist on equitable treatment – i.e., offered material pricing with the same cost markups, or be given a valid reason why they must pay a premium.

The Product Assembler's Material Management Cost as a Function of Product Assembly Location (Low Labor Rate vs. High Labor Rate Regions)

The product assembler can control some of the material management cost variables. Others cannot be controlled. In addition, the material management cost of some of the variables is dependent on the labor rate in the assembly location.

1. Cost of Carrying the Inventory – This is largely a controllable cost that should not dependent on product assembly location.
2. Material purchasing – This is a controllable cost that has traditionally been tied to labor rates.
3. Material planning – This is a controllable cost that has traditionally been tied to labor rates.
4. Incoming inspection – This is a controllable cost that has traditionally been tied to labor rates.
5. Attrition rate – This is largely a controllable variable that is not dependent on product assembly location.
6. Scrap rate – This is largely a controllable variable that is not dependent on product assembly location.
7. Shipping – This is a controllable cost to the extent that good MRP can minimize the need for expedited shipping charges. There can be a cost premium for high labor rate regions since most material is manufactured in geographically distant locations. Either the product assembler's cost is F.O.B. the ODM (Original Device Manufacturer) location, or this cost is embedded in the distributor's price. Higher material delivery cost is often given as the reason, or one of the reasons, for the higher material cost in high labor rate locations.
8. Distributor loaded cost – This is not controllable by the product assembler, but does affect the material price the assembler pays. The potential dependence of this variable on assembly location will be discussed later in this paper.
9. Distributor Markup – Historically, this has not been controllable by the product assembler, but does affect the material price the assembler pays. The potential dependence of this variable on assembly location will be discussed later in this paper.
10. Purchase price variance (PPV) – This is controllable to a limited extent by the product assembler, and does affect the material price the assembler pays. It is not dependent on product assembly location.
11. Currency exchange rate – This is an uncontrollable variable. However, material that is manufactured in China is priced in Yuan. Because of the unvalued nature of this currency, buying this material in the United States with US dollars is cheaper than buying it in China with the Yuan. This result is a net advantage for the product assembler in a high labor rate region like the U.S. buying material from China.

12. Trade policy / Corporate tax rates – The duties and tariff policy between importing and exporting countries and the taxes imposed on earnings. These costs are uncontrollable because they are government imposed.

Minimizing the controllable material management costs listed above is a critical success factor in assuring the competitiveness and profitability of a product assembly operation. Some of the elements required to establish a good material management strategy are:

a. The ability of the suppliers to provide the material in a just-in-time fashion. Maximizing inventory turns.
b. For EMS providers, relatively stable customer delivery schedules. For OPDs, a relatively accurate market forecast.
c. Minimal indirect labor (Manufacturing management cost variables 2, 3 and 4).
d. Statistically capable and in-control assembly processes – low yield loss.
e. A proactive process control strategy that identifies, in real time, processes that begin to vary in a non-random way. This will help avoid the all too common policy of including rework as part of the labor standard estimate.

Clearly, properly addressing these variables is more critical in high labor rate environments. However, except for variables 2, 3 and 4, the influence of these 12 variables on the total manufacturing management cost has little to do with the location of the assembly. And, organizational rethinking can mitigate the labor rate effect on variables 2, 3 and 4 [22].

During the product assembly quoting process, some of the controllable material management costs, such as variables 1 through 6, can be added as a small percentage of the BOM cost. The degree to which this can be done is largely a function of what the customer will tolerate. Those material management costs that are not embedded in the material cost, are added to the non-product-specific labor and facility overhead cost and become part of the loaded labor rate.

This brings us to the final element of material cost, the cost of the material itself, raw "M."

PROCURING THE MATERIAL

Companies in Tiers 1, 2 or 3 may have manufacturing operations in both high and low labor rate regions. This can provide material purchasing leverage (See paragraph below: *The Big Guy Versus The Little Guy*). Regardless of the product assembly locality there are four principal categories of sources for purchasing electronic material:

1. Direct from the Original Device Manufacturer (ODM) – Purchasing directly from the component manufacturer should result in the lowest material pricing. Usually an electronic product assembler must offer the component manufacturer an annual business opportunity in the $5-10 million USD range. This type of purchasing power is generally only possible for Tier 1 and 2 assembly companies.

2. Component distributors – Distributors provide the most common source of electronic components for lower volume product assemblers who don't have the purchasing power to buy directly from the ODM. There are two basic types of distributors:
   a. Franchised – These companies are contracted by ODMs to distribute their components.
   b. Independent (Brokers) – These companies have no formal contracts with ODMs. They buy and sell components on the open market.

   As part of a supply chain, distributors can add significant value by providing the logistical advantage of *one-stop shopping* to fulfill a product assembler's BOM, as well as the opportunity to approach or achieve just-in-time component deliveries. This service can reduce the cost of inventory and purchasing for a product assembler, but increases the purchased material cost since the distributor must load his price from the ODM to absorb these additional costs.

3. Component catalog distributors – This source is used primarily for prototype and low volume assembly applications, as well as short lead-time situations.

4. Third party after-market sellers – These are companies that buy up excess inventory, discontinued stock and other after-market material. The *reward* for the product assembler is that the material is generally offered at a discounted price, or it may be the only source of discontinued components. The *risk* is in the material's *gray market* nature (counterfeit potential) and the material's history as it affects the component's quality (e.g., solderability).

Material Cost for the Electronic Product Assembler: A Case Study

In 2006, an OPD in the United States who had been manufacturing all of their own products decided to test the offshore manufacturing waters with a new product they had just designed. They first took their BOM for the new product and had the material quoted stateside: The circuit board components from a local distributor, the circuit board and mechanical parts from sources they had often used in the past. The initial quantity quoted was enough material to build 3000 units. The total costed BOM came to $7.44 per unit. They sent the same BOM and a set of product assembly drawings to a contract manufacturer in China for pricing. The quote received was $7.30 – for the completely assembled and tested product! There was nothing the OPM had to do but ship the completed product to their customers. In addition, the associated material management costs that would have been incurred by the OPM if they built the product themselves were eliminated.

No matter what the material/labor cost split was for the product, no reduction in labor rate would reconcile the total assembled price difference. A significant portion of the assembly price disparity between the two quotes *had* to be in

an inflated cost of the stateside quoted BOM. This fact suggests that an investigation in material pricing as a function of assembly location is in order.

Establishing the Cost Models:
Sources of Material Manufacturing Cost Data

The most significant challenge in trying to reconcile any global electronic material cost disparity between low and high labor rate assembly regions is unearthing the material manufacturer and distributor cost data. It is a difference in the cost of doing business that is often used as the explanation for the difference in price for the same material.

Unlike relative labor cost data that are widely available through government and other labor tracking sources, cost structures for material manufacturing companies and their distributors are very difficult to obtain. The competitive reasons for this are obvious. Many sources were contacted to acquire cost data for this paper. Those that contributed did so only with the understanding that everything they said was off the record. And the *data* provided were anecdotal, since there were no sources that could be referenced. Therefore, a different strategy was needed to establish a detailed relative cost data set that could be analyzed. The financial disclosure data that public companies are required to file provide an oblique way to back into a model. It is these data combined with the off the record information that supplied the pieces to this puzzle.

Material Manufacturing Versus Product Assembly:
Cost and Location

The same market forces that affect the product assembly industry drive the material manufacturer. However, there is a significant distinction. Electronic product assembly can fall back on manual labor for their standard and rework processes much more frequently than electronic component manufacturing. For example:

- SOT 23 transistors can be hand placed on a circuit board, but you can't *hand place* the thin film transistors on a silicon wafer.
- It is much easier to manually touchup a SOIC solder joint that connects it to a circuit board, than to manually rework the solder-bumped or wire bonded connection between a component's silicon die and lead frame.

This means that the degree of automation at the component level must be at least as high as that at the circuit board level. Typically it is much higher. Direct labor *content* on a manufacturing process basis is always smaller when manufacturing electronic components than when assembling a circuit board, regardless of location. This tends to lessen the effect of a location's labor rate. It also means that because the workforce skill level must be higher to manage the required automation, it has taken longer for component manufacturing to gravitate to low labor rate areas. However, it also means that once the skills are available and the shift in location has taken place, expenses such as facility costs weigh more heavily and favor the lower cost environments.

When an assembled circuit board fails in-circuit test it usually will be troubleshot, reworked and retested. Why? Because the value of the material (components and circuit board) and value-added labor applied up to this point in the assembly offset the rework costs. Or worse, the quantity of product that must be shipped requires this board. If neither of these is true, the board should be scrapped to cut losses.

Whether the inability to develop a capable process that can be kept in control is caused by lack of skill, lack of proactive process control, or an inherently poor board design, the same production mentality often occurs: *We need to ship 1000. The kit size needs to be 1300.*

When a silicon wafer is probe tested, the dice that fail are marked and discarded. No rework. Of course, the probe test results are statistically treated in the spirit of continuous wafer fab process improvement.

In other words, a significant level of statistical yield loss in wafer fab is accepted. Assuming the circuit board design is robust from an assembly point of view, a small statistical yield loss due to random defects is acceptable (<0.5%). However, larger ICT yield losses caused by manufacturing process defects are not. It's a matter of what is controllable in the respective wafer fab and circuit board assembly environments.

The shift of electronic component manufacturing to low labor rate areas, mirroring the shift in product assembly, seems to make sense logistically – to provide the material close to the point of its assembly. But, the decision to move solely because of the labor rate differential is belied by the same reasoning that challenges the decision to move the assembly. [23]

THE MATERIAL MANUFACTURING COST MODEL

If the material pricing to product assemblers is dependent on the assembler's location, either:

1. the ODM is charging their distributors more in high labor rate regions, or
2. the distributor is marking up the material more in high labor rate regions, or
3. some combination of 1 and 2.

The Material Manufacturing Cost Model:
The ODM

If the product assembler has enough purchasing power to buy directly from the material manufacturer, the pricing received should be independent of the assembler's location, regardless of the ODM's cost model. The only cost variables related to product assembly location that could have a cost impact on raw "M" are:

1. Shipping
2. Currency exchange rate
3. Trade policy.

What follows is a brief analysis of each ODM cost variable:

1. Shipping – Any adverse shipping cost differential between the material manufacturer and the location of the high volume product assembler is negligible. For example, just divide the difference in shipping cost by the number of SOIC-8 components (2500 on a 330mm diameter reel), or the number of 0402 components (10,000 on a 178mm diameter reel) and the increase is normally less than a few percent of the component cost. Whatever the difference in shipping, this increased cost will be reduced or offset by the difference in cost to ship the finished product to the assembler's primary consumer markets – in many cases, high labor rate markets.

2. Currency Exchange Rate – If the components are being manufactured in China (or, any country with an undervalued currency), a product assembler buying those components with a currency valued correctly will, effectively, be buying at a discount when compared to an assembler operating in the same country where the components are being manufactured.

3. Trade Policy – Wide variation exists in international trade policies. They are dependent on the countries or the trading blocks that are conducting the trade. For example, it is very difficult for any other country to sell electronic products into the Brazilian market. This is because of the tariff policy applied to importing finished products. Product assembly companies (both OPMs and EMS providers) who have wanted to sell into this market have had to establish a product assembly capability in Brazil. Items 2 and 3 are part of the manufacturing business environment a country establishes. It is discussed in the final section of this paper entitled, *Cost Bucket 3: NOP (National Out-Bordering Predisposition)*.

Therefore, the product assembler in a high labor rate market should pay no significant premium for material if buying directly from the ODM.

The Material Manufacturing Cost Model:
The Distributor

The only justified reason for a material cost difference based on a product assembler's location would be because a distributor's indirect or overhead costs are more in high labor rate regions.

Using a composite of Security and Exchange Commission (SEC) financial filing data for a number of component distributors with multiple global locations, the following generalized cost model can be developed:

Distributor Location Average (USD/year)
Sales = $50 million
Gross Profit = $4.65 million
Net Profit = $0.50 million
Indirect and Overhead Costs = $4.15 million

The average facility indirect and overhead cost as a percent of sales = $4.15M/$50M = 8.3%

Assume the following difference in indirect and overhead cost as a percent of sales/year:

High labor rate region = 12% ($6 million)
Low labor rate region = 4% ($2 million)
i.e., $4 million (200%) more per year to do business in high labor rate regions.

Using this model, a BOM that is priced by distributors at $100 in a high labor rate location would be priced at $92 by distributors in a low labor rate region.

The Material Manufacturing Cost Model: Summary

The analysis suggests if buying directly from the ODM the product assembler in a high labor rate market should pay no significant premium for material.

Conservatively speaking, the analysis indicates about an 8% increase to buy components from a distributor in a high labor rate area is justified because of a higher indirect and overhead cost base. While not an insignificant amount, 8% is certainly not the 20 to 50% or even higher premiums reported by certain product assemblers since the exodus of both material and product manufacturing from the U.S. began.

The Big Guy Versus the Little Guy

There is one other consideration when discussing the material cost disparity – the size and the geographic mix of the product assembler's locations. To provide flexibility and to maximize competitiveness, many large product assemblers have established a global presence – assembly operations in both high and low labor rate regions. In addition, many of these large OPMs and EMS providers have established a central or corporate group to buy material for all their assembly facilities. This, along with the volume purchasing power they offer, provide significant pricing leverage with material distributors. It often gives them the ability to receive low labor rate distributor pricing for their high labor rate assembly locations. The smaller, lower volume assemblers in only high labor rate regions are at a disadvantage – they are bound to the distributor pricing in that region.

Material Other Than Electronic Components

The material discussed to this point has been associated with the electronic functioning of the product. But the mechanical parts used for the *box build* or *higher level assembly* must be considered as another principal category of material for electronic products. These parts are unique to the specific product. They consist of custom machined parts, castings, molded parts and standard hardware. For these parts, the non-recurring labor the supplier needs to develop the tooling – an injection mold, for example – is strongly tied to labor rate. It is significant, giving the low labor rate source an advantage. However, if the production quantities of the finished product are high, this cost advantage is reduced as the savings are amortized over a large number – similar to the shipping cost discussion.

M – SUMMARY OF RESULTS

Market economies depend on the freedom to contract, a free pricing system, and the principle of supply and demand, for their vitality. This means that companies have the right to do business and set prices without the interference of government. Without interference, as Adam Smith said, the *invisible hand* of the free market would naturally establish the price of goods and services. [24] So, it should be clear that material manufacturers and their distributors have the right to charge whatever prices they want. The check and balance against this pricing is competition and the customer.
The rapid change in manufacturing locations brought about by access to the global marketplace has caused pricing *tradition*, *expectation* and *willingness to pay* in the high labor rate regions to be additional price setting factors for material manufacturers and distributors.

Over time, the effect of this perturbation has dampened out and the market's invisible hand has begun once again to be the predominant force in establishing material pricing.

Is there a non-cost-justified, inflated material price paid by product assemblers in high labor rate manufacturing regions? Anecdotal information and case studies indicate there is. However, it also appears the disparity in material pricing between these high and low labor rate locations has declined significantly over the last ten years as the pricing policy continues to migrate toward free market principles.

Where does all of this leave the product assembler in high labor rate environments? Attention must be paid! Material pricing challenges must be made. The following questions must be asked:

1. Do low labor rate assemblers receive material pricing favoritism that is *not* related exclusively to component manufacturers and distributors lower indirect and overhead cost bases?
2. Why do material manufacturers charge *more* for the same components when the product assembler or franchised distributor is in a high labor rate environment whether selling to a distributor or directly to the product assembler?
3. Do the distributors in high labor rate regions mark up the material more than their 8% higher operating costs?
4. Is 8% an accurate number? If not, what is the increased cost of doing business in high labor rate environments, and how should this cost be reflected in material pricing?

The elements of the two controllable cost buckets, labor and material, as they affect assembling products in high and low labor rate regions have now been analyzed. The final section of this paper addresses how the degree of friendliness (or, animosity) a region presents to business in general, and manufacturing specifically, affects high tech product assembly cost.

COST BUCKET 3: NOP
(NATIONAL OUT-BOARDERING PREDISPOSITION)
There are several national conditions that must exist to create a climate for an electronic product manufacturing business to compete on a national basis successfully:
1. Good access to capital: Bank policy
2. A reliable supply chain
3. A trained workforce
4. A reliable source of utilities
5. A competitive regulatory environment
6. Competitive corporate tax laws

The degree to which a country or state excels at creating a positive environment that includes these attributes is a good predictor of whether manufacturing will thrive or wither.

These conditions are grouped under the banner NOP (National Out-Bordering Predisposition): the tendency of the business climate created by a country to cause a company to either manufacture or seek manufacturing outside of its borders. It represents the least controllable of the three assembly cost buckets this papers addresses.

An OPD or EMS that builds high tech electronic products normally does not consciously seek out the country that provides the least government and financial burden. However, since national policies in areas such as regulation, taxation and access to capital, affect the cost of doing business wherever the production takes place, the NOP costs are embedded in the total overhead cost and, hence the product cost or service.

Small companies usually confine manufacturing and sales to the country in which they reside. However, if they are located within countries like the United States, Switzerland or China the potential exists for an intra-border variety economic climate. In a country like the United States, this variety manifests itself in another level of economic competition that occurs between the states. America's founders called it *federalism*. In Switzerland, the states are called cantons and are relatively autonomous.

Centrally controlled economies such as China have much less opportunity for business competition within their boundaries. Hong Kong is an exception. The central government in Beijing has largely left what historically has been the bastion of free markets and capitalism from the end of the Second World War under English rule through its return to China in 1997 largely unencumbered. (It is called a *special administrative area*, or SAR, within China.) It continues to be a cash-generating dynamo for the country.

The Economic Freedom of the World (EFW) index [25] is a much broader measure of a nation's economic freedom than the six conditions listed above. However, they are proportional, and the six conditions of NOP can be thought to be a subset of the EFW. Both contribute to a healthy manufacturing climate. The higher the EFW index of a country, the less predisposed a company located in that country is to out-border its manufacturing. The EFW index uses 24 metrics that are organized into 5 major categories to rate each country:

1. Size of Government
   a. Government consumption
   b. Transfers and subsidies
   c. Government enterprises and investment
   d. Top marginal tax rate
      - Top marginal income tax rate
      - Top marginal income and payroll tax rate
2. Legal System and Property Rights
   a. Judicial independence
   b. Impartial courts
   c. Protection of property rights
   d. Military interference in rule of law
   e. Integrity of the legal system
   f. Legal enforcement of contracts
   g. Regulatory restrictions on the sale of real property
   h. Reliability of police
   i. Business costs of crime
3. Sound Money
   a. Money growth
   b. Standard deviation of inflation
   c. Inflation: most recent year
   d. Freedom to own foreign currency bank accounts
4. Freedom to Trade Internationally
   a. Tariffs
      - Revenue from trade taxes (% of trade sector)
      - Mean tariff rate
      - Standard deviation of tariff rates
   b. Regulatory trade barriers
      - Non-tariff trade barriers
      - Compliance costs of importing and exporting
   c. Black-market exchange rates
   d. Controls of the movement of capital and people
      - Foreign ownership/investment restrictions
      - Capital controls
      - Freedom of foreigners to visit
5. Regulations
   a. Credit market regulations
      - Ownership of banks
      - Private sector credit
      - Interest rate controls//negative real interest rates
   b. Labor market regulations
      - Hiring regulations and minimum wage
      - Hiring and firing regulations
      - Centralized collective bargaining
      - Hours regulations
      - Mandated cost of worker dismissal
      - Conscription
   c. Business regulations
      - Administrative requirements
      - Bureaucracy costs
      - Starting a business
      - Extra payments/bribes/favoritism
      - Licensing restrictions

What is the current EFW landscape, how has it changed over the last 25-years and do the changes track with global manufacturing activity? Table 8 ranks countries by the

Table 8. 2010 EWF Index Ranking By Country (source: Economic Freedom of the World: 2012 Annual Report) [16]

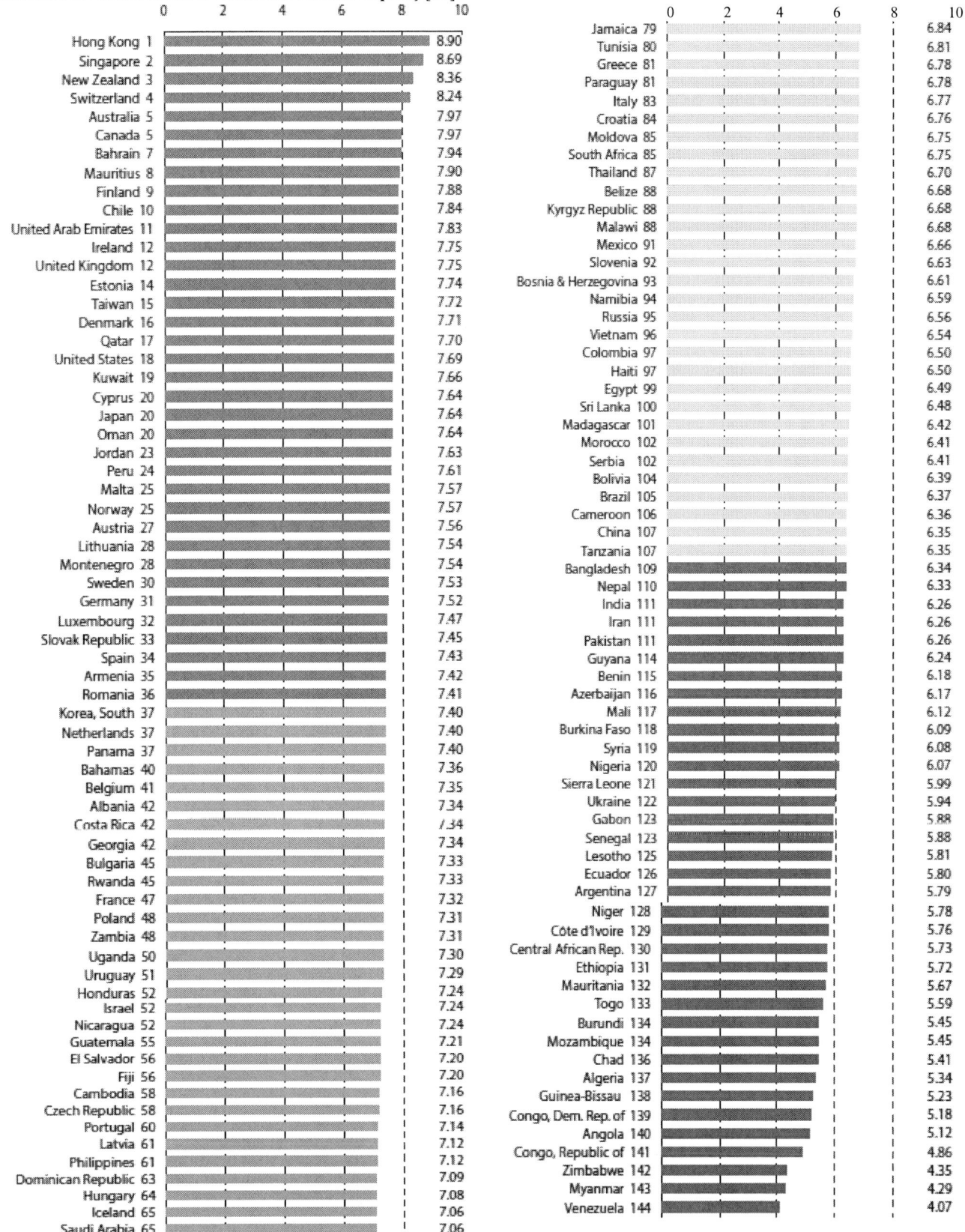

Economic Freedom Index, a compilation of 24 factors in 5 areas that indicate the relative level countries shackle private business development primarily through regulation, monetary policy, trade barriers, and tax rates. As described in the report, the most significant increases in economic freedom have been in former Communist nations that were characterized by centrally controlled economies to economies that are more free market based. This occurred in Europe and Asia in 1990 with the collapse of the Soviet Union, and in formerly Communist countries in Africa over the last 10 years.

Based on the EWF index, since 2000 Venezuela, Argentina, Iceland and the United States have shown the biggest declines in economic freedom [26]. Table 9 compares the changes in the Economic Freedom Index to manufacturing output for several selected countries. In the U.S., the EWF index dropped from 3rd best in 2000 to ranking 18th in 2010. At the same time, the U.S. manufacturing output has continued on a precipitous slide as a percentage of its total GDP. It has declined from being ranked 16th in 1975 to 75th

in the world in 2004. The opposite effect was experienced in the former Soviet bloc country of Slovakia. As its EWF index continued to improve from the fall of the Soviet Union in 1989 to its independence as part of the dissolution of Czechoslovakia in 1993, its manufacturing output has improved as well. The Slovakian world ranking for manufacturing GDP as a percentage of total GDP has improved from 77th in 1975 to 30th in 2005.

These examples seem to show a correlation between a country's economic freedom index and manufacturing activity – but is there causation? The data for China belie the causation hypothesis. In China, the meteoric rise in manufacturing ranking from 30th in 1975 to 2nd in 2005 has occurred without any significant change in the EWF index. Logic suggests that the relatively low EWF index is overwhelmed by the manufacturing activity spawned in 1978 when the government began to convert its 1.4 billion-person population from a centrally planed economy to a quasi-market economy. To prove causation would require a statistical analysis that is beyond the scope of this paper.

Table 9. Selected Country Comparisons for Changes in Nominal GDP, EWF and Manufacturing Output (Current USD)

Country	Nominal GDP per Capita / Ranking, 1990 Source: World Bank	Nominal GDP per Capita / Ranking, 2011 Source: World Bank	EWF Index / Ranking. 2000 [28]	EWF Index / Ranking, 2010 [27]	Manufacturing GDP (per $1000 of GDP) / Ranking, 1975 [29]	Manufacturing GDP (per $1000 of GDP) / Ranking, (Year) [29]
United States	23,198 / 10	48,442 / 16	8.65 / 3	7.69 / 18	547.17 / 16	132.32 / 75 (2004)
China	341 / 125	5,445 / 91	5.75 / 101	6.35 / 107	381.23 / 30	334.82 / 2 (2005)
Singapore	12,387 / 27	46,241 / 19	8.61 / 2	8.69 / 2	223.54 / 57	268.21 / 7 (2005)
Venezuela	2,482 / 56	10,810 / 63	5.83 / 82	4.07 / 144	151.54 / 94	171.08 / 39 (2003)
Slovakia	2,527 (1993) / 159	17,646 / 44	6.20 / 83	7.45 / 33	184.47 / 77	187.83 / 30 (2005)
Greece	9,073 / 31	26,427 / 34	6.91 / 45	6.78 / 81	562.23 / 14	99.15 / 102 (2004)

NOP – SUMMARY OF RESULTS

Is the manufacturing robustness of a country influenced by government size, regulation, monetary policy, and availability of capital? Any condition or policy that adds cost to a manufacturing operation must be absorbed in its price structure. Therefore, countries or states that create a climate that is averse to manufacturing produce a competitive disadvantage to manufacturing operations.

Outliers like China suggest that although there is a significant relationship between EWF index and manufacturing activity in a region, it is not a totally predictable indicator.

CONCLUSION

Can an electronic product assembler in a high labor rate market that sells labor for $37.93/hour compete with an assembler selling labor at $12.00/hour? This paper has demonstrated that in many applications it can. However, as this paper also demonstrates, the critical success factors rely on addressing many subjects on many fronts. Exploiting the automation, dismantling the traditional organization, changing the nature of the direct workforce, and challenging material cost are some of the areas requiring action. Over the

last ten years, the dramatic increase in manufacturing activity in low labor rate areas has been accompanied by the development of a middle class. This, along with worker unrest and currency inflation, has produced upward pressure on labor rates in these regions as the middle classes develop. Consequently, the cost gap between high and low labor rate regions has narrowed. It will continue to close, but there will be other low labor rate areas to seduce production. Cambodia and Vietnam are the latest countries to attract disciples from the church of the low labor rate. [30] High yield automation is the counterweight to low labor rate manufacturing. However, following the automation route is more difficult than seeking low labor rates, but more prudent. Finally, governments can play a positive role in encouraging manufacturing in high labor rate regions. Reducing NOP, challenging unfair labor and material practices, assisting in the formation of a world class workforce, and developing a high EWF index are all ways in which governments can create a positive manufacturing environment. Things are not always the way they seem. This is certainly the case when making the decision on where to assemble electronic products.

REFERENCES

[1] T. Borkes, "… Like Holding "the Wolf by the Ears …" The Key to Regaining Electronic Production Market Share: Breaking Free of the Division of Labor Manufacturing Model in High Labor Cost Global Regions, Proceedings of SMTA International, 2008, Orlando, Florida.

[2] T. Borkes, "Paper or Plastic? Choosing to Move Offshore," SMT Magazine, April 2006.

[3] T. Borkes, P. McDonough, The Economical Development of a Lead-Free Assembly Process: A Practical Case Study That Minimized Conversion and Operational Costs, P. 1, Proceedings of SMTA International, 2007, Orlando, Florida.

[4] Borkes, op. cit., "… Like Holding the Wolf by The Ears"

[5] NationMaster.com - Industry Statistics / Manufacturing, value added > in current US$ (per $ GDP), by Country http://www.nationmaster.com/graph/ind_man_val_add_cur_us_pergdp-added-current-us-per-gdp

[6] UN Classifications Registry, International Standard Industrial Classification of All Economic Activities, Rev.4, Detailed Structure and Explanatory Notes. http://unstats.un.org/unsd/cr/registry/regcst.asp?Cl=27

[7] T. Borkes, Producing the Design: Designing Electronic Products in the Year 2000 and Beyond, Proceedings of SMTA International, September 2000, Chicago, Illinois.

[8] Borkes, op. cit., "… Like Holding the Wolf by The Ears" P. 11.

[9] Borkes, op. cit., "Paper or Plastic?"

[10] U.S. Department of Labor, VentureOutsource.com, February 2008.

[11] U.S. Bureau of Labor Statistics, International Comparisons Of Hourly Labor Costs In Manufacturing, 2010, Press Release, p. 5, December 21, 2011

[12] Ibid.

[13] Borkes, op. cit., "… Like Holding the Wolf by The Ears" P. 7.

[14] U.S. Bureau of Labor Statistics, Hourly Compensation Costs For Computer and Electronic Product Manufacturing: Earnings by Occupation: Earnings and Hours of Production and Nonsupervisory Employees, NAICS 334, http://www.bls.gov/iag/tgs/iag334.htm

[15] U.S. Bureau of Labor Statistics, Hourly Compensation Costs For Computer and Electronic Product Manufacturing: NAICS 334, Earnings by Occupation: Electrical and Electronic Equipment Assemblers, November 2012.

[16] U.S. Bureau of Labor Statistics, International Comparisons Of Hourly Labor Costs In Manufacturing, 2010, Economic News Release, International comparisons of hourly compensation costs in manufacturing, Table 1, December 21, 2011

[17] U.S. Bureau of Labor Statistics, op.cit., Hourly Compensation Costs For Computer and Electronic Product Manufacturing: NAICS 334, Earnings by Occupation: Electrical and Electronic Equipment Assemblers

[18] Mojonnier, Tim, "China's Supply Chain Rocked by 13.6% Labor Cost Increase," businesstheory.com, February 14, 2012

[19] Borkes, op. cit., "Paper or Plastic?"

[20] T. Borkes, "Concurrent Education: A Learning Approach to Serve Electronic Product Manufacturing," Proceedings of SMTA International, 1990, San Jose.

[21] Payscale, Inc., payscale.com/research/ US/Job=Project_Engineer

[22] Borkes, op. cit., "… Like Holding the Wolf by The Ears," P. 9.

[23] Borkes, op. cit., "… Like Holding the Wolf by The Ears"

[24] Adam Smith, An Inquiry into the Nature and Causes of the Wealth of Nations, Ed. Edwin Cannan, 5th ed. London: Methuen and Co., Ltd., 1904.

[25] James Gwartney, Robert Lawson, and Joshua Hall, 2012 Economic Freedom Dataset, published in Economic Freedom of the World: 2012 Annual Report, Economic Freedom Network, 2012 http://www.freetheworld.com/datasets_efw.html

[26] James Gwartney, Robert Lawson, with Walter Park, Smita Wagh, Chris Edwards and Veronique de Rugy, Economic Freedom of the World: 2002 Annual Report, Vancouver, BC, The Frasier Institute, 2002, www.thefreeworld.com

[27] Ibid

[28] James Gwartney, et al., op. cit., Economic Freedom of the World: 2012 Annual Report

[29] NationMaster.com, op. cit., Industry Statistics / Manufacturing, value added > in current US$ (per $ GDP), by Country

[30] S. Montlake, "Vietnam Seeks Gains as China Labor Costs Rise," Christian Science Monitor, September 12, 2010

TAMPER PROOF, TAMPER EVIDENT ENCRYPTION TECHNOLOGY

Phil Isaacs,[1] Thomas Morris Jr.,[2] and Michael J. Fisher[2]
IBM Corporation
[1]Rochester, MN, USA
[2]Poughkeepsie, NY, USA
pisaacs@us.ibm.com

Keith Cuthbert[3]
W. L. Gore & Associates
Dundee, Scotland

ABSTRACT

Hardly a week goes by where there isn't a report of cyber-crime having occurred. So much so that there is a special branch of the FBI established to address the many forms of Cyber-Crime. While the internet is convenient for many regular on-line activities, for example: Information searches, goods purchasing, sales, airline and hotel reservations, banking, bill-pay, driving directions and telephone/address look-up. It is the ease at which this information is so readily available that makes it vulnerable to attack.

One solution would be to completely isolate the computer applications. However, most applications cannot perform their function in isolation. In order to prevent cyber crimes from occurring on server level products, IBM and Gore[1] have collaborated on a state-of-the-art physical security package to protect the hardware components of a cryptographic coprocessor module.[2][3] This package meets the highest level of physical security requirements contained in the U.S. Government Federal Information Processing Standard (FIPS) 140-2 Security Requirements for Cryptographic Modules- (Level 4), and supports the overall attainment of FIPS 140-2 (Level 4) for the cryptographic coprocessor. This is the highest level of encryption technology which is allowed outside of the government or military. The packaging technology includes tamper response where any attempt at physically gaining access would render the cryptographic module useless.

This paper will provide an overview of the protection features of the assembly and the manufacturing processes used to manufacture the product.

Key words: Encryption, Tamper proof, Tamper evident, FIPS 140-2

INTRODUCTION

Encryption alone is no longer adequate to protect sensitive data. Imagine having your encryption keys breached without knowing it happened. Storing data in electronic form may be convenient, but it is also susceptible to stealth attacks. In order to perform cryptographic functions (encrypt, decrypt, sign, authentication) a computer system requires access to cryptographic keys and other security relevant data in a clear format. It is evident by getting access to such security relevant data in clear format a hacker can easily get access to the data being protected and also impersonate other authorities. Continuous technology improvements are affording unscrupulous individuals with opportunities to unravel encrypted data. The way to prevent this is to generate and never expose the most important cryptographic key outside an enclosure capable of detecting and responding to any type of physical tamper.

One can find many approaches to tamper prevention, tamper detection and appropriate reaction. Often these approaches are concepts. Some of the concepts are as simple as a passive circuit pattern with no electrical connection, which can be monitored for any physical intrusion by change or distortion in the monitored electro-magnetic field.[4] Others use capacitive networks or fringe capacitance to create a sensor device.[5] There are also quite sophisticated approaches such as using quantum mechanics to create a non-repeatable encryption key.[6] Sensors can be made from a variety of materials: Semi-conductors, metallic traces, organic traces, etc.[7] The approach selected for this product is a sensor with an organic trace network constantly monitored for any attempted intrusion into the package. This technology has been proven successful in previous products.[8] This paper focuses on the hardware design and the manufacturing process used to make this product.

Tamper Detection and Response

Data stored in electronic form such as electronic components when left unprotected can be susceptible to access without detection. Simply wrapping the components in a physical enclosure hampers component access, but does not prevent data retrieval. The use of a thick impenetrable envelope is not practical and can not be considered secure by itself. The enclosure needs to do more than visually signify that an intrusion has taken place. An after-the-fact indicator of data breach means the data has already been compromised. The enclosure must detect and respond at the time of the intrusion.

Our two companies have partnered in the development of a secure environment that supports the physical security requirements of Federal Information Processing Standard 140-2 (FIPS 140-2) certification.[9] This solution is currently in use in the PCI express Cryptographic Coprocessor.[10]

The methodology behind this secure solution is the use of a multilayered random pattern mesh sensor incorporated with response circuitry. Tamper Respondent Sensor[11] [12] technology is wrapped around the security sensitive components (i.e. secure module). This wrapping shields against physical intrusion such as puncture, chemical attack, and laser penetration.

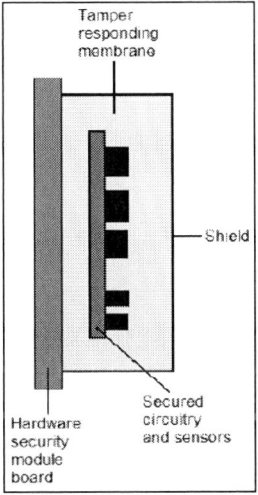

Figure 1: Secure Module Diagram

Utilizing the concepts of a Wheatstone Bridge and comparator logic, the resistance of each "leg" of the sensor mesh is constantly monitored for deviation against a known base value. The tamper sensors, control electronics and small key memory (part of BBRAM) are highly integrated in a small tamper detection and response module (DS3645[13]). This tamper detection module provides higher overall reliability including more reliable tamper validation thus preventing false tamper incidents. The employment of a tamper module versus discrete electronics results in a lower battery back up current drain. It also better enables the housing of the tamper subsystem within the confines of the secure enclosure.

The IBM 4765 coprocessor is shipped from the factory with a certified device key which is stored in the card's battery back up protected memory. The electronic key digitally signs test messages to confirm that the coprocessor is genuine and that no tampering has occurred. The coprocessor cannot operate without this device key. If any of the secure module's tamper sensors are triggered by tampering or accident, the coprocessor erases (zeroizes) all

data in the protected memory destroying the device key. This renders the coprocessor permanently inoperable with no recovery.

A change in the mesh sensor characteristics triggers an imbalance in the tamper circuitry. When a physical or laser penetration is attempted the resistance of the sensor mesh conductive ink track changes the resistance. The response module senses this imbalance and invokes the immediate erasure of the high speed erase battery backed up memory (3KB BBRAM) eliminating all security sensitive data (i.e. coprocessor critical keys and certification). The 3KB BBRAM hardware controller embeds a function that offloads the firmware task of flipping the data in order to avoid imprinting data. A chemical attack (reagents and solvents) causes the conductive ink track to "dissolve" changing the "leg" resistance resulting in a similar detected imbalance. Attempts to unwrap the adhesive mesh sensor causes permanent changes to the characteristics of the ink tracks also resulting in a detected imbalance.

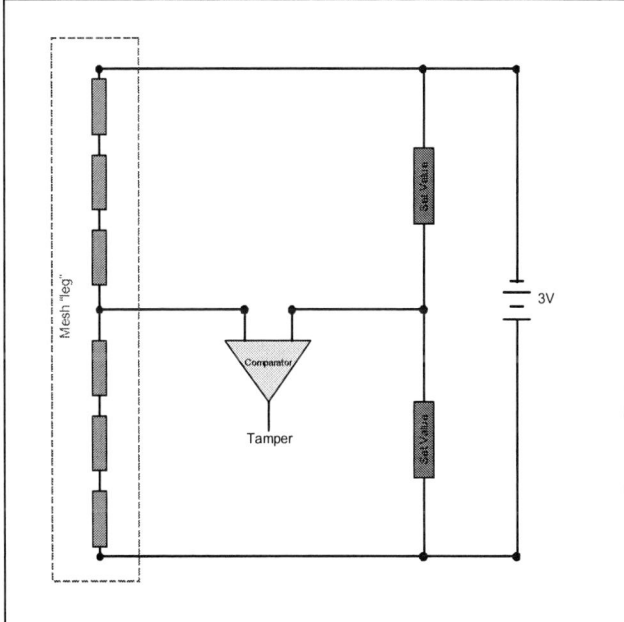

Figure 2: Tamper Circuit Schematic

The data stored in this secure memory is encrypted for added security with a key stored in the small key memory of the DS3645 Security Manager. Once the sensitive data is erased, the IBM 4765 is placed into diagnostic mode and left in a permanently inoperable state.

A pair of batteries mounted on the coprocessor board ensures the tamper subsystem is always active even when the IBM 4765 is not in a powered on machine. Removal of these batteries outside the authorized battery replacement process will trigger a tamper event.

Tamper Respondent Technology
The current Cryptographic card uses the Tamper Respondent Technology. This technology defends the

physical security boundary of the module by creating a "tamper respondent" envelope. Protection is provided by an organic, flexible sheet sensor which enfolds the electronic package creating an enclosure with no direct entry points.

Conductive ink traces are deposited onto an organic substrate. The electrical state of the sensor changes if an ink trace is broken, triggering tamper respondent mechanisms, such as zeroing encryption key memory. An opaque outer resin coating prevents attackers from optically seeing the traces. The traces are also invisible to X-rays, further thwarting analysis. In its finished form, entry into the module without circuit damage and detection is extremely improbable.

Figure 3: Tamper Respondent Sensors

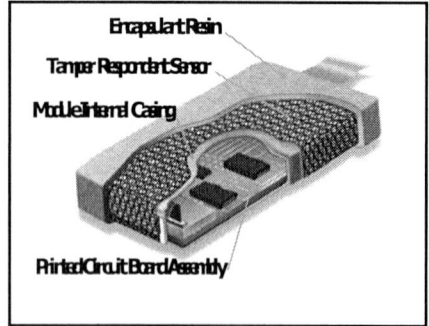

Figure 4: Tamper Respondent Secure Encapsulated Module

Electrically, the sensor consists of a resistive network which is constantly monitored by a detector circuit inside of the package. When a trace breaks, it triggers a fast and unrecoverable change in electrical state.

This sensor network is validated in the Cryptographic Module to FIPS 140-2 (Level 4) physical security. Tamper

Respondent Technology in this application has undergone a number of validations to FIPS 140-2 (Level 4).

Manufacturing Encapsulation Process
In the following section we will describe the manufacturing process and key process controls used to assemble the tamper proof hardware assembly.[14]

Manufacturing Process and Storage Environments
All processing operations must be performed in a temperature and humidity controlled environment.

Encapsulation Process Steps
1. Electronic Card Assembly and Test, ECAT Card Primary Enclosure
2. Tamper Proof Sensor Folding and Cure
3. Resin Encapsulation

ECAT Card Primary Enclosure
Insert the signal flex cable assembly and the Power Flex cable assembly into the daughter card mating connectors.

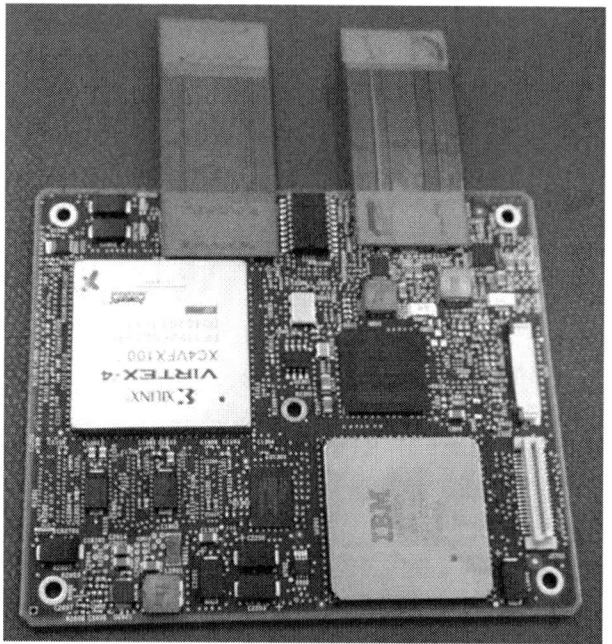

Figure 5: Flex Cable Plug

The two flex cables must be pre-folded creating an upward right angle with respect to the daughter card plane. This operation is meant to facilitate the enclosure of the daughter card and flex cable assemblies into the inner cover while sliding the cables through slots in the top cover.

Figure 6: Flex Cable Pre-Bending

Remove the blue plastic release sheets from the 2 thermal pads in the inner bottom cover and lay the daughter card down in this cover with the flex cables up, aligning the 5 card holes with the 5 rivets in the cover.

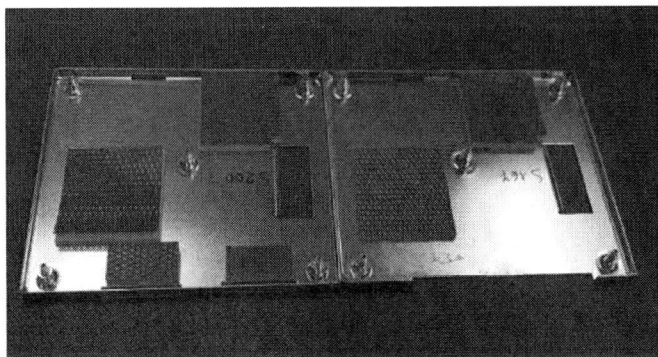

Figure 7: Inner Covers with Thermal Pad Release Sheets.

Figure 8: Inner Covers without Thermal Pad Protectors

Remove the blue plastic release sheets from the 4 thermal pads in the inner top cover and place the top cover aligning to the rivets below while passing the 2 Flex cables through the cover cable slots to fully enclose the daughter card inside the inner cover.

Figure 9: Card Assembly placed into Inner Cover

The resulting assemblies are placed inside the pre-riveting holding tools to avoid inner covers becoming loose while moving pre-riveted assemblies around the manufacturing floor.

Figure 10: Pre-Riveting Holding Tools

Customized tooling is used to form the rivets on the inner cover assembly. The package is placed into a pre-load fixture and automatically shuttled under a press head with rivet forming punches.

Figure 11: Riveting Alignment Aids

Critical to function measurements of the inner can assembly are as follows:

Rivet Attributes
- Max assembly thickness
- Minimum diameter of rivet head
- Lack of cracking, breaking and burrs

Cover Attributes
- Top and bottom inner covers outline alignment.
- Cover surfaces are free of burrs and sharp edges
- Cover assembly planarity
- Total covers thickness
- Good Flex cable to daughter card interconnection by electrically testing connection.

Figure 12: Riveting Pre-load Fixture

Figure 13: Rivet Forming Punches

The inner cover assembly is cleaned with IPA, handled with low ionic gloves and should be processed immediately after cleaning.

Figure 14: Completed Inner Assembly

TAMPER PROOF SENSOR FOLDING AND CURE
Tamper Sensor Preparation
The tamper sensor is pre-tested prior to application for critical function electrical and mechanical measurements.
Tamper Respondent Sensor Folding
1. Blow off the tamper respondent sensor with nitrogen under ionized flow.

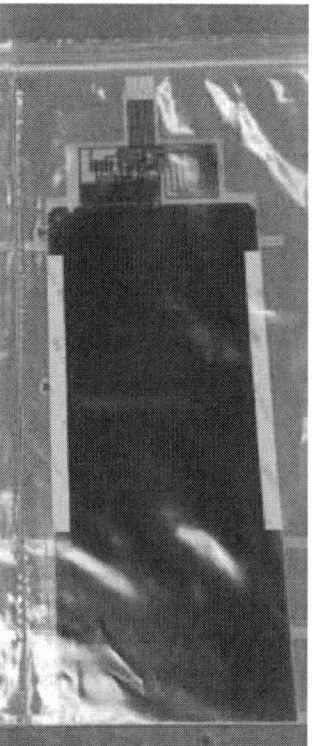

Figure 15: Tamper Proof Sensor

2. Remove the release layer
3. Align on the folding tool with the adhesive side up

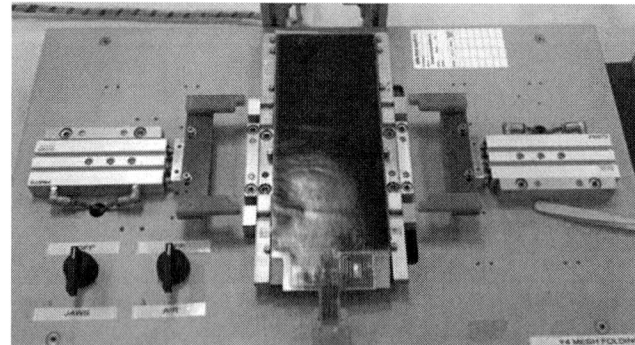

Figure 16: Sensor Folding Equipment

4. Place the inner can onto the tamper respondent sensor using the guides.
5. Press the inner can onto the tamper respondent sensor to activate the pressure sensitive adhesive.
6. Fold the first fold which contains the tamper respondent sensor leads using the folding apparatus.
7. Insert the tamper respondent sensor I/O cable into the PCBA sensor connector.

Figure 17: Inner Assembly Placement

Figure 18: Sensor Plug into ECAT Card

Figure 19: Sensor Alignment into ECAT Card

8. Fold the tamper respondent sensor around the inner cover to complete the first folding.

Figure 20: First Fold of Tamper respondent sensor on Inner Cover

9. Tamper respondent sensor folding continues on the two package sides where there are the flex cables on one side and the vent on the opposite side

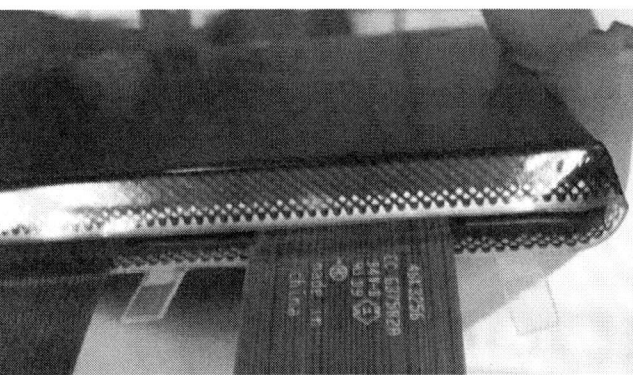

Figure 21: Edge Fold on Ribbon Cable

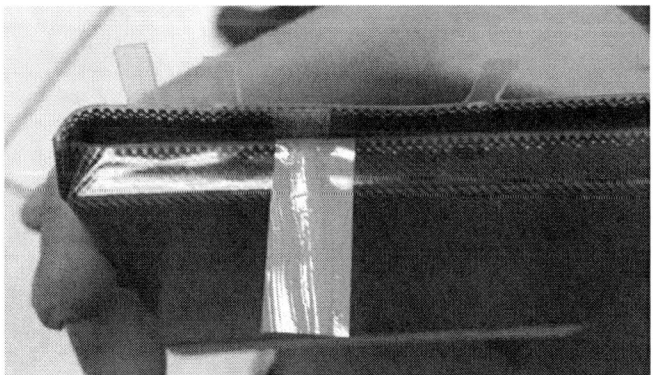

Figure 22: Edge Fold on Vent Side

Figure 23: Completed Fold on Vent Side

10. Inspect for cracks, scratches, creases, bubbles and any gaps in the sensor.
11. Place the folded assembly into the holding tool.

Figure 24: Holding Tool

Tamper Respondent Sensor Holding/Curing
The assembly is held in the holding fixture just after the folding operation. The package, retained in the fixture, must be cured for 1 hour at 60°C. The package is encapsulated in resin within 24 hours after curing.

RESIN ENCAPSULATION
Process Indicators and Process Controls
Water in the polyol component can lead to bubbles or foaming in the polyurethane, PU, resin. Such bubbles can also occur when the curing PU mixture is allowed to pick up water during processing. Thus water content must be controlled. Transfer of polyol needs to be done under nitrogen with controlled pressure (to avoid over pressurizing the shipping container).

Transfer of isocyanate also needs to be done under nitrogen with controlled pressure to avoid over pressurizing the shipping container. The environment in which the dispense operation occurs must have an RH maximum of 30%. Properties of the fully cured PU must be determined after change of either polyol or isocyanate and periodically during manufacturing. Whether in the tool production or back up tanks, the Polyol and the Isocyanate must be stirred constantly.

The mixing ratio of Isocyanate to Polyol is from 0.91 to 0.99:1 by weight. Defects in the cured resin such as swirls, areas of inhomogeneity, wet or soft spots are not allowed. When starting a new encapsulation lot, a sample of the PU resin will be taken from the dispense tool for the purpose of obtaining a time to gel point determination, G'/G". The gel point must fall between the values of 70 to 125 minutes.

First Resin Dispense
Dispense 4 shots of mixed Resin Polyurethane into the outer cover using a long plastic mixing nozzle. This operation must be performed in a dry environment such as a Dry Hood through which either dry nitrogen or dry air flows to keep the humidity level below 30%. The polyurethane should not be allowed to be stationary in the static mixer nozzle for longer than 2 minutes. The nozzle must be replaced frequently in order to assure good polyurethane mixing. The resin must fully cover the bottom of the outer cover.

Assembly Package Positioning With Template Alignment
After dispense insert the Crypto card package in the outer cover with the cables oriented toward the longer outer cover side. The assembly is manually centered into the outer cover. The thickness of the resin around the enclosure must be maintained.

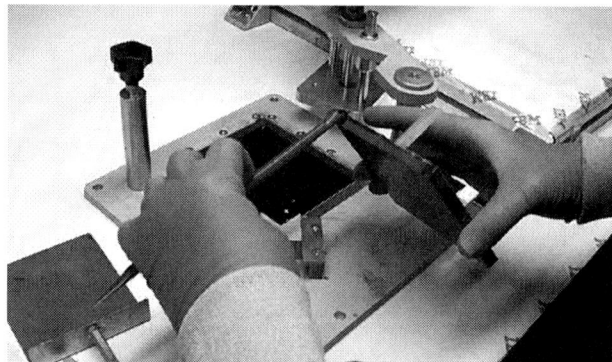

Figure 25: Applying Resin to Package Corners

Figure 26: Positioning of Package within Outer Cover

Second Resin Dispense

Dispense 4 shots of mixed Resin Polyurethane over the assembly. Put the part in a Dry Hood. No part of the assembly should be visible after this resin dispense.

Figure 27: Fully Encapsulated Module

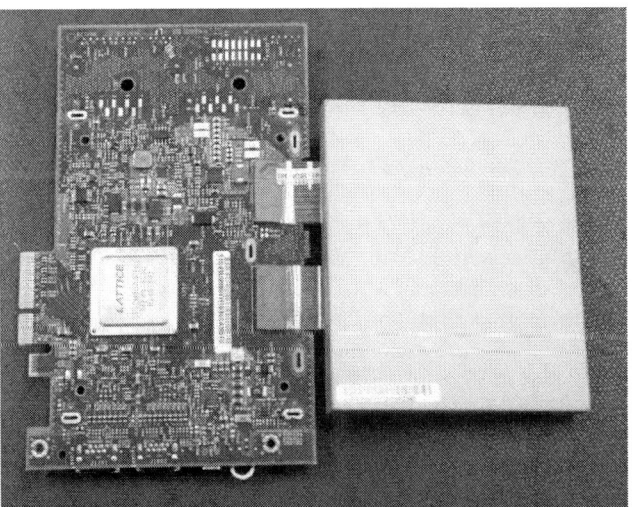

Figure 28: Encapsulated Module Connected with PCIe Assembly

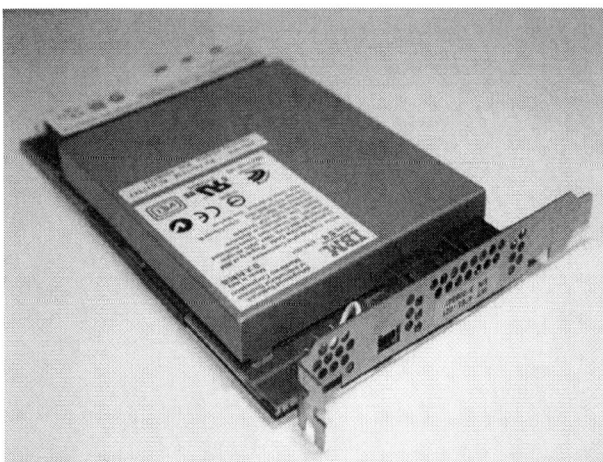

Figure 29: Finished Cryptographic Module assembled to PCIe Card

Second Polymerization (Cure)

The resin polyurethane must be cured in a nitrogen oven using the following parameters:
- Cure PU for one hour at 25 C.
- Ramp to 50 C and cure for 70 minutes.

Vent Trim

Trim the vent to within 0.5 mm above the level of the resin.

Secure Module Encapsulation Visual Inspection

The resin must completely cover the folded sensor, no uncovered area is allowed. Separation between resin and cover is not allowed. The resin must be below or equal to 0.5 mm below the cover edge. The cured resin must be shiny and show only minimal bubbles.

Quality and Reliability

There are several process steps included as package verification quality gates. They can be found in Table 1. In addition to the in-line quality tests the package has passed a series of stress tests to assure the package will last a lifetime consistent with the requirements of high end, mission critical server products.

Table 1: Quality Gates

Item	Test	Criteria
1	Receiving Inspection of the Sensor and circuit verification	Dimensions
2	Resin	Chemical analysis
3	Visual Insp. of the Sensor prior to folding	Damage & contamination
4	VI Sensor after folding	Proper folding
5	Electrical verification of sensor after folding	Sensor circuits
6	Electrical verification after sensor cure	Current circuits
7	PU Material Properties	Stoichiometry & cure
8	VI after PU dispense	Physical appearance
9	Electrical verification after resin	Sensor circuits
10	VI after Crypto to PCI merge	Solder defects
11	PCI compliance	Thickness gage
12	Burn-in	Functional Test

SUMMARY

The combination of a sensor mesh and monitoring circuitry provides an environment that protects sensitive data from cyber theft. It is manufactureable and reliable both in preventing undesired access and its longevity in the field.

ACKNOWLEDGEMENTS

The authors gratefully acknowledge the following contributors: Assistance provided by Vincenzo Condorelli, Nihad Hadzic, and William Santiago Fernandez of IBM Poughkeepsie, NY toward the content of this paper. We would also like to acknowledge the team who have worked on this project: Dave Allan, Ed White, Frank Orapello, Jason Wertz, Jim Wilcox, Jing Zhang, Mitch Ferrill, Nandu Ranadive, Norm Curry, Stu Lake and Tim Donahue.

REFERENCES

1. Gore is a trademark of the W. L. Gore & Associates, Inc. Newark, DE.

2. T. W. Arnold, C. Buscaglia, F. Chan, V. Conderelli, J. Dayka, W. Santiago-Fernandez, N. Hadzic, M. D. Hocker, M. Jordan, T. E. Morris, Jr. and K. Werner, " IBM 4765 Cryptographic Co-Processor," IBM Journal of Research and Development, Vol. 56 No. ½, pp. 10:1-10:13.

3. Arnold, T., Dames, A., Hocker, M. D., Marik, N., Pellicciotti, A. and Werner, K., "Cryptographic system enhancements for the IBM System z9", IBM Journal of research and Development, Volume 51 Number 1.2. web site:
http://ieeexplore.ieee.org/xpl/tocresult.jsp?isnumber=53886 99&punumber=5288520

4. Woodard, S. E.., Functional Electrical Sensors as Single Component Electrically Open Circuits Having No Electrical Connections, IEEE Transactions on Instrumentation and Measurement, Vol. 59. No. 12, December 2010, pp. 3206-3213.

5. Eren, H. and Sandor, Lucas D., "Fringe-Effect Capacitive Proximity Sensors for Tamper Proof Enclosures," Proceedings of Sensors for Industry Conference, Feb. 8, 2005, pp. 22-26.

6. Liebholz, Stephen W., "Solutions for the grand Challenges of Information Security: Protection Against Rogue Insiders, Dynamic Compartmentalization and True Quantum Encryption." Proceedings from 2007 IEEE Conference on Technologies for Homeland Security, pp. 129-132.

7. Paul, P., Moore, S. and Tam, S., "Tamper Protection for Security Devices." from the proceedings of the 2008 Symposium on Bio-inspired Learning and Intelligent Systems for Security, pp. 92-96.

8. Isaacs, P., Buscaglia, C., Feger, C., Pearsall, K., Wolf, H., Cesana, M., Moscheni, G., Cuthbert, K. and Hunter, S., "Packaging and Processing of a State-of-the-Art Encryption Technology." From proceedings of 2007 IMAPS Symposium, San Jose.

9. National Institute of Standards and Technology Cryptographic Module Validation Program (CMVP). website: http:/csrc.nist.gov/groups/STM/cmvp/index/html.

10. IBM 4765 PCIe Cryptographic Coprocessor Installation Manual. web site: http://www-03.ibm.com/security/ cryptogards/pciecc/pdf/4765install.pdf.

11. Tamper Respondent Mesh is a trademark of the W. L. Gore &Associates, Inc. Newark, DE.

12 Gore Anti-Tamper Physical Security for Electronic Hardware. web site: http://www.gore.com/en_xx/products/ electronic/anti-tamper/anti-tamper-respondent.html.

13. Maxim Integrated DeepCover[TM] Security Manager (DS3645). web site: http://www.maximintegrated.com /datasheet/index.mvp/id/5424.

14. IBM Engineering Specification, PCIe Cryptographic Coprocessor Secure Module Assembly Requirements, written by IBM and SEM, Services for Electronic Manufacturing, Milano, Italy.

IMPACT OF LEAD-FREE COMPONENTS AND TECHNOLOGY SCALING FOR HIGH RELIABILITY APPLICATIONS

Chris Bailey, Ph.D.
University of Greenwich
London, United Kingdom
c.bailey@gre.ac.uk

ABSTRACT

Semiconductor technology is increasingly meeting the demands and challenges posed by roadmaps such as ITRS in terms of technology scaling and packaging, where the interconnect size is decreasing to ever smaller dimensions. The drive for this is increasing functionality of the devices and is governed primarily by the consumer electronics markets.

For applications in high reliability sectors such as aerospace, automotive, oil and gas, etc, this scaling in technology, and the use of lead-free solders to satisfy ROHS legislation, is posing a number of challenges. This paper discusses current status in using lead-free components for high reliability applications and developments in modelling that can aid organisations in assessing different design options before using lead-free COTS components in their applications.

Key words: Lead-free, refinishing, technology scaling, metal migration and reliability

INTRODUCTION

Semiconductor companies are now focused on the consumer electronics sector where market size is now significantly greater than the high reliability sector. For example the high reliability sector now only accounts for <1% of electronics components matket. The life-times required for components in this sector are much shorter that that required for the high reliability sectors.

Microclectronics components are increasingly manufactured and packaged as "lead-free" as a result of legislations such as the EU RoHS directive. This directive restricts the use of lead and other hazardous substances.

As the manufacturing of defense, aerospace and other high-reliability equipment relies more and more on the use of commercial-of-the-shelf (COTS) components, the use of electronics parts with pure or tin-rich alloy finishes on their terminations becomes a major reliability issue due to the problem of tin whisker growth (see figure 1).

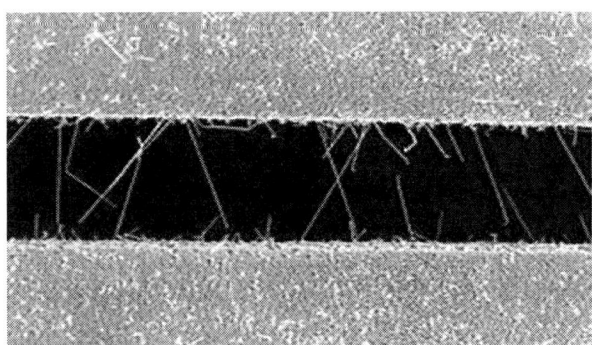

Figure 1: Tin Whiskers

For Hi-Reliability equipment manufacturers, one possible response to the tin whisker risk is to 're-finish' component terminations, replacing tin-rich termination materials with tin-lead by a process known as hot-solder dipping. This approach, originally developed as a manual 'hand dipping' technique, has migrated to a robotically controlled system (see figure 2) and a more repeatable process available from a limited number of suppliers.

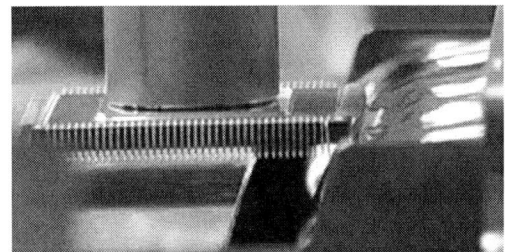

Figure 2: Robot Controlled Solder Refinishing (Courtesy: Micross Components Limited)

A major concern related to hot solder dipping is that the process temperatures may result in thermally induced damage in the electronic component parts. It is also likely that parts with different design and construction will have different thermal behaviour during the refinishing process, hence there is a need to optimise the hot solder dipping process in terms of rates of change in temperature and resulting stress magnitudes imposed on the parts.

Limited research on hot solder dipping has been conducted and reported to date. One of the main studies in this area has adopted an experimental approach to investigate the effect of hot solder dipping, and has identified potential damage issues in some types of packages [1].

Another area of concern particularly with regards technology scaling and the use of lead-free solders is metal migration. Metal migration can occur within a metal structure within the die (e.g. copper or aluminium interconnects) or within the packaging (e.g. lead-free solder joints). Figure 3 illustrates metal migration in electronic materials, which can be due to a combination of thermal, electrical and/or stress effects.

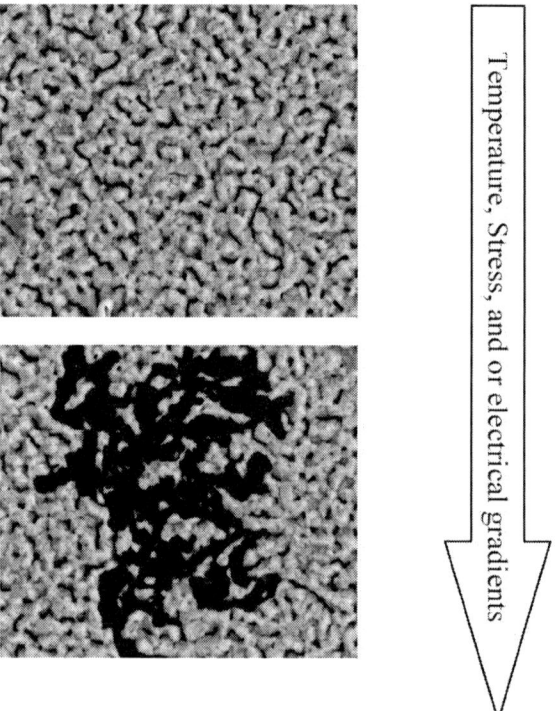

Figure 3: Metal migration and formation of voids in interconnects

One of the failure mechanisms that is causing considerable concern is Electro-migration (EM). EM is due to metal transportation at the atomic level caused by high current density which is an inevitable consequence of miniaturization. EM is known to cause voids and hill-locks in metal conductors and in the worst cases, this leads to open or short circuits. Moreover, higher current density and complexity of interconnect structures also generates high temperature and stress gradients which result in void formation due to thermo-migration and

stress-migration respectively. As a result, the cause of metal migration is governed by multi-physical cross coupling relationships. For example, in flip-chip interconnects the ever decreasing size of solder joints can lead to current densities reaching 104 A/cm^2, these will promote electro-migration but also result in high temperature and stress gradients which need to be understood, particularly when aiming to develop qualification tests for this phenomena.

Another adoption being used for assembling COTS components for high reliability applications is to use underfills to support the second level interconnects between the package and PCB.

This paper details some of the challenges in adopting lead-free COTS components for high reliability applications. The following will discuss (i) refinishing processes to remove lead-free materials, (ii) metal migration and (iii) use of underfills to support second level interconnects.

REFINISHING PROCESSES
The refinishing process detailed in figure 4 is a double-dip hot solder dip. It is a fully automated process where a robot arm with a vacuum sucker holds the component and takes the package though a complex sequence of process steps. The package subjected to HSD is first picked with the robot arm and assessed for positioning. Then the part is taken to a flux bath and the leads at each side are fluxed in a sequence. Package is then moved to a pre-heater and heated from ambient enclosure temperature (38-42C) to 140C in a close-loop temperature control using an IR sensor. The IR sensor is integrated with the pre-heater and measures the temperature at the package bottom surface. Based on the IR readings, the heat is controlled so that the ramp rate of pre-heating does not exceed 3C/sec. Then the package is moved to the solder bath. In a sequence, the leads on each of the four sides of the package are dipped in the molten tin-lead solder wave. This step is undertaken under nitrogen blanket for the solder bath. The solder is at 250C and the time of dipping each package side is 3 seconds.

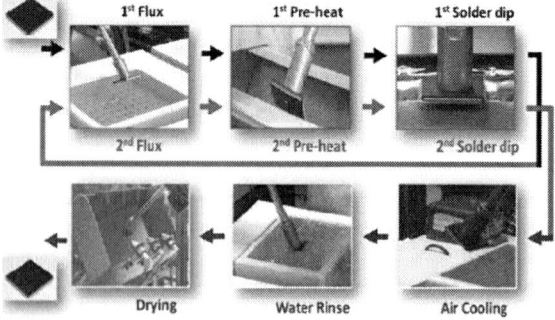

Figure 4: Schematic of the double-dip HSD process steps

The package is then taken for second time to the flux bath where in a similar way and in sequence the leads at each side are fluxed. During these steps package cools down, hence second pre-heating is required to heat the component from its current thermal state to 140C. Second solder dip of the package leads follows. The double-dip approach is utilized to ensure the quality of the re-finishing. The next steps in the process involve air cooling followed by water wash. Drying the package and placement complete the process.

A thermal modelling methodology for analysis of the effects of double-dip hot solder dip process on re-finished leaded components has been utilised, demonstrated and validated (2). This exploits the use of finite element analysis which solves the temperature equations over the domain of the package. Figure 5 details such a model representation of the HSD process.

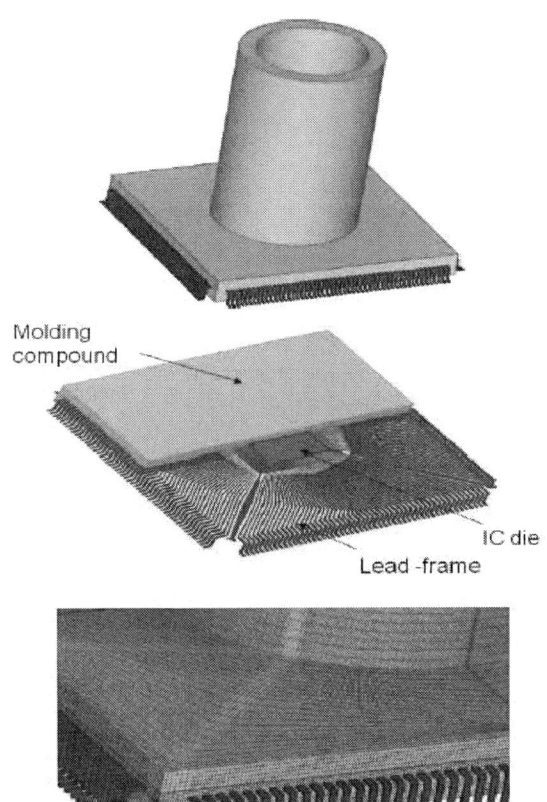

Figure 5: Finite element model of the HSD process

The above modelling approach has been applied to a particular Quad Flat Package where transient temperature changes in the package during the HSD process have been predicted and compared against thermocouple data. Figure 5 details the temperature calculations from such as model where we can clearly see the effect of the dipping process on the temperatures in the package.

The model predictions were validated against thermo-couple data (see figure 5). The modelling results on the QFP part in the reported study have been extremely beneficial in identifying issues related to the re-finishing process, process instrumentation, and the accuracy of obtained thermocouple measurements. In addition tmodel has supported also the assessment of the performance of cooling rates in the HSD e process and for characterisation the effects of tooling (component mounting) on the component temperatures.

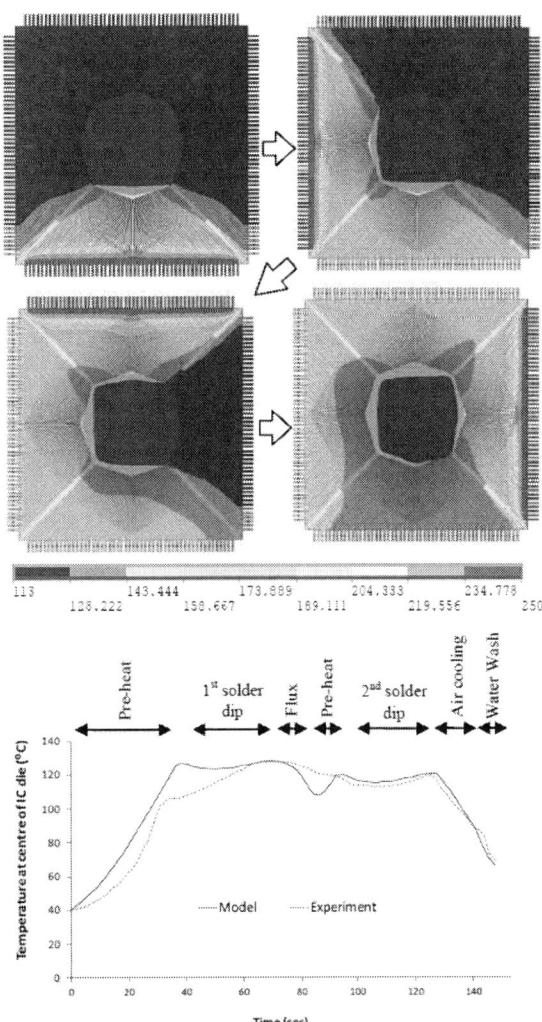

Figure 6: Thermal modeling results and comparisons with thermocouple data

USE OF UNDERFIILLS FOR SECOND LEVEL INTERCONNECTS

To help increase the reliability and lifetime of COTS components a number of organisations are using underfills at the second level interconnects between package and PCB. The following modelling analysis, although using lead-based solders illustrates that care needs to be undertaken when choosing an underfill for this type of application.

In this study a PBGA packaged is analysed. It is 35x35 mm in size and has a full array (1152) of solder joints at 1 mm pitch size. A copper stiffener at the peripheral area of the package provides structural integrity between the substrate and copper heat spreader. Figure 7 illustrates the PBGA slice model with colours representing different materials. A detailed view at the solder joint level and the mesh across the joint is also provided. The package is modelled in detail so that all layers and materials in the vertical build up of the package are represented. Appropriate boundary conditions for the slice modelled are applied.

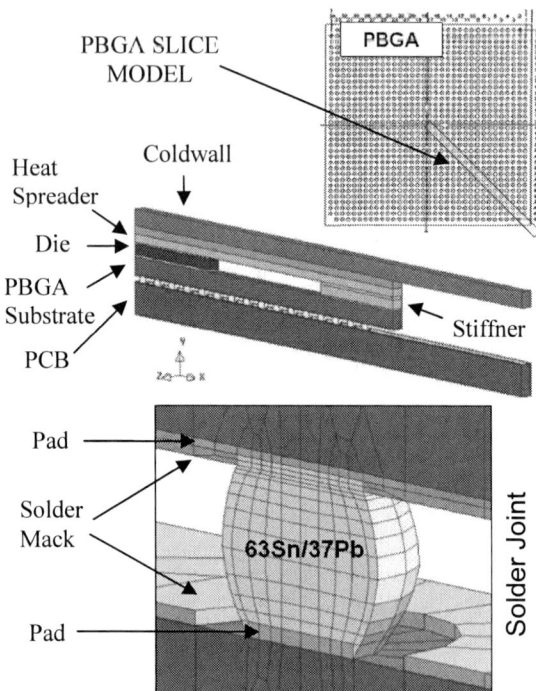

Figure 7: Finite element model for PBGA

The solder material (63Sn/37Pb) is assumed to behave in visco-plastic manner according to an inelastic strain rate relationship. The rest of the materials are elastic with temperature dependent properties. The board is orthotropic. Transient stress and inelastic strain analysis of PBGA finite element model under temperature cycling is used to identify the most damaged solder joints and to calculate a damage parameter based on creep strain energy. It was found that the corner solder joint has the highest value of damage. The two solder joints beneath the edge of the silicon die follow closely. Figure 8 illustrates a typical distribution of accumulated inelastic strain energy density in the solder joints at the end of the thermal cycle (a certain field profile); the location of highest damage in solder is identified near the package interface where the crack is expected to occur.

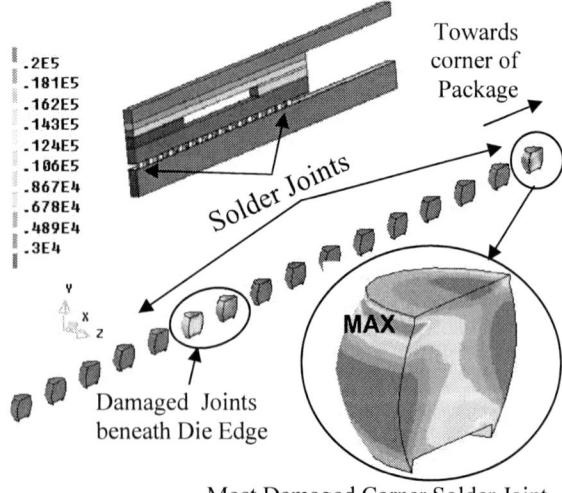

Figure 8: Energy density [Pa] in PBGA solder joints.

Figure 9 illustrates the accumulated inelastic strain energy density after a thermal cycle in the most damaged solder joint as function of the underfill selection. It was found that underfills A and B both improve solder joint reliability providing a reduction in damage (crack growth rate) up to 60%. However, underfills C and D unfortunately increase damage by up to 4X. Therefore care must be taken in choosing the correct underfill.

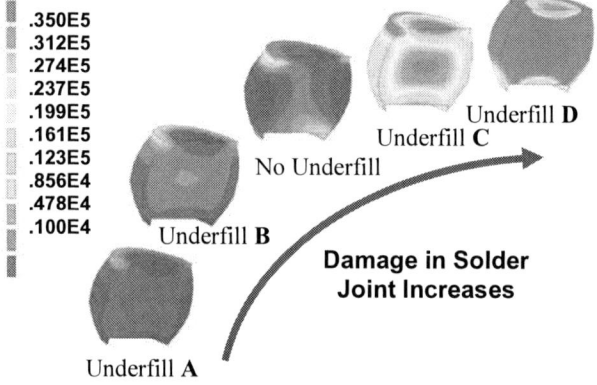

Figure 9: Impact of underfill properties.

TECHNOLOGY SCALING
Trends in semiconductor packaging and the use of lead-free materials is causing a number of concerns particularly in the high reliability sectors (aerospace, oil&gas, etc) where increasingly the use of commercial off the shelf components are used. One of these concerns is metal migration due to stress, electrical and thermal gradients. At the die level the move towards smaller technology nodes (e.g. down to 32nm and below) and for solder interconnect of much finer pitch and smaller size the risk of metal migration is a concern in particular for components that are required to survive in the field for 10 or more years.

The TTF model [3] for stress migration failure mechanism provided in JEDEC standard is exponentially related to the hydrostatic stresses in the metal line, as shown in the following equation

$$TTF = Bo * (\sigma)–N * \exp(Eaa / kT) \quad (7)$$

Where Bo = pre-factor, σ = constant stress load, N = 2 to 3 for ductile metals, Eaa = apparent activation energy, 0.5 to 0.6 eV for grain-boundary diffusion; ~1 eV forsingle-grain (bamboo-like) diffusion, k = Boltzmann's constant, T = temperature in kelvins.

It is interesting to note that in the JEDEC standard the time to failure is a function of stress magnitude and not stress gradient. Stress migration (SM) or stress induced voiding (SIV) is a failure mechanism often occurs in IC metallization such as the aluminium interconnects. Hydrostatic stress and hydrostatic stress gradient are considered as the driving forces for the void nucleation and growth respectively (4). The void is likely to form at the sites where the stresses are high, but voids would not grow without the presentence of the stress gradients.

66

As mentioned above, metal-migration is affected by many factors that are difficult to decouple. Recent work (5) has attempted to provide a modelling framework that provides a multi-physics/scale approach to predicting metal migration. In this work, a closely coupled multi-physics modelling method has been proposed (see Fig. 10). It can be used predict atomic concentration and void formation in metals where electro-migration is affected by electrical, thermal, stress, and geometry factors.

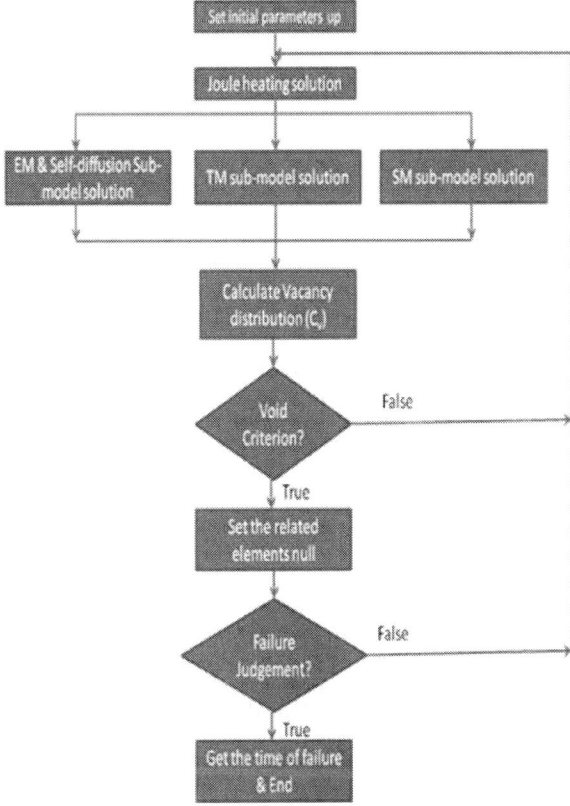

Figure 10: Integrated modeling approach for metal migration

To test the model, a one dimensional analysis was used. The computational domain has a length of 100 μm and an electric potential difference was applied at the two ends of the domain generating a uniform electric filed and a constant current.

In order to test the effect of temperature gradient, a linear temperature profile was imposed. Similarly, a linear hydrostatic stress profile was created to test the stress gradient effect. The values of the temperature and stress gradients are +0.107 C°/m and +0.107 MPa/m. Fig. 11 shows the vacancy concentration at t=800 s for simulations with electric effect only, electric and thermal effects, and all three effects respectively. The results show that the temperature and stress gradient have both had an impact on metal migration.

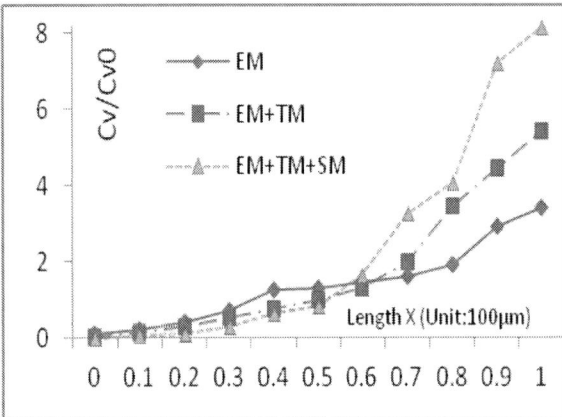

Figure 11: Influences for electro, thermal and stress migration on void formation in metal tracks.

The above numerical values correspond with the analytical solution given by R.L. de Orio and his colleagues (6) for metal migration when only electric effects are considered. It is interesting to see that the numerical results show that including thermal and stress gradients does indeed show a greater degree of metal migration demonstrating that both effects should be considered. Although these are just numerical results, experimental studies are ongoing to try and validate the above.

CONCLUSIONS

This paper has discussed some of the issues faced by organizations in the high reliability sector when adopting COTS and lead-free components. The paper also details some of the developments in modeling technologies that can aid organizations understand the impact of refinishing lead-free components, using underfills for second level interconnects and identify the impact of technology scaling on failure mechanisms such as metal migration.

REFERENCES

1. Sengupta, S et al., "Assessment of Thermomechanical Damage of Electronic Parts Due to Solder Dipping as a Postmanufacturing Process", *IEEE Transactions on Electronics Packaging Manufacturing*, 30(2), 2007, pp. 128-137
2. Stoyanov et-al, Thermal Modelling and Optimisation of Hot Solder Dip Process, Proc IEEE Eurosime conference, pp 1-8, (2012)
3. Failure mechanisms and models for semiconductor devices, JEDEC Standard JEP 122F, Nov 2010
4. D. Ang, R.V.Ramanujan, "Hydrostatic Stress and Hydrostatic Stress Gradient in Passivated Copper Interconnects", Materials Sciences and Engineering A 423 (2006), pp. 157-165
5. Bailey et-al, Modelling metal migration for high reliability components when subjected to thermo-mechanical loading, Proc IEEE EPTC conference, pp1-5 (2012)
6. R.L. de Orio, H. Ceric, S. Selberherr, "Physically based models of electromigration: From Black's equation to modern TCAD models", P775-789, *Microelecronics Reliability*,2010

PANEL LEVEL PACKAGING – A MANUFACTURING SOLUTION FOR COST-EFFECTIVE SYSTEMS

R. Aschenbrenner, K.-F. Becker, T. Braun, and A. Ostmann
Fraunhofer Institute for Reliability and Microintegration
Berlin, Germany
aschenbrenner@izm.fraunhofer.de

ABSTRACT

Developing demands and the market show two main trends helping to shape the ongoing development of system integration technologies. First of all is an ongoing increase in the number of functions directly included in a system — such as electrical, optical, mechanical, biological and chemical processes — combined with the demand for higher reliability and longer system lifetime. Second is the increasingly seamless merging of products and electronics, which necessitates adapting electronics to predefined materials, forms and application environments. Only by these means systems sensors — which are often installed in extremely harsh environments — and signal processing can be implemented near to the point where signals are occurring.

Large area mold embedding technologies and embedding of active components into printed circuit boards (Chip-in-Polymer) are two major packaging trends in this area. This paper describes the potential of heterogeneous integration technologies researched at Fraunhofer IZM with a strong focus on embedding in printed circuit boards and embedding in molded reconfigured wafers with an outlook of advanced large area encapsulation processes for multi chip embedding in combination with large area and low cost redistribution technology derived from printed circuit board manufacturing.

INTRODUCTION

Most electronic systems available today are realized through an organic printed wiring board, on which the individual components are placed. The wiring board is exclusively used with regard to electrical and mechanical function. However there are numerous attempts and necessity in the development of modern electronic products, which have to lead to the integration of further system functions into the board.

Future board and substrate technologies have to ensure a cost efficiently integration of highly complex systems, with a high degree of miniaturization and sufficient flexibility in adaptation to different applications. Their functionality will be considerably enlarged by integration of non-electronic functions such as MEMS, antennas or optical components. New production methods will ensure a high throughput at very low cost.
To ensure high data transmission and processing rates new cost effective cooling technologies and 3D-

Packaging concepts will ensure a stable operation mode. The following priorities are seen for multi functional board and substrate technologies:

- Embedded devices technologies
  MEMS, passives, antennas, IC´s
- Low cost finer line & smaller via substrate and interposer
- Impedance controlled wiring
- Flexible substrates (reel to reel manufacturing)
- Integrated optical interconnects

To reach these priorities new materials for embedding and encapsulation have to be developed:

- High K and low K dielectrics
- High Tg polymers
- CTE matching between dies and substrate

On the following pages an overview will be given on the recent technological developments towards heterogeneously integrated SiPs, focusing on embedding technologies.

EMBEDDING INTO PCB

Integration technologies for electronic systems or subsystems have attracted research and developments efforts in recent years. Device miniaturization and increasing functional density are pushing not only CMOS fabrication technologies to decrease feature sizes on the semiconductor devices. Also packaging, interconnection, and circuit board technologies have to keep pace with miniaturization requirements.

The EU-funded project "HERMES" [i] has initiated, with wide participation of European industries and research institutes, advancing the embedding technology borders at R&D level and more importantly of bringing embedding technology in real manufacturing PCB production [ii].

Embedding technologies offer the advantages of direct contact to the chips without use of long wires or solder bumps and thus improved electrical performance, capability of 3D-stacking, reduced package thickness if thin components are available, and enhanced thermal performance for components assembled on thermal interfaces and heat sinks, as in the case of embedding power components. The emergence of embedding packages steadily changes the packaging value chain and consequently sets new roles for all players in the whole packaging value system. The initial value share of substrate suppliers for production of FCBGA devices was about 20% but now with embedded technologies the substrate suppliers can also perform the assembly of components before embedding and thus count for 55% of the embedded package value [iii]. However, it should be underlined, that the shift to embedding technologies marks also necessary adaptations to the supply chain which will potentially burden the value system. For instance, the necessity of RDL layer for chip pitch enlargement that makes chip components compatible with existing embedding capabilities or copper pad deposition should be definitely accounted for before the shift in embedded packages. Initial applications for embedded packages will be low cost, low pin counts applications such as analog and power devices (DC/DC converters, Power MOSFETS etc.). There are forecasts for a half billion dollars extended market by 2015 [iii].

In a number of European cooperation projects with partners from industry and research, embedding of power chips, like IGBTS and power MOSFET, is of high interest [iv]. In this paper current achievements of these projects will be shown, especially examples of realized devices and their characteristics. The dominating technology for power chip embedding is a face-up technology. Chips are bonded with their backside (drain contact) to a Cu substrate using highly conductive adhesive or solder. Using the face up assembly, a direct contact to the backside of the die is possible, allowing a lot of benefits for driving high currents and applying an efficient thermal management for the power devices. Then the chips are embedded by vacuum lamination of prepreg or RCC (resin coated copper) layers. Via holes to the top contacts (gate and source) are formed by laser drilling. The vias are metalized using conventional Cu plating. Finally conductor structures are etched in the top Cu layer, finalizing the circuit.

On the following pages the technology developed in the EU-funded project "HERMES" is described whose scope is to further develop all embedding technologies at advanced R&D level and more importantly to bring embedding technology in real manufacturing PCB production with a concrete goal of embedding components in 18"x24" PCBs.

Power System-in-Package Demonstrator
The combination of two or more embedded dies e.g. MOSFET or diodes, but also controller chips, results in embedded Power System-in-Packages (*Figure 1*). Here the embedding technology offers a variable technology platform for the realization of a large variety of packages on the same process line. One big advantage of such packages is their short interconnects, resulting in low inductances which allow faster switching speed.

Figure 1. Embedded Power SiP

The project goal is to embed the power devices into the PCB and to place the application-specific logic devices and passives on top of the PCB.

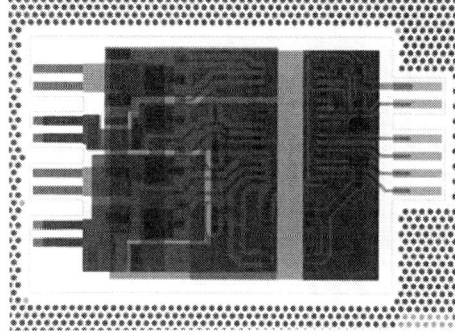

Figure 2. X-ray of a power module with four embedded power dies

This concept provides high flexibility for the logic parts combined with a careful power removal design for the power devices embedded in the PCB and isolated from the heat sink. The challenges for this solution are particularly great since in addition to the embedding process the high thermal conduction under electrical isolation has to be implemented. Therefore suitable materials for this isolation layer were investigated, which provide the needed thermal conductivity and fulfill the electrical isolation specifications. In *Figure 2* an x-ray picture of such an embedded power module, without the SMD components which will be assembled on top finally, is shown. The embedded MOSFET die has a thickness of 120 µm and is attached to a 70 µm thick copper foil using highly thermal conductive silver glue. Then the die is embedded into epoxy prepreg layers. The backside of the die is isolated to the thermal pad at heat sink location by a thermal laminate layer which is able to provide the needed thermal conductivity. In total the module is a four layer construction, containing four embedded power chips. All needed SMD components for the logic part are assembled on top of the module. These modules are currently under a fully qualification.

EMBEDDING INTO MOLDING COMPOUNDS
There are two main approaches for embedded die technologies:
First, there is the ChiP technology, the embedding of active dies into PCB as described above. Second, there is wafer level integration, where dies are embedded into polymer encapsulants and 3D vertical integration, where dies are embedded into the substrate. For wafer level integration a lot of activities are running worldwide. Main drivers are here the Embedded Wafer Level Ball Grid Array (eWLB) by Infineon [v] and the Redistributed Chip Package (RCP) by Freescale [vi]. Singulated dies are assembled on an intermediate carrier and encapsulated by compression molding, forming a polymer wafer with embedded silicon dies. This "reconfigurated" wafer is then released from the carrier. Using thin film technology, an electrical redistribution layer is routed on the wafer.

Finally, the wafer is singulated by sawing into single packages. One trend in eWLB technology is at the moment a double sided eWLB packaging with integration of vias through the encapsulant by integration of preformed PCB based vias allowing the stacking of eWLB packages [vii].

The combination of both concepts embedding into polymer by molding and redistribution by PCB technologies has the potential for highly integrated low cost packages and was successfully demonstrated for a 2-chip LGA package [viii]. The direct integration of Through Mold Vias (TMVs) can be easily integrated in such packages as vias are a standard feature in the PCB manufacturing process and can be adapted for the proposed concepts of embedding into polymer by molding and redistribution by PCB technologies. A principle draft of a Package-on-Package assembly (PoP) based on a wafer level embedded package with PCB based redistribution technology is shown in Figure 3. Within this section the development and evaluation of such a packaging technology with TMVs is described.

Process Flow
The general process flow starts with the lamination of an adhesive film to a carrier. This special adhesive film has one pressure adhesive side and one thermo-release side, i.e. by heating up the tape above a certain temperature, the thermo-release side of the tape loses its adhesion strength. On this carrier-adhesive film sandwich dies are precisely placed, the active side facing down towards the carrier. High accuracy is needed as die pads have to match with the redistribution layer. Molding is done by large area compression molding.

For chip redistribution, low cost PCB based technology with RCC has been selected. After lamination of the RCC film on both wafer sides in one step, µvias are drilled to the die pads and through mold vias in the same process step to connect to and bottom side.

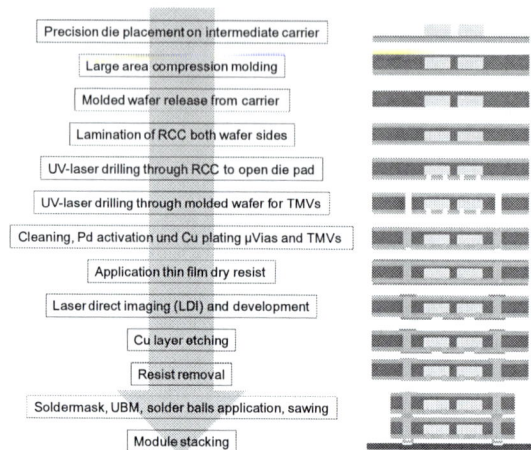

Figure 4. Process flow: PoP assembly based on wafer level embedded package with PCB based redistribution technology

Next process steps are cleaning, palladium activation and copper plating. By plating both, via filling and die pad connection to the copper layer and the top copper layer to the bottom copper layer are achieved. Conductor line formation is done by laser direct imaging (LDI) in combination with a dry film resist and copper etching. Finally, a solder mask and solderable surface finish as NiAu and solder balls can be applied. After package singulation by sawing the package can be stacked and connected by reflow soldering. The process steps described above are summarized in Figure 4.

Demonstrator Manufacturing
Reconfigured wafer assembly and compression molding was done on 6" wafer size. For compression molding a liquid compound with a maximum filler size of 55 µm has been selected. Wafer thickness was set to 670 µm due to sensor thickness of 550 µm to allow homogeneous overmolding of the components without damaging the stress sensitive MEMS components by the filler particles.

The embedded sensor and ASIC dies need to be prepared with an under bump metallization (UBM) to avoid incompatibilities with the used PCB processing. Typically, NiPd pad reinforcement is applied to the silicon wafer or to the reconfigured mold wafer.

Double sided redistribution has been processed accordingly to the process described above. A filled RCC material and process parameters as for the TMV evaluation have been used. Laser drilled via diameter was set to 150 µm which gave homogeneous and reproducible via drill and metallization results in the 670 µm thick mold wafer. *Figure 5* depicts the final acceleration sensor package with pads on the package top side for stacking of the pressure sensor package and TMV for 3D routing to the package bottom to connect ASIC and acceleration sensor as well as the substrate.

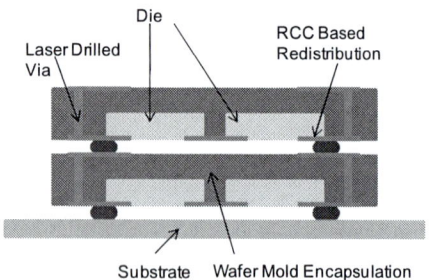

Figure 3. Schematic of a Package-on-Package assembly based on wafer level embedded package with PCB based redistribution technology

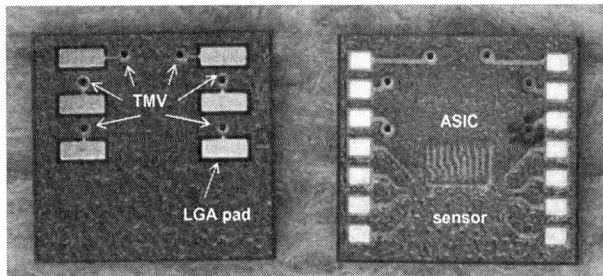

Figure 5. Photograph of an acceleration sensor with ASIC package, package top (left) and bottom (right)

ASIC and sensor show a minimum contact pitch of 110 µm and pad size of 80x80 µm². Die positions have been measured after molding and automatically used to adapt the µvia drill position to the die pads and the wiring of the conductor lines to compensate the die shift during assembly and molding. Therewith, a good alignment between µvias and die pads could be achieved without shorts or off target positions (s. *Figure 6*).

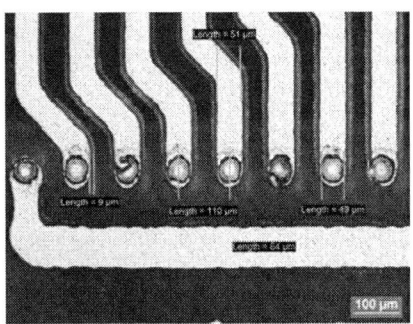

Figure 6. Optical micrograph of an RCC based RDL µvia interconnection to ASIC pads with 110 µm pitch

Figure 7 depicts an X-Ray image of a manufactured demonstrator package showing the interconnection of the top and bottom package metallization by the through mold vias.

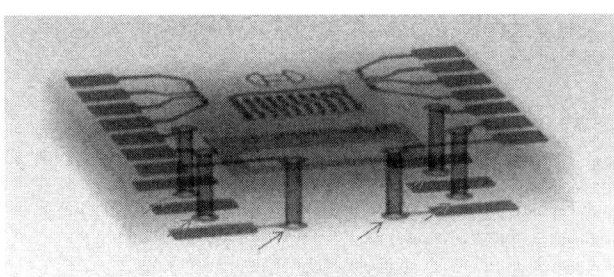

Figure 7. X-Ray image of an acceleration sensor and ASIC package with through mold vias (TMV) marked by red arrows?

Cross sections has been done to investigate µvia interconnects as well as through mold vias. *Figure 8*

shows the interconnection between ASIC and acceleration sensor with the µvia connection the conductor line with the dies. µvia are well aligned and Cu-filled without any air entrapments.

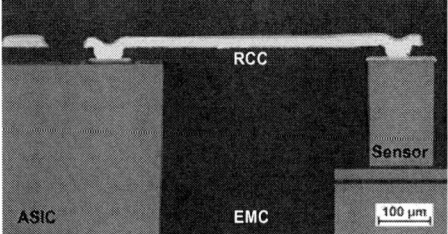

Figure 8. Cross section of a mold embedded acceleration sensor and ASIC package with RCC based redistribution

Through mold vias are homogeneously metallized with Cu with a thickness of around 15 µm depending TMV wall roughness (s. *Figure 9*). Via edges are also well connected to the top and bottom metallization. Solderable surface finish NiAu was also applied in the TMVs resulting in a constant layer thickness of 5 µm as TMVs were not plugged or covered during the electroless metallization process.

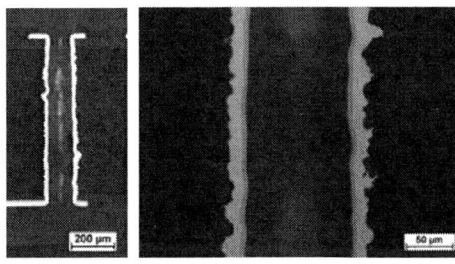

Figure 9. Cross section of a metallized through mold via; left: TMV overview, right: detail via metallization

Pressure and acceleration sensor packages were stacked and assembled on board by soldering. Demonstrator as shown in Figure 10 allows now the functional testing of the entire sensor stack and therewith the proof of technology.

Figure 10. Photograph of a demonstrator stack with acceleration and pressure sensor package

Fan out wafer level acceleration sensor and ASIC package with RCC based redistribution technology and through mold vias as a single package as well as a stacked package have been successfully tested on their functionality. The acceleration sensor performed properly within its specification range. Hence, it can be derived that the above described technology is well suited also for stress sensitive components as MEMS devices and sensors.

PANEL LEVEL PACKAGING

Cost is a major driver of technological developments in microelectronics packaging, so the increase in size found with wafer diameters having evolved from 2" in the 70s to 300 mm today is mimicked by embedding technology. Currently, wafer form factor is mandatory for the manufacturing of mold embedded components, as cost effective thin film RDL technology is only available for circular shapes. But with the PCB-derived redistribution layer application as a feasible fine pitch alternative, a cost effective upscaling of wafer level mold embedding to panel level molding will be a promising path.

Technologically speaking, there is a variety of compression molding compounds for embedded wafer level molding from different suppliers on the market available at the moment. For mold embedding technology materials should have low chemical shrinkage, low cure temperature and match thermo-mechanical properties for low warpage of the molded wafer and low die shift after molding. Flow properties should allow homogeneously filling of large cavities. Basically, state of the art materials can be divided in liquid and granular compounds. From the processing point of view, the liquid materials are dispensed in the middle of the cavity and flow during closing and compression of the tooling to fill the entire cavity. In opposition to the liquid, the granular compound is distributed nearly homogeneously all over the cavity. The compound melts and the droplets have to fuse during closing and compression of the tooling. With a view to large area encapsulation up to 610 x 457 mm² granular compound is the material of choice as here processing is not influenced by cavity size where for liquid compounds larger cavities mean longer flow lengths or application specific dispense pattern with a risk for incompletely or inhomogeneously filling of the cavity.

Sheet or roll lamination under vacuum are alternative process options for large area encapsulation especially as state of the art compression molding machines at the moment only allow encapsulation of areas up to 300 mm in diameter or side length respectively. For encapsulation by lamination, highly filled epoxy resin films are used with material properties comparable to liquid or granular compounds and available film thicknesses range from 50 µm up 1000 µm defining also final product thickness.

Standard PCB lamination presses allow the processing of full format sheets up to 610 x 457 mm² under pressure, heat and vacuum. Here, drawback is the quite long cycle time due to heating and cooling of the entire press during lamination. However, there is the possibility to process more than one sheet in one cycle, what also leads to a higher yield of the process. Short cycle presses even eliminate the heating and cooling time during lamination cycle as these machines work with preheated plates making the process much faster.

There is also potential for using PCB lamination presses for low cost, large area encapsulation of reconfigured wafer sheets with well-known wafer level molding compounds as described above. At the same time tooling costs for lamination are expected to be much lower as for compression molding. With this technology combination already today the manufacturing of large area mold embedded components is feasible.

CONCLUSIONS

Heterogeneous integration bridges the gap between microelectronics and its derived applications. Two main forces drive progress in this area — emerging device technologies and new application requirements. New technologies and architectures are arising to bring the progress made in microelectronics, microsystem technologies, and bio-electronic or photonic component technologies into application. The future belongs to integration technologies that combine several components into a highly integrated assembly in one package.

One target application was a multi sensor device for indoor navigation purposes integrating magnetic sensors, an acceleration sensor and a pressure sensor – based on sensor development within the MST-SmartSense project. For demonstration purposes the wafer level approach has been applied to pressure sensor / ASIC and to acceleration sensor combinations. For each combination the best suited technology variant was chosen and package stacking could be demonstrated. Successful functional testing of this stacked device also proofs that this packaging technology is suited for stress sensitive sensor ICs.

Apart from showing the overall suitability of embedding technology the special focus was put on z-axis routing possibilities, where through mold vias have been evaluated regarding process and reliability.

In this paper a wide variety of technologies has been described that allow the generation of maximum miniaturized microsystems or SiPs, consisting of at least two components – all technologies bear the possibility to integrate multiple heterogeneous components, fulfilling this demand to heterogeneous integration.

REFERENCES

i   For information on HERMES refer to:
http://www.hermes-ect.net/

ii   A. Ostmann, D. Manessis, H. Stahr, M. Beesley, J. De Baets, M. Cauwe, "Industrial and technical Aspects of chip embedding", Proceedings in the 2nd ESTC 2008 Greenwich, September 1-5, 2008, pp. 315-320

iii   Yole Development report, "Embedded Wafer-Level-Packages", 2010

iv   D. Manessis, L. Boettcher, A. Ostmann, K.-D. Lang, "Embedded Power Dies for System in Package", International Workshop on Power Supply On Chip October 13-15, Cork, Ireland; http://www.powersoc.org/index.php

[v]   T. Meyer, G. Ofner, S. Bradl, M. Brunnbauer, R. Hagen; Embedded Wafer Level Ball Grid Array (eWLB); Proceedings of EPTC 2008, Singapore.

[vi]   B. Keser, C. Amrine, T. Duong, O. Fay, S. Hayes, G. Leal, W. Lytle, D. Mitchell, R. Wenzel; The Redistributed Chip Package: A Breakthrough for Advanced Packaging, Proceedings of ECTC 2007, Reno/Nevada, USA.

[vii]   Y. Jin, X. Baraton, S. W. Yoon, Y. Lin, P. C. Marimuthu, V. P. Ganesh, T. Meyer, A. Bahr; Next Generation eWLB (embedded Wafer Level BGA) Packaging; Proceedings of EPTC 2010, Singapore.

[viii]T. Braun, K.-F. Becker, L. Böttcher, J. Bauer, T. Thomas, M. Koch, R. Kahle, A. Ostmann, R. Aschenbrenner, H. Reichl, M. Bründel, J.F. Haag, U. Scholz; Large Area Embedding for Heterogeneous System Integration; Proceedings of ECTC 2010, Las Vegas, USA.

3D-TSV VERTICAL INTERCONNECTION USING Cu/SnAg DOUBLE BUMPS AND NON-CONDUCTIVE FILMS (NCFs)

Kyung-Wook Paik*, Yongwon Choi, and Jiwon Shin
Department of Materials Science and Engineering
KAIST
Daejon, Korea
kwpaik@kaist.ac.kr

ABSTRACT

In this study, the chip to chip or wafer copper/SnAg eutectic solder bonding method using NCFs for 3-D TSV stacking was investigated as an alternative 3D-TSV interconnection method. The specially formulated non-conductive epoxy polymer adhesive was applied at 3d-TSV wafers as a B-stage film format before Cu/SnAg eutectic solder bonding resulting in easier processing and no extra underfill process. In addition, the electrical interconnections between Cu/SnAg double bumps for 3D-TSVs of the stacked chips and joint morphology were investigated. Excellent electrical joint properties through the arrays of the Cu/SnAg double bumps between two stacked chips were obtained and the bump joint resistances showed no change even after the reliability tests. And excellent solder joint structure was obtained by the flux function added NCF material. Therefore, excellent 3D-TSV vertical interconnection method with Cu/SnAg double bumps using NCFs bonding was demostrated.

Key words 3D-TSV, Vertical interconnection, Cu/solder double bump, NCFs

INTRODUCTION

Packaging density is defined by total chip area per total package area. To achieve the most number of circuits to be packaged in the least amount of space, packaging density has increased by many packaging techniques such as CSPs and flip chip. However, as hand-held electronic equipments have been widely used, higher packaging density was required to package more chips or circuits on smaller board areas. For this reason, new ways of packaging techniques have been emerged. Such as 3-D stacking including stacking dies, packages, or modules in a z-direction. By 3-D stacking, the board area can be substantially saved and interconnection length can be shortened resulting in extreme high packaging density. Therefore, 3-D stacking has been widely adapted in in various chips such memory and micro-processor chips packaging.

In 3-D stacking, the die stacking is widely used than packages or modules stacking. Die stacking is usually achieved using wire bonding, solder bump, and through silicon via (TSV). Among these ways, TSV is the most advanced state-of-art method. Broad researches have been done in the 3D-TSV stacking areas.

Nowadays, many researches have been reported on the 3D-TSV processing, vertical interconnection and stacking of multiple TSV chips. Among various vertical interconnection methods, Cu/SnAg eutectic solder double bum is one of the promising bonding method. After the 3D-TSV chips are bonded with double bumps, the gap between stacked chips should be filled with underfill materials. However, underfill between 3D-TSV stacked chips become very difficult because of increased chip/wafer area, decreased gap height, and multiple chip/wafer stacking.

Therefore, new method to bond 3D-TSV chips with Cu/SnAg double bumps and underfill the gap of 3D-TSV chips is introduced using specially formulated NCFs. NCFs were specially formulated to give flux functions, and can be also wafer-level laminated on a bumped wafer acts as both bonding and underfill material.

In this paper, the NCFs bonding mechanism and the effects of material properties of NCFs such as curing behavior and viscosity on the Cu/SnAg double bumps joint formation were investigated.

EXPERIMENTS

Several kinds of NCFs were formulated using various kind of curing agents, epoxies, and thermoplastic resins. Each ingredient was mixed together with solvents in a bottle. They became a homogeneous resin by milling for several days. The resins were formed into a film by coating on a releasing film using a manual doctor-blade coater. The thickness of the adhesive films was precisely controlled.

The test vehicles were designed and fabricated to investigate the bumps joint formation and joint stability as shown in Fig. 1. Two patterns for daisy chain resistance and single bump joint resistance were adapted at the chips and substrates. 10 um of Cu pillar and 10 um of SnAg were electroplated for the bumps on a wafer. Al and Au were electrodeposited on another wafer along the pattern to measure the contact and daisy chain resistance after the bonding of a chip and substrate.

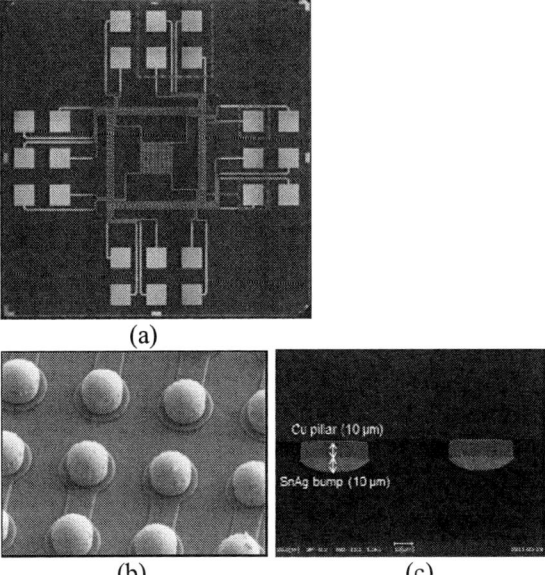

(a)

(b)   (c)

Figure 1. Test chip and substrate
(a) Substrate with daisy chain and single bump resistance patterns, (b) test chip with Cu/SnAg bumps, (c) cross-section view of Cu/SnAg bumps.

NCFs were laminated on a test chip before the bonding process. Lamination temperature was below the onset temperature of NCF to prohibit the curing. A flip chip bonder was used for the aligning and bonding of a chip to a substrate. Bonding temperature was set enough to ensure the melting of SnAg bump and therefore the metallurgical reaction could happen. Temperature was elevated after the alignment and attachment under a given pressure.

Effects of viscosity and degree of cure of NCFs on the electrical interconnection of the bumps were investigated at various temperatures during the bonding process. The effects of the NCF curing behaviors were investigated by controlling the curing temperature and curing time of NCFs. The effects of the viscosity of NCAs were also investigated with two kinds of NCAs having different viscosities.

A differential scanning calorimeter (DSC) was used to measure the curing behaviors of each NCF. The dynamic scanning was performed from 30 °C to 250 °C, to measure the onset and the peak temperatures. Isothermal scanning was performed to measure the curing time. A rheometer was used to measure the viscosity variation of NCFs according to temperature ranging from 40 °C to 250 °C. Daisy chain resistance and bump contact resistance were measured for every test conditions. SEM observations were also performed to see the IMC formation on the joint surface after the shear test .

RESULTS AND DISCUSSION
NCFs with High Viscosity
Two types of NCFs having different viscosities were formulated. Figure 2 shows the viscosity changes of two NCFs at various temperatures. It showed that the minimum values of two NCFs differ from 400 to 12000 Pascal second.

Two kinds of viscosity have their own pros and cons. Low viscosity gives easier gap filling and no polymer residues inside the bump joint interface even after the melting of SnAg solder. However, NCF with low viscosity easily overflowed along the edge of upper chips causing the contamination of bonder tools and potential malfunction of equipment.

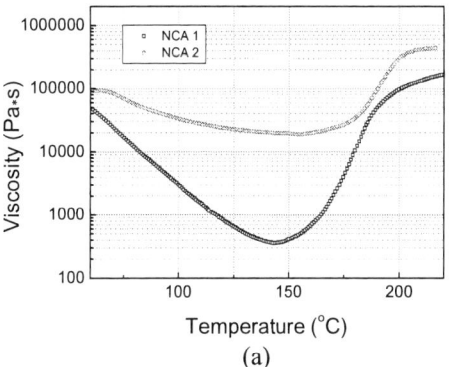

(a)

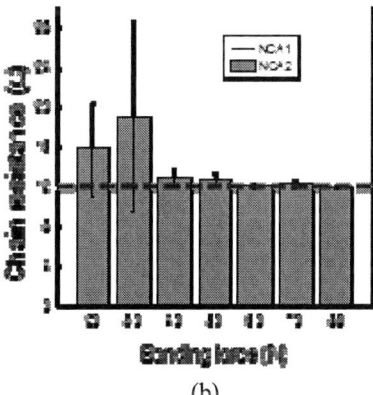

(b)

Figure 2. (a) Viscosity of two NCFs used in this experiment, (b) Daisy chain resistances of two NCFs at various bonding forces.

On the other hand, high viscosity needs more sensitive control of bonding conditions such as bonding force and temperature. However, high viscosity NCFs prevents NCF from overflowing over the thin chips. Therefore, NCF with high viscosity is more favored. Therefore, the bonding force with high viscosity NCF was optimized to give good electrical interconnection by measuring the chain resistance after bonding. At more than 60 N bonding forces, daisy chain resistance became stabilized.

Mechanism of the Joint Formation
The bump joints were formed with the complicated interactions between the properties of SnAg and NCFs. Once the pressure was applied and the temperature was raised, SnAg and NCFs mutually interacted according to their material properties at the certain temperature.

The mechanism of the bump joint formation was experimentally identified. Once the pressure and heat were applied, NCFs which is laminated on the bump side started to adhere on the substrate side and filled the cavity between bumps and the gap between stacked chips. While pressure is still applied, elevated temperature allow the NCF flow easily so that NCF could fill the gap between the chip and substrate. However, bumps from upper and electrodes from lower chips did not physically contact each other because the temperature was not high enough for NCF to flow out completely to remove the gap between upper bumps and lower electrodes. As temperature increases with constant applied pressure, NCF is forced to flow out of the area of bump surface and fill the gap between near bumps. And finally, bumps and electrodes are physically touched each other and electrically interconnected. At the same time, high temperature start the curing reaction of NCA with gradual increase of viscosity. After the temperature reached the melting point of SnAg, SnAg bump wetted on the pad surface and formed the metallurgical joint. During this solder wetting process, specially added flux function removes the oxide of SnAg. Fig. 3 shows the changes of viscosity and the daisy chain resistance as a function of temperature, and the cross-section of the bumps and electrode at various temperatures.

Figure 4 (a) and (b) showed the joint surface of both upper chip and lower substrate after the die shear test.

From the pictures, the remained parts of IMC on both sides were detected by EDS analysis. On the both of sides, the IMC composed of Sn and Cu was found showing the evidence of the formation of eutectic solder joints. Therefore, Cu/SnAg double bumps for the TSV interconnection were well established using Cu/SnAg double bumps with NCFs. After the bonding, ten samples having four chain resistance patterns were put into the Pre-con reliability test, which is the reflowing after the moisture absorption. From the Figure 5, all the 40 patterns showed no open failure with the stable resistance value after the Pre-con reliability test.

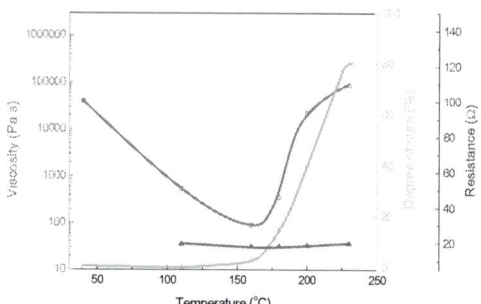

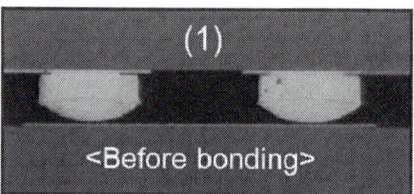

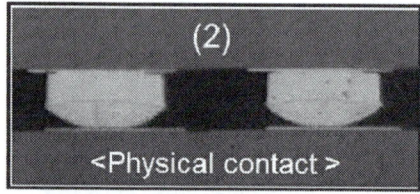

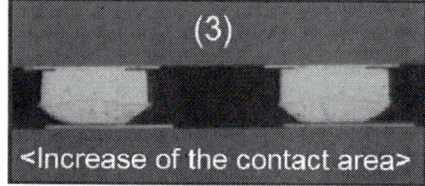

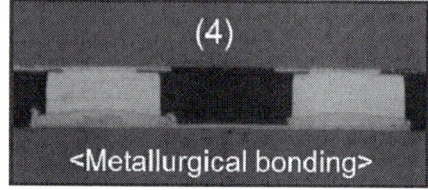

Figure 3. The changes of viscosity and the daisy chain resistance as a function of temperature, and the cross-section of the bumps and electrode as bonding progress at various temperatures.

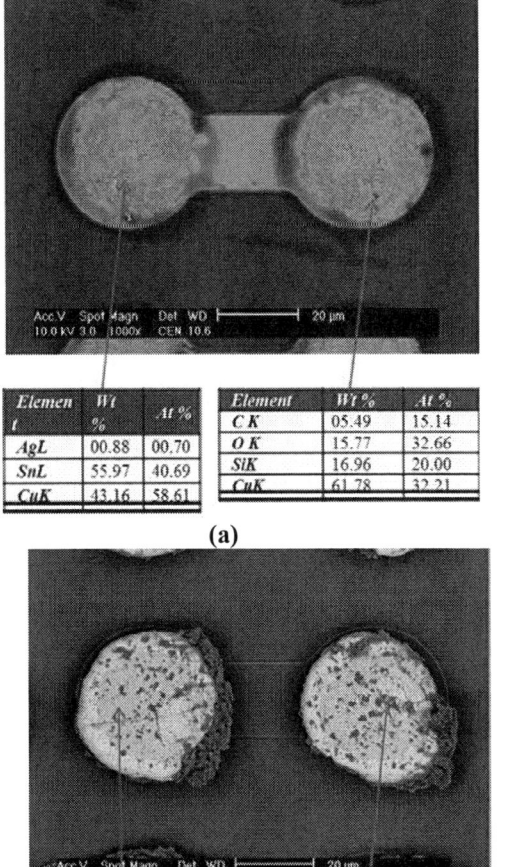

Element	Wt %	At %
AgL	00.88	00.70
SnL	55.97	40.69
CuK	43.16	58.61

Element	Wt %	At %
C K	05.49	15.14
O K	15.77	32.66
SiK	16.96	20.00
CuK	61.78	32.21

(a)

Element	Wt %	At %
AgL	66.69	66.71
SnL	29.48	26.80
CuK	03.82	06.49

Element	Wt %	At %
C K	06.00	16.11
O K	22.25	44.88
SiK	16.46	18.92
AgL	01.88	00.56
SnL	32.16	08.74
CuK	21.25	10.79

(b)

Figure 4. The joint surface of both upper chip and lower substrate after the die shear test. EDS analysis showing the presence of Sn and Cu at the both side of chip and substrate.

Fig. 6 shows the excellent bump joint structure after optimizing the NCFs and bonding condition. Well defined solder joints between Cu bumps and electrodes were obtained at the 40 micro pitch 3D-TSV test samples.

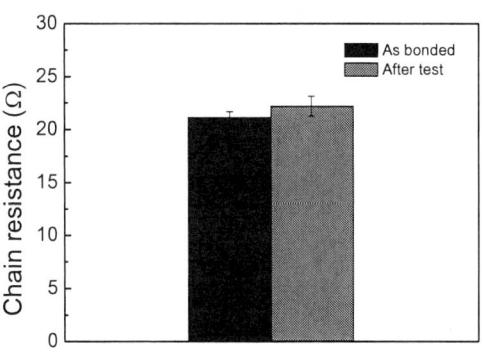

Figure 5. Daisy chain resistance changes after the Pre-con reliability test.

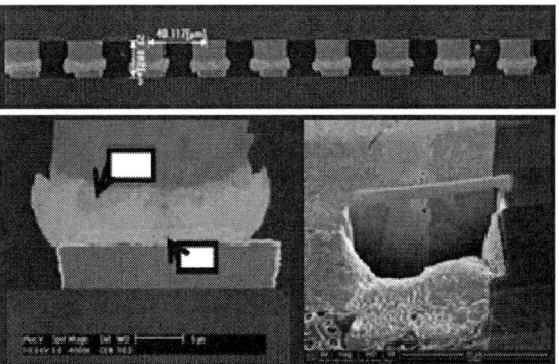

Figure 6. Cross-section of 40 micron pitch 3D-TSV assembled samples using NCFs.

CONCLUSION

The new bonding processes using specially formulated NCFs which is a non-conductive polymer adhesive for the 3D-TSV chip stacking was demonstrated. The mechanism of Cu/SnAg double bumps with NCFs polymer hybrid joint formation was well determined. In addition, the effects of NCFs viscosity on the bump joint formation were investigated. SnAg bump joints were successfully formed with stable electrical interconnection as well as the reliability test results.

REFERENCES

1. R. R. Tummala, Fundamentals of Microsystems Packaging, McGRAW-HILL, 2001.
2. R. R. Tummala, Microelectronics Packaging Handbook, Chapman & Hall, New York, 1997.
3. Karnezos, M., "3D packaging: where all technologies come together", Electronics Manufacturing Technology Symposium, 2004 IEEE/CPMT/SEMI 29th International, 2004, p.64-67.

MICROSTRUCTURE CONTROL OF UNI-DIRECTIONAL GROWTH OF η-CU_6SN_5 IN MICROBUMPS ON (111) ORIENTED AND NANOTWINNED CU

Han-wen Lin[1], Jia-ling Lu[1], Chien-min Liu[1], Chih Chen[1,*], King-ning Tu[2], Delphic Chen[3], and
Jui-Chao Kuo[3]
[1]Department of Materials Science & Engineering, National Chiao Tung University
Hsinchu, Taiwan, R.O.C.
[2]Department of Materials Science and Engineering, University of California Los Angeles
Los Angeles, CA, USA
[3]Department of Materials Science & Engineering, National Cheng Kung University
Tainan, Taiwan, R.O.C.
*chih@mail.nctu.edu.tw

ABSTRACT

The growth of η and η'-Cu_6Sn_5 has been proven as a preferential growth behavior on single crystal copper. However, a layer of single crystal copper is not possible to be electroplated. It can not be utilized in the electronic industry. In this paper, we electroplated an array of (111) uni-directional Cu pad followed by electroplating SnAg2.3. After being reflowed at 260°C for 1 minute, the η-Cu_6Sn_5 showed a preferential growth to (0001) plane. As reflow time extended, the preferential growth behavior would change. The intensity of (0001) decreased while that of ($2\bar{1}\bar{1}3$) increased. It means the preferential growth of η-Cu_6Sn_5 would change during reflow. Eventually, the preferred orientation of η-Cu_6Sn_5 changed to ($2\bar{1}\bar{1}3$) after 5 minutes of reflowing. It is also found that this preferential growth behavior of Cu_6Sn_5 would be affected by the quality of (111) uni-directional Cu.

Key words: Intermetallic compounds; Soldering; Copper;

INTRODUCTION

The use of conventional Sn-Pb solder alloy in consumer products has been forbidden since 2006 due to RoHS (Restriction of Hazardous Substances Directive). Some new alloy system such as Sn-Ag, Sn-Cu, Sn-Ag-Cu are therefore developed and studied. These alloy systems are all massive tin matrix with minor addition, typically less than 5 wt%, of other metal elements. For this reason, the study of metallurgical reaction and mechanical behavior between tin and copper or nickel substrates are carefully concerned in recent years. Among all the factors, growth of Sn-Cu and Sn-Ni intermetallics draw many attentions since the intermetallic would act not only as a mechanical joint but also a brittle interface at the same time due to its nature.

As the trend goes to minimize the packaging size, the volume fraction of tin in solder joints is decreased. In micro-bumps, the solder height is only 5-20μm. It turns out that the tin might be consumed during reflow process. In that way, the Cu_6Sn_5 would control the properties of

solder joints and prevent the early failure caused by Sn orientation.[1] Also, Cu_6Sn_5 can be employed in Li-ion batteries as anode materials due to its favorable properties such as lower cost and storage capacity.[2][3] So, it is beneficial to control the properties of Cu_6Sn_5 by controlling its orientations.

The growth of η and η'-Cu_6Sn_5 has been proven as a preferential reaction with single-crystal Cu substrates. Suh et al. found that the preferential growth behavior between η'-Cu_6Sn_5 and (001) single crystal Cu.[4][5] Zou et al. found the η'-Cu_6Sn_5 a very strong texture on (001), (011), (111) and (123) single crystal Cu substrate.[6][7]. However, for the electroplating technology nowadays, it is unlikely to electroplating single crystal metals on silicon wafers. On poly-crystalline Cu, Kumar et al have reported that the orientations of Cu_6Sn_5 grains formed on were random and had no preferential relationship. [8] So, the preferred orientation behavior can not be used in electronic industries. In this paper, we produce a layer of (111) uni-directional Cu by electroplating. The SnAg 2.3 was used as solder materials. The orientations relationship between uni-directional Cu and Cu_6Sn_5 were examined. Also, the effect of quality of uni-directional Cu on the orientation of Cu_6Sn_5 was discussed.

EXPERIMENTAL PROCEDURE

To examine the relationship between the Cu_6Sn_5 and (111) uni-directional Cu pad, which can be utilized in 3D-IC packaging, Cu pads were electroplated by various current densities of 1 ASD (amps per square decimeter) and 8 ASD. Then, the solder alloy of SnAg2.3 is electroplated.

In the first part, the bump die was reflowed at 260°C for 1 and 5 minutes to grow Cu_6Sn_5. Then, after cooling in the air, the sample was grinded by abrasive papers of #1000, #2000, and #4000 followed by polishing with Al2O3 powder of 1.0μm and 0.3μm. Finally, we use colloidal silicas to remove the surface layer, which might be damaged during the process of polishing. We observe the morphology and the orientation image map from cross section view and top view after grinding and polishing.

In the second part, we jointed two single bumps together to make a complete solder joint and then made them reflowed for several minutes. Samples were grinded by using abrasive papers #400, #1000, #2500 and #4000 after air cooling, and then also polished by Al_2O_3 of 1 and 0.3 µm and colloidal silica. Focus ion beam technique was also adopted in cross-sectional observation. The images of solder joints were taken by JOEL FE-SEM 7001. Orientation maps, inversed pole figures and pole figures were collected by EDAX EBSD system. The XRD and TEM would also adopted to justify the experimental results.

The monoclinic η'-Cu_6Sn_5 has been found by A.K. Larsson et al. [9] in 1994. Ever since that, the crystal structure of Cu_6Sn_5 has become controversial. Larsson has reported the transformation temperature of Cu_6Sn_5 from η' to η is 186 °C. However, since the process of making solder joints includes reflow at up to 260 °C and cooling down to room temperature. The real stable phase after the joints were made is not sure. Ghosh reported on the η↔η'-Cu_6Sn_5 kinetics and transformation energy.[10] Laurila et al. showed that time for the η to η' transformation was insufficient with typical cooling rates after soldering.[11][12] Therefore, the η-Cu_6Sn_5 would be remained as a metastable phase. Nogita et al. also make a TTT diagram about the phase transformation of Cu_6Sn_5. They conclude that under 70 °C, the η-Cu_6Sn_5 in the joint could hardly transform into monoclinic structure.[13] Nogita have also found that the Ni has a effect to stabilize the structure of η-Cu_6Sn_5. [14][15] And Schwingenschlögl has further proven it with first principles calculations.[16] The most common combination used as under bump metallization layer in packaging industry is Cu and Ni. Therefore, after reflowing, the solder must contain some Ni atoms, making the intermetallics $(Cu,Ni)_6Sn_5$, which is more stable as hexagonal crystals. So, the orientation relationships between η-Cu_6Sn_5 and Cu are more important.

EXPERIMENTAL RESULT
1. The (111) Uni-Directional Cu
To make a preferential relationship between intermetallics and copper useful in packaging industry, the copper must be able to be electroplated. In that way, the single crystal copper is so far unpractical to be adopted. The (111) uni-directional Cu with nano-twins were first found by Li et al. and they also did lots of research about the mechanical and electrical properties.[17][18] In our experimental, we made several arrays of Cu pads by electroplating with some particular combinations of Cu seed layer and electroplating setups. With these configurations, we made an array of Cu pads with (111) uni-directional Cu. The diameter and thickness of copper pad are 100 µm and 20

µm respectively.

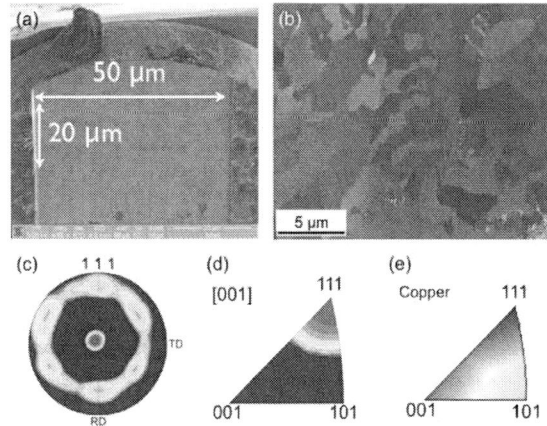

Figure 1. a) The plane-view of (111) uni-directional Cu Pads; **b)** the orientation image map, **c)** the (111) pole figure and **d)** the inversed pole figure of Cu pads; (e) the reference figure.

To examine the consistence of the orientations of copper, we created a flat surface on the copper pads by focused ion beam. The created area with length of 50 µm and width of 20 µm is shown in figure 1(a). By using EBSD, the colors of copper grains were all blue or blue purple in the orientation image map of figure 1(b). With the reference figure 1(e) about the orientation of copper, the orientations of copper on the pads were confirmed to be either (111) or close to (111). It means that the surface of copper is (111). Since the copper has a crystal structure of face-centered cubic, the direction of [111] is therefore normal to the surface. The grain size of electroplated copper is about 2 - 5 µm. The (111) pole figure of figure 1(c) proves that the pole is along the normal direction of surface. And by this figure, it is suggested that the (111) pole of Cu is not exactly normal to the surface. However, the angle difference between (111) pole of Cu and the normal direction of Cu surface is less than 10 degrees. Also by the inversed pole figure, the orientation was confirmed again.

In addition, a TEM sample was prepared by FIB. In figure 2(a), the bright field image was taken and the grain of copper was observed. The diffraction patterns of 3 adjacent grains were taken separately and then superimposed on each other in figure 2(b). It shows that these grains were all with a pole of (111) and only few rotating degrees exist between different grains. Although the Cu is still polycrystalline, the orientation is preferred to (111). And it is possible to electroplate (111) uni-directional Cu.

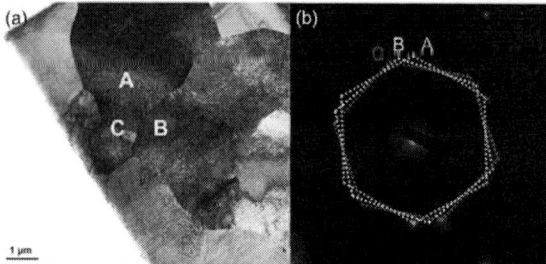

Figure 2. a) The TEM image of Cu pads from top-view; **b)** the diffraction patterns of three adjacent Cu grains.

The Preferential Orientation Relationship In Bump-Die Structure.

The SnAg2.3 was electroplated on (111) uni-directional Cu pad. The reflowing temperature was set to be 260 $^{\circ}$C. After reflowing for 1 minute, the chip was carefully mounted by low-temperature epoxy. To show the cross-sectional area of single side bump, the sample was gridded and polished. The focused ion beam was adopted to perform a final cut. Since there were Cu, Cu_6Sn_5 and Sn coexisting, the advantage of adopting FIB is to make a cut precisely at the desired position and to a definite thickness. Also, by performing the FIB technique, the strain-free surface can be revealed for both 3 phases.

Figure 3(a) shows the cross-section of a bump die after final cut by focus ion beam. The void inside Cu pad was damaged during grinding and those inside Sn was because of the flux. The red rectangle area was carefully observed by EBSD. Figure 3(b), (c), (d) show the orientation image maps of Sn, Cu_6Sn_5 and Cu. The orientation map shows the orientation of these phases in the direction perpendicular to silicon chip. In other words, the orientation relationships along the direction of growth of these phases are shown in figures. And again the color of orientation image maps would represent the orientation of phases.

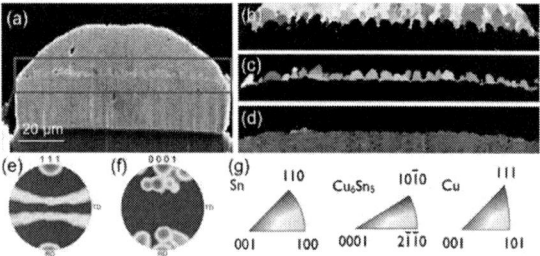

Figure 3. a) The cross-sectional area observed in bump die structure. The orientation image maps of **b)** Sn, **c)** Cu_6Sn_5 and **d)** uni-directional Cu. **e)** the (111) pole figure of Cu. **f)** the (0001) pole figure of η-Cu_6Sn_5; **g)** the reference figure.

As displayed in figure 3(d), the grain of Cu is columnar and the color of Cu is blue with some interlaced bands on

it. Corresponding to figure 3(g), the orientations of (111) Cu was assured again. The interlaced bands are confirmed as nano-twins with only 5 - 20 nm of twin spacing. The grain size of intermetallics is about 8-μm widths and 2-μm height. Each scallop has different colors, indicating that each scallop is a single grain of Cu_6Sn_5. These intermetallics are all in the color of red and orange. It suggested that the intermetallic have certain preferred orientation on the (111) Cu pads. By reference of figure 3(g), the orientation of intermetallics is close to (0001). To assure this result, we examine the pole figure of (0001) of intermetallics and that of (111) Cu. In figure 3(e), the (111) pole figure of Cu showed the (111) pole was majorly align with the rolling direction of Cu. Combined with figure 3(f), the (0001) pole figure of Cu_6Sn_5, the (0001) pole were also along the rolling direction. As mentioned in section 3.1, [111] direction of Cu is not exactly normal to the surface, so the pole figure of both Cu and Cu_6Sn_5 showed some leaning behavior. The degree of leaning is few. With these results, the orientation relationship between (111) Cu and η-Cu_6Sn_5 can be clearly justified: $\{111\}_{Cu} \parallel \{0001\}_{Cu6Sn5}$. Since the Cu/Sn couple was reflowed at 260 $^{\circ}$C for only 1 minute, the Cu_3Sn may form little. These orientation relationships are credible because the intermetallics contacted with Cu directly.

Effect of Reflowing Time On The Orientation Relationship

To obtain a general view on the orientation of every intermetallic on one Cu pad, the surface of intermetallic are revealed by grinding and polishing from top surface. Since the intermetallics in section 3.2 were too small and hard to be reached while grinded by human hands, the sample was further reflowed for 4 minutes more to make intermetallics larger.

Figure 4(a) shows the SEM image of Cu pad with Sn and Cu_6Sn_5 on it. The islands on the Cu pad were Cu_6Sn_5 formed during reflow. Between them are the residue tin phase. With the orientation image map of figure 4(b), the colors of intermetallics were red, orange and yellow. It should be noticed that some noisy signals were shown on the figure. These noises at center of the pad are mainly because of the residues left while polishing. And because we revealed the plan-view of intermetallic by grinding from the top of solder ball, the shape of hemisphere makes the peripheral area of pad bumpy. That also makes some noises around the pad.

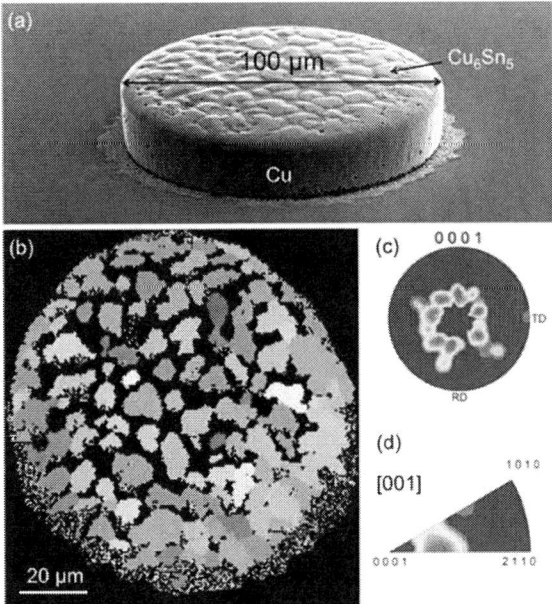

Figure 4. a) The SEM image from plane-view of Cu_6Sn_5; **b)** the orientation image map of Cu_6Sn_5 from top-view; **c)** the (0001) pole figure and **d)** the inversed pole figure of Cu_6Sn_5

There are something different from those reflowed for only 1 minute. First, more Cu_6Sn_5 grains appears in yellow in figure 4(b). It indicates that the orientation of intermetallics had been changed from (0001) toward (20). The (0001) pole was dispersed from the center of normal direction in figure 4(c). It also means that the (0001) plane of η-Cu_6Sn_5 were not parallel to the (111) plane of copper substrate. By the accumulated result of figure 4(d), the inversed pole figure showed that the major orientations of intermetallics have been changed. Many researches have already shown the morphology of Cu_6Sn_5 would change after reflowed at higher temperature or for longer time. The other intermetallics of Cu_3Sn start to grow at the interface between copper and Cu_6Sn_5 then. The formation of Cu_3Sn would break the preferential growth relationship because the Cu_6Sn_5 would then grow on Cu_3Sn, rather than Cu.

The Preferential Orientation Relationship in a Solder Joint

The preferential relationship between Cu_6Sn_5 and Cu could be used in real packaging joints. The solder joints were fabricated by joining two bump-dies together. We flipped one chip with uni-directional Cu pads and SnAg2.3 and then stacked it on another one. The joints were then reflowed at 260 °C for 3 minutes in total to make the joints compact. The sample was then grinded and polished. And the focused ion beam was used to make the final cut.

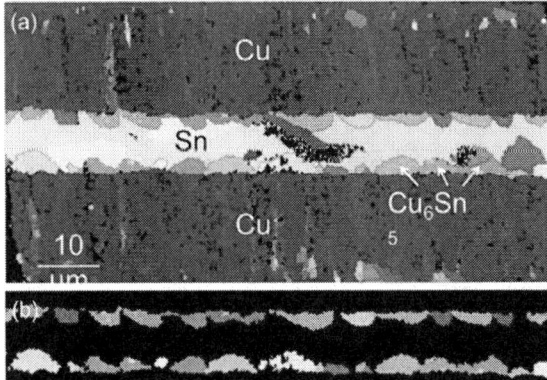

Figure 5. The orientation image maps of **a)**the solder joints reflowed at 260 °C for 3 minutes; **b)** the Cu_6Sn_5.

Figure 5(a) shows the orientation image maps of the joints. The Cu pad was blue, because it was a uni-directional structure from the bottom to the top. By the reference figure 3(g), it can be sure that the whole column was all (111)-oriented and with nano-twin on each grain. There were some grains showing different colors rather than blue on the bottom side. It might be a transition region from the seed-layer and the uni-directional Cu. The mechanism is not clear yet. The scallop-type Cu_6Sn_5 appeared in the color of red, orange and yellow. Figure 5(b) shows the layer of Cu_6Sn_5 separately. The middle part was remaining SnAg2.3. As noted in previous section, the orientation of Cu along the direction perpendicular to the silicon chip was (111) and that of Cu6Sn5 was close to (0001).

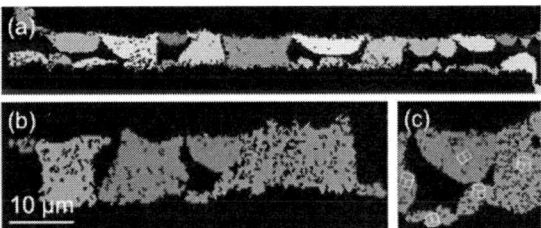

Figure 6. The orientation image maps of η-Cu_6Sn_5 after the solder joints were reflowed at 260 °C for **a)** 4 minutes and **b)** 5 minutes; **c)** the magnified image for figure 6(b).

As we reflowed the other joint for 4 minute, some of the intermetallics contacted each other and seemed to be merged together in figure 6(a). It should be noted that the unequal growth rate of Cu_6Sn_5 on each side of Cu pad. It is mainly because of the unequal Cu flux during reflowing. This phenomenon has been reported in our previous study [19]. And the orientation map of Cu_6Sn_5 in this figure shows that the two intermetallics became one single grain as soon as they were jointed. It could express some natural properties of Cu_6Sn_5. Although each grain has its own orientation, these grains would merge into one single grain as one reached another. Mind that the intermetallics were small after 3 minutes of reflowing. So they needed to grow first and then they can contact with each other. The reaction must be such fast that it can be finished in less than one minute at 260 °C.

The reflow time was further increased for 1 more minutes for other joints. The total reflow time was 5 minutes. The results are shown in figure 6(b). Most of the intermetallics were merge together. The orientations of these intermetallics were still close to (0001). By the magnified image of figure 6(c), the white hexagons represent the exact orientations of each -Cu_6Sn_5. As illustrated by this figure, the (0001) planes of hexagons were almost parallel to the Cu pads whether they were at the top or bottom chip. All the c-axis of hexagons lied along the direction perpendicular to the Cu pads. The two grains started to merge together, and their orientations seemed to be similar. However, the mechanism is still under research.

DISCUSSION
Coherence Between η-Cu_6Sn_5 and (111) Uni-Directional Cu

There have been lots of researches about the preferential growth of η'-Cu_6Sn_5 on Cu. The crystal structure of η'-Cu_6Sn_5 is monoclinic (C2c, a = 11.022 Å, b = 7.282 Å, c = 9.827 Å, β = 98.84o). However, in the introduction part, we have provided lots of study showing that the η-Cu_6Sn_5 would be mainly presented in solder joints. In in this study, we mainly focused on the preferential growth behavior between η-Cu_6Sn_5 (P63/mmc, a = 4.2032 Å, c = 5.1107 Å) and the (111) uni-directional Cu. The crystal structure of Cu are face-centered Cubic with a = b = c = 3.610 Å.

By our experimental results in section 3.2, the (0001) plane of η-Cu_6Sn_5 was parallel to (111) plane of Cu at the early stage of reflowing. After superimposing the Cu atoms of these two planes, we could barely found any directions that are of lower lattice mismatch through whole plane. So it is difficult to explain this coherent relationship by matching the Cu lattice. Nevertheless, in this study, although the Cu pads were made uni-directional and also the surface of Cu pad was totally (111)-planed, it was still a poly-crystalline structure. Figure 2(b) has already proven that there were rotating behaviors between adjacent copper grains. In this way, it is not a totally ordered arrangement of Cu atoms on the surface. Also, the grain size of Cu_6Sn_5 was larger than that of columnar Cu, indicating that the η-Cu_6Sn_5 must grow on several grain of Cu. So it is not possible to have a whole low lattice mismatch through entire area.

The spontaneous reaction between Cu and Sn makes the formation of Cu_6Sn_5. It is obvious that the total energy between Cu-Sn must be lower than Cu-Cu.[11] We should discuss the coherence between the Sn inside η-Cu_6Sn_5 and

Cu. However, it's hard to achieve since Sn atoms are hard to be located in the hexagonal Cu_6Sn_5. Moreover, the tin atoms in the hexagonal Cu_6Sn_5 lie on the plane close ($2\overline{11}3$). By now, we can only conclude that the (0001) plane of η-Cu_6Sn_5 must be parallel to (111) plane of uni-directional Cu due to its favorable lower energy state.

CONCLUSION
The uni-directional Cu with surface covered by (111) plane can be made by electroplating. The shape of Cu grain was columnar. The diameters of these columnar grains were 2 – 5 μm. After electroplating SnAg2.3 on the Cu pad and then reflowed at 260 oC, the η-Cu_6Sn_5s have shown a preferential growth relationship on the uni-directional Cu. At the early stage of reflowing, the orientations of Cu_6Sn_5 were preferred at (0001). As the time of reflow extended, the orientations of Cu_6Sn_5 would change to be preferred at ($2\overline{11}3$). Since the uni-directional Cu was still poly-crystal metal, the coherence must be achieved by Cu-Sn bonding at the interface between Cu pads and the intermetallics. Electroplating parameters would affect the quality of uni-directional Cu and therefore affecting the preferential behavior of Cu_6Sn_5. With the technique of electroplating (111) uni-directional Cu, it is possible to control the orientations of intermetallics in the solder joints.

REFERENCES
[1] Lu M, Shih DY, Lauro P, Goldsmith C, Henderson DW. Appl Phys Lett 2008;92:211909
[2] Sarakonsri T, Apirattanawan T, Tungprasurt S and Tunkasiri T. J Mater Sci 2006;41:4749
[3] Ju SH, Jang HC, Kang YC. J Power Sour 2009;189:163
[4] Suh JO, Tu KN, Tamura N. Appl Phys Lett 2007;91:051907
[5] Suh JO, Tu KN, Tamura N. J Appl Phys 2007;102:063511
[6] Zou HF, Yang HJ, Zhang ZF. Acta Mater 2008;56:2649
[7] Zou HF, Yang HJ, Zhang ZF. J Appl Phys 2009;106:113512
[8] Kumar V, Fang ZZ, Liang J, Dariavach N. Metall Mater Trans A 2006;37:2505
[9] Larsson AK, Stenberg L, Lidin S. Acta Cryst 1994;B50:636
[10] Gosh G, Asta M. J Mater Res 2005;20:3102
[11] Laurila T, Vuorinen V, Paulasto-Kröckel M. Mater Sci Eng R 2010;68:1
[12] Laurila T, Vuorinen V, Kivilahti JK. Mater Sci Eng R 2005;49:1
[13] Nogita K, Gourlay CM, McDonald SD, Wu YQ, Read J, Gu QF. Scripta Mater 2011;65:922
[14] Nogita K. Intermetallics 2010; 18:145
[15] Nogita K, Nishimura T. Scripta Mater 2008;59:191

[16] Schwingenschlögl U, Paola CD, Nogita K, Gourlay CM. Appl Phys Lett 2010;96:061908

[17] Lu L, Shen Y, Chen X, Qian L, Lu K. Science 2004;304:422

[18] Lu L, Chen X, Huang X, Lu K. Science 2009:323:607

[19] Kuo MY, Lin CK, Chen C, Tu KN. Intermetallics 2012;29:155

3D INTEGRATION
A THERMAL-ELECTRICAL-MECHANICAL-RELIABILITY STUDY

K. Weide-Zaage[1], J. Schlobohm[1], H. Frémont[2], A. Farajzadeh[1], J. Kludt[1]

[1]Information Technology Laboratory, Leibniz University Hannover
Hannover, Germany
[2]Laboratoire IMS, Université Bordeaux I
Talence Cedex, France
weide-zaage@lfi.uni-hannover.de

ABSTRACT

Increasing demand, regarding to advanced 3D-packages and high performance applications, accelerates the development of 3D-silicon integrated circuit, with the aim to miniaturize and to reduce the cost. Due to drastic dimension mismatches between interconnects, through-silicon-vias (TSV), and landing pads, the reliability of the systems and components are affected by thermal and thermal-electrical loads due to high temperature as well as high applied currents. This stress leads to degradation effects like electro- and thermomigration (EM, TM). Mechanical or thermal stress due to coefficient of thermal expansion (CTE) mismatch of the different materials on one hand and induced stress during the flip-chip-packaging process on the other hand can lead to delamination and cracking on the packing side or in the IC's.

Investigations of electro- and thermomigration as well as the mechanical stress concerning the reliability of the through silicon vias, BGA-PoP-Packages as well as μ-bumps which are the most critical areas for the emergence of failure, remains a major concern in reliability studies. Generally measurements are time consuming and expensive and the time-to-market cycle is in the focus of interest too. Due to this, simulations offer a possibility for a fast analysis of weak links and problematical areas in the investigated structures and avoid re-design.

Key words: μ-bump, PoP, migration, TSV, delamination, cracking, reliability

INTRODUCTION

Due to the exponential growth in device density the miniaturization down to ultra large scale integration (ULSI) reached physical and economical limits. As one consequence the development of electronic devices was made with an increasing number of ICs, which have to be placed in a constant or shrinking amount of space. The width of interconnects, as well as the dimensions of solder balls and bumps or copper pillars for flip-chip application and therefore the distance (pitch) between them decreases. As a result of the finer pitch, the density of solder bumps in flip-chip designs increases. Solutions for the vertical assembly of ICs and complete packages have been developed. This for instance allows the usage of small 3-D assembly technologies like package-on-package or TSV in compact applications (figure 1). The reduction in the geometrical dimensions leads to an increase in the carried current density in the solder bumps, pillars, interconnects or TSV's. Especially TSV has a critical role in 3-D applications [21]. The reliability in the frame of migration or thermal stress effects is determined by aging test. To shorten these time-consuming long-term tests, highly accelerated stress tests of bump and metallization systems under high current and temperature loads are used for reliability characterization. To understand the failure mechanism and physical background simulations can assist the tests. The prediction of local weak spots in interconnects contacts as well as TSV and solders bumps by finite element simulations is described as a helpful procedure [5, 20]. Beside this the modern 3-D integration leads to more complex material compositions in the systems concerning the different physical material properties leading for instance to intrinsic stress during the processing. Therefore the process induced stress should be considered in the simulations [2]. Higher applied currents on interconnect, contacts and bumps result in Joule heating as well as high temperature gradients in the bump and metallization systems. The exponential temperature dependence of the diffusion coefficient influences the temperature acceleration and the mass flux becomes more important compared to the influence of the applied current on the mass flux. Thus TM effects can not anymore be neglected [1].

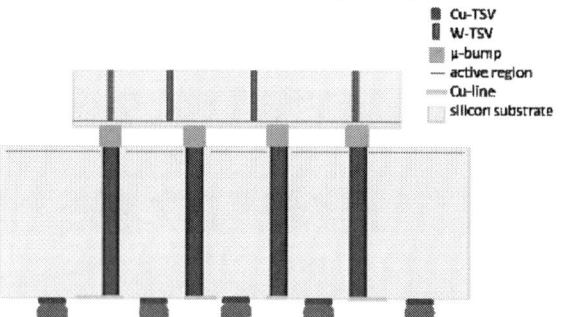

Figure 1: Simplified illustration of 3-D integration.

Simulation results of accumulated stress and plastic strain show that interface stress between copper and silicon is an indicator for a potential failure such as delamination and die cracking. The stress in the through silicon via also depends on the filling material, on the size of holes and on the thickness of the wafers. Increasing via diameter increases the stress in the through silicon via and the effect of thermal expansion mismatch between copper, silicon and silica [20]. Also electromigration (EM) in TSV was detected, although the dimensions in comparison to interconnects of for instance 65nm technology node are quite big. Simulation investigations of BGA-bumps are described in detail in [5]. Also the mechanical behavior of TSV was investigated by simulations [20].

Concerning the simulation in 3-D integration all possible effects of the interconnect system and the packaging have to be in view, and have to be investigated in detail.

GOAL, WORKFLOW AND MODELING

The degradation in interconnects, BGA- and μ-bumps, pillars and TSV under high current and temperature load is investigated and the current, temperature and mechanical stress distribution calculated. As a result the weakest spots in the structure can be determined. The finite element analyses and the mass flux divergence calculation of these phenomena will show the suitability of the method by comparison with experimental results.

Beside the electromigration due to the joule heating, temperature gradient driven thermomigration can occur in the bumps. Also accelerated intermetallic compound (IMC) growth due to electromigration (EM) is found after stress testing. Out of this the reliability prediction due to the different migration mechanisms like electro- and thermomigration (TM) become more and more important.

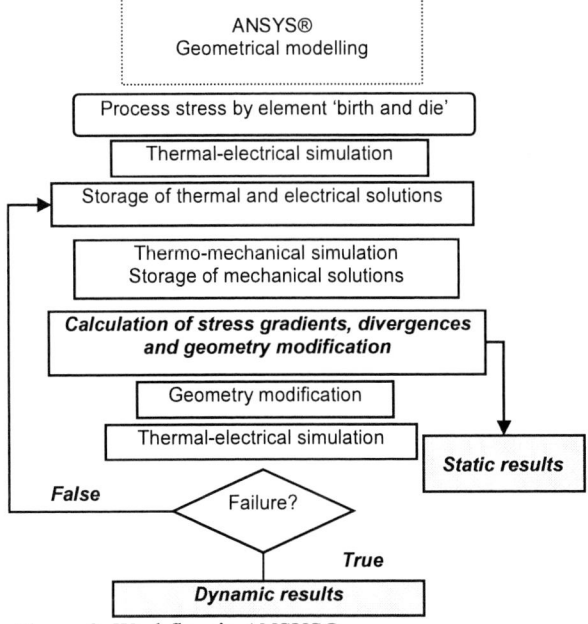

Figure 2: Workflow in ANSYS®.

The geometrical dimensions used in the simulations are normally taken from layouts, SEM or TEM pictures as well as literature. Based on the parameters a finite element model with an adequate mesh concerning electrical or mechanical simulation has to be constructed with all

boundaries. For packages heat conductivity and radiation have to be considered. In the case of interconnect structures this effects can be neglected. The main heat distribution there is heat conductivity. A typical metallization scheme with power metal in layer 6 is shown in figure 3. In the next step the material properties for the simulations are taken from literature or measurements. If the simulations are carried out with ANSYS® the element 'birth and die' capability can be used to determine the process induced stress [13]. This facility virtually removes (or adds) materials by management of the stiffness. In a first step the whole finite element model is made including all material properties and boundaries. Afterwards all elements not used for the first process step are removed. The removed materials are still present in the model, but they have only an insignificant contribution to the matrix of stiffness. When the elements of a material are removed, their strain is set to zero. When the elements of a material are reactivated, the original physical properties are set back.

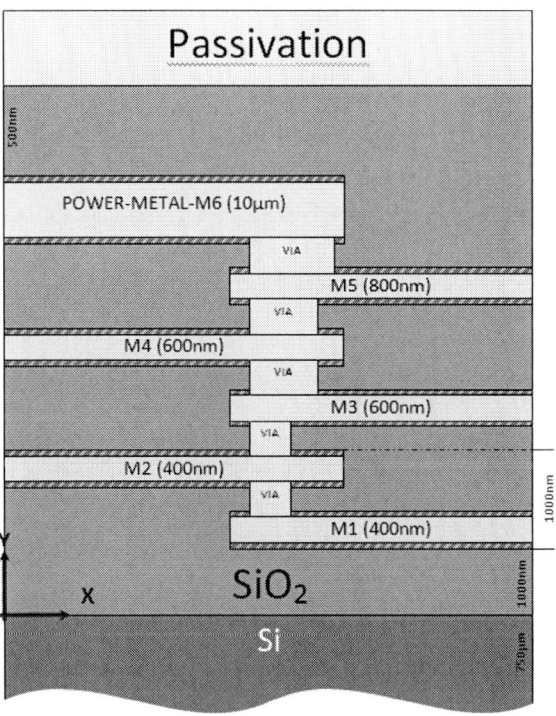

Figure 3: Example of a metallization scheme with power metal in layer 6.

Out of this the pre-stress out of the specific temperatures of every single process step is calculated for the whole metallization system. After doing the finite element analysis all results are stored in files for a further processing with an external program. The stress gradients, the distribution of hydrostatic stress and the local mass flux and mass flux divergence distribution of the different migration mechanisms are calculated and the results restored for graphical interpretation. The workflow in ANSYS® is shown in figure 2. The influence of grain boundaries or interface migration is represented by the measured activation energy. The calculation gives the result for each node in the finite element mesh for a worst case scenario of the coincidence of maximum stress, maximum mass flux divergence, triple point or interface. Out of this the weakest point in the structure is identified and the different migration mechanisms under applied current can be calculated.

MASS FLUX AND MASS FLUX DIVERGENCE

EM is highly temperature dependent. The divergence of the mass flux density is a scalar product of the local current density and the local temperature gradients. The mass flux divergence depends on the amount of these two vectors and the angle between them. A positive mass flux divergence leads to a possible void growth and a negative mass flux divergence to a hillock formation.

An external program is used for the determination of the stress gradients as well as hydrostatic stress, the mass flux and the mass flux divergences [2]. The electromigration mass flux is defined by

$$\overrightarrow{J_{EM}} = \frac{N}{k_B T} e Z^* \left(\vec{j} - \vec{j}_{th} \right) \rho \, D_0 \exp\left(-\frac{E_A}{k_B T} \right) \quad (1)$$

and the thermomigration mass flux is defined by

$$\overrightarrow{J_{TM}} = -\frac{N Q^*}{k_B T^2} D_0 \exp\left(-\frac{E_A}{k_B T} \right) grad \, T \quad (2)$$

the electromigration mass flux divergence is defined by

$$div \, \overrightarrow{J_{EM}} = \left(\frac{E_A}{k_B T^2} + \frac{\alpha_0}{1 + \alpha_0 (T - T_0)} - \frac{1}{T} \right) \cdot \overrightarrow{J_{EM}} \cdot grad \, T \quad (3)$$

and the thermomigration mass flux divergence is defined by

$$div \, \overrightarrow{J_{TM}} = \left(\frac{E_A}{k_B T^2} - \frac{2}{T} \right) \cdot \overrightarrow{J_{TM}} \cdot grad \, T - \frac{QND}{k_B T^2} \cdot \Delta T \quad (4)$$

N is the atomic concentration, k_B the Boltzmann constant, T the local temperature, ρ the resistivity, Ω the atomic volume, $Q^*/k_B T^2$ is referred as Soret coefficient. A value of $Q^* > 0$ means a heat flux is generated to keep the solute atoms isothermal, which takes place towards the dissolved flux. Is $Q^* < 0$ the flux of dissolved particles and the heat flux are counter set. It follows that in an isothermal system, a density gradient produces a thermal flux and vice versa a temperature gradient leads to a material flux. The activation energy E_A was taken from measurements or literature. In the case of interconnect system the effect of stress migration has to be analyzed.

SIMULATION EXAMPLES

Migration effects like electro- and thermomigration can act as a failure mechanism in ball grid arrays (BGA) used in power modules [4]. An investigation of the current and temperature load on the thermal electrical mechanical behavior of BGA bumps is described in [5]. The failure location after EM stress test and a mass flux divergence analysis of a PoP-BGA will be compared in concern of the thermal-electrical results. Furthermore the influence of the stress distribution on the possible local fracture appearances will be shown through an example exemplary in a Package on Package (PoP). Using the FE-analyses a massive shear load was found. This can lead to local fractures in the bump.

As an illustration a flip-chip bump is investigated concerning its thermal-electrical behavior. Especially the thickness of the aluminum metallization in the die was in the focus here.

For a simplified assembly, a smaller pitch between the bumps CuSn-pillars, bumps with a thin layer of Sn on the top, can be used. These layer thicknesses can vary and the influence on the thermal-electrical behavior was investigated [15, 16].

A comparison between BGA-bumps and μ-bumps, flip-chip bumps as well as Cu-Sn pillar bumps will be investigated under the same applied current and the mass flux divergence will be calculated. It is assumed that the reliability of the μ-bump as well as the Cu-Sn pillars have an electro- and thermomigration risk.

For high power applications with a high performance, a low resistance of the TSV material is necessary. The resistivity of doped poly-silicon is too high. The resistivity of copper is about three times lower compared to tungsten. Nevertheless the thermal expansion coefficient, of copper is 6.5 times and of tungsten 1.7 times higher than that of silicon. The reliability of W-filled vias under high-current stress is reported in [21]. In this study it was found that compared to the W-TSV the bumps have the reliability risk [21]. On the other hand EM failures could be found depending on the current flow direction above and under the TSV [22, 23].

MATERIAL PROPERTIES

The mechanical and electrical material parameters of the solder and interconnect used in the investigations are given in table 1. The material parameters for the calculation of the mass flux and mass flux divergence are given in table 2.

Material	α [1/K]	E [GPa]	ρ [Ωμm]	K [W/(mK)]
SnAgCu (SAC)	2.8 10⁻³	41.4	0.132	58.7
Cu	1.6 10⁻³	125	0.0174	395
Al	2.3 10⁻³	68.9	0.0316	237

Table 1: Mechanical and electrical parameters of the solder and interconnect.

Material	D₀ [m²/s]	Z*	Q [eV]	E_A [eV]
SnAgCu (SAC)	27x10⁻³	-23	-0.0084	0.792
Cu	17x10⁻⁵	-4	0.217	0.9
Al	78x10⁻⁶	10	-0.104	0.7

Table 2: Material parameters of the solder and interconnect [7-9].

BGA POP BUMP

In previous investigations was shown that current crowding is the major reason for the electromigration induced void formation at the edges between the cooper traces and the SAC bumps [9, 10]. High electric field strength leads to an increased current density and joule heating at those edges. Due to the high current density the mass flux due to electromigration is increased and in combination with the raised temperature gradients an accelerated void formation appears. Additionally the EM induced mass flux accelerates the IMC growth at the interfaces. In figure 4 (right) the failure location after the EM stress test was determined by SEM inspection after micro-section of the bump. The mass flux divergence distribution in the bump (left) shows high values on the top of the bump. At this location void formation occurs after EM stress load.

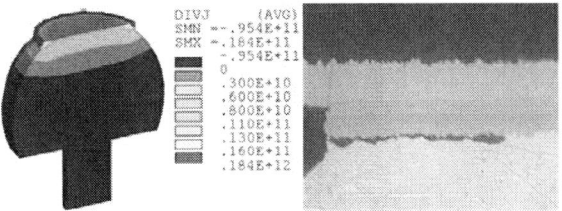

Figure 4: Mass flux divergence (left) and SEM inspection of the failure location on top of the bump (right).

DETERMINATION OF CRACK EVOLUTION IN A BGA BUMP

The influence of the stress distribution at the bump surface will be discussed particularly on the possible local fracture appearances at this position. As a follow-up step of the stress analysis the distribution of the local stress components was carried out by finite element analyses. Different principles for a cracking modeling are described in [11]. One way to indicate the location of cracking appearances is the first principal stress criterion. The material cracks whenever one of the principal stress components exceeds the ultimate tensile strength (UTS). So the principal stresses $\sigma 1$, $\sigma 2$ and $\sigma 3$ have to be calculated and compared with the UTS. According to the first principal stress criterion cracking is expected to appear along the surface of the bump. The influence of the stress distribution on the possible local fracture appearances is illustrated in a 400µm diameter bump. By the FE-analyses a shear load was found. This can lead to local fractures in the bump. In figure 5 the indication of the cracking appearance on top of the bump is shown as an example. The circles indicate an open crack and the circles with crosses indicate a closed crack. The left side of figure 5 shows the hydrostatic stress distribution with maximum values at the top corners of the bump.

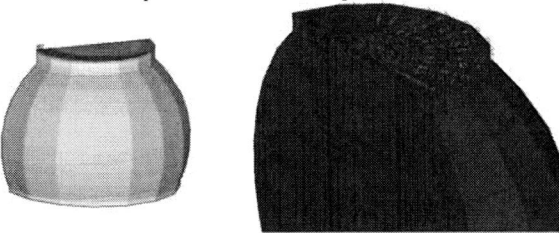

Figure 5: Hydrostatic stress distribution (left) and possible cracking evolution on top of the bump.

FLIP-CHIP SAC BUMP

The accelerated trend to smaller and lighter electronics has accentuated many efforts towards size reduction and increased performance in electronic products. Moreover, RF performances are limited by parasitic effects due to the

RLC network between the wirebond from the dies to the leadframe. The use of flip-chip bonding technology employing micro bumps for very fine pitch packaging permits high integration and limits parasitic inductances. However, both electromigration (EM) and thermomigration (TM) may have serious reliability issues for fine-pitch Pb-free solder bumps in the flip-chip technology used in consumer electronic products.

In this investigation the thermal-electrical behavior of a flip-chip SAC-bump and a die containing a power metallization (figure 3) was determined. The model is shown in figure 6 right. High current crowding is found at the bond pad opening. Due to this at this position the highest mass flux divergences occur (figure 6, left). The maximum temperature was found in the aluminum metallization of the die.

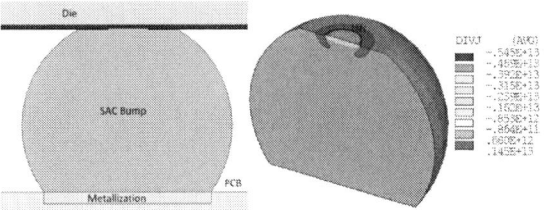

Figure 6: Flip-Chip SAC bump and mass flux divergence distribution.

In figure 7 the mass flux divergence values depending on the applied current are given for an aluminum metallization with a thickness of 1µm and a power metallization of 5µm. In the case of thicker aluminum the current crowding at the bond pad opening decreases. Due to this the mass flux and the mass flux divergence decrease as well.

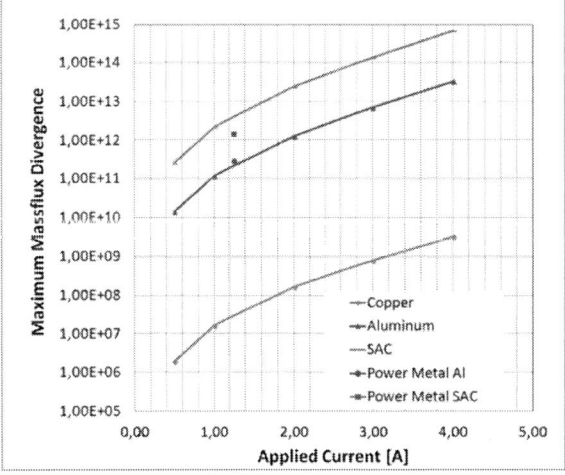

Figure 7: Maximum mass flux divergence depending of the applied current in a flip chip bump with die.

COPPER TIN PILLAR

The evolution of the solder joints lead to smaller bump diameters as well as thinner traces, expecting the same power consumption for the mounted ICs with a higher amount of current crowding. For a smaller pitch between the bumps and a low interconnect inductance as well as a simplified assembly, CuSn-pillars can be used. Due to this copper pillar bumps are the next generation flip chip interconnects with a thin layer of Sn on the top. The benefit of copper pillars are low costs, lead free concerning green package solutions and a better electromigration performance for high current power applications. The CuSn-pillars can be formed under pressure and a temperature load. This leads to the formation of Cu₃Sn and Cu₆Sn₅ phases. In this section CuSn pillars with different Sn thickness and location in the bump like described in [12] were investigated. In [12] the CC was determined by simulations. Beside CC also the temperature gradients influence the reliability of the pillar.

The finite element mesh of the investigated CuSn-pillars is shown in figure 8. The bumps had a diameter of 24μm and a complete high of 45μm. Above the bumps a Cu trace and below the bumps a TSV (figure 2) were placed in the model.

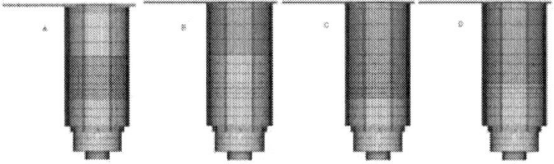

Figure 8: Mesh of the CuSn Pillar with Cu in magenta and Sn in yellow with different Sn thickness.

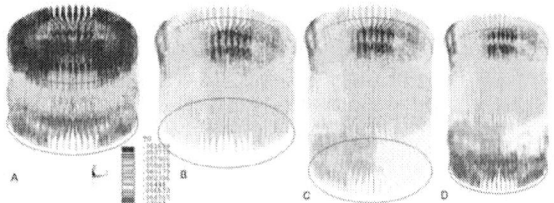

Figure 9: Temperature gradient distribution for the different pillar models.

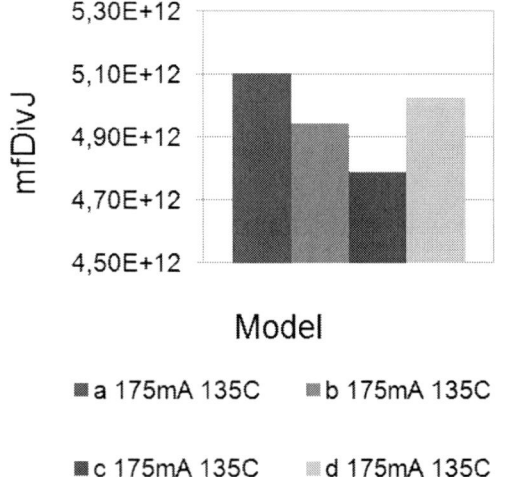

Figure 10: Mass flux divergence electromigration in copper tin pillar bumps.

The applied current was set to 175mA, the substrate temperature was varied from 135 to 150°C and the stress free temperature was set to 150°C. In figure 9 the temperature gradient distribution in the different bumps is shown. The homogenous temperature distribution the

bumps can be achieved by a placement of the Sn in the middle (model A). In the case of model B-C high temperature gradients are found at the corner of the Sn and the Cu metallization and beneath the Sn in the pillar. At this position also strong current crowding occurs. Both can lead to a weak link at this position. Depending on the current flow direction the flux will be increased or decreased.

In figure 10 the maximum mass flux divergence due to electromigration is shown for the different pillar models for an applied current of 175mA and a stress temperature of 135°C.The lowest values are found in model C were the Sn is in the middle of the pillar.

μ-BUMP AND COMPARISION OF THE SOLDER JOINTS

A variation of the applied current in a Package-on-Package (PoP) bumps and μ-bump was carried out and the mass flux divergence distribution was determined [10, 18]. The simulations were carried out with anisotropic and temperature depending material parameters. The dimensions of the μ-bumps are similar to the test structures used in [17]. The diameter of the μ- bump is 25μm and the height is 10μm. Over and under the μ-bump a 100μm silicon layer resp. a 50μm thick silicon layer is representing the ICs of a CoC (Chip-on-Chip) structure. The ICs are covered with a 1μm thick Si3N4 passivation layer. The copper traces at the upper and lower contact surface have a height of 0.5μm and a width of 32μm with a pitch of 40μm. The Fe-Model of the μ-bump is shown in figure 11.

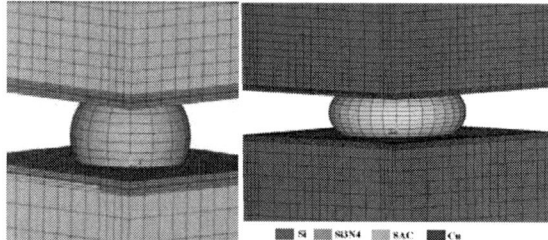

Figure 11: Flip-Chip SAC bump and μ-bump.

The μ-bumps show a strong current crowding as well as high temperature gradients. Due to this EM and TM may occur for smaller applied currents in comparison to BGA bumps. In figure 12 the maximum mass flux divergence for the SAC BGA bump as well as μ-bump depending on the applied current is shown.

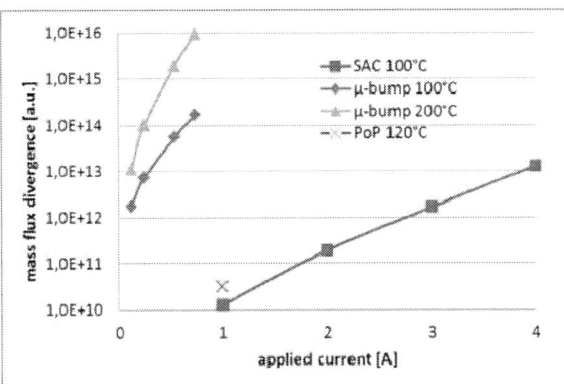

Figure 12: The different materials are indicated by colors. The mass flux divergence vs. the applied current for SAC bump and μ-bump.

TSV AND CIRCUIT DESIGN

3-D IC technology, an advanced IC package architecture, has drawn much interest in semiconductor manufacturing. A 3-D integrated circuit is a chip in which two or more layers of active electronic components are integrated both vertically and horizontally into a single circuit. For this kind of package through silicon vias (TSVs) provide high wiring density interconnection, thus improve electrical performance due to shorter interconnection from the chip to the substrate. However, TSV technology is still facing severe challenges as the physical design problems due to the existence of the copper vias remain resolved. Apart from thermal expansion mismatch, the problems are due in part to many factors including design parameters such as via radius, via aspect ratio, via pitch, chip thickness, and underfill thickness [24].

W-TSVs may outperform Cu-TSVs not only due to its ability to form sub-micron vias, but also W is not a potent diffusant in Si substrate as Cu and it only leaves very minimum stress in active Si owing to smaller difference in coefficient of thermal expansion (CTE) between W and Si. Therefore W-TSV is preferable for high density and high speed TSVs with small diameter and small capacitance for signal lines. However, W-TSV is not suitable for power/ground (GND) lines because of its higher resistance. Cu-TSV with larger diameter and lower resistance should be employed for TSVs for power/GND lines. Cu-TSVs with larger diameter are more preferable to suppress Cu diffusion since a barrier metal such as Ta can be conformal and uniformly formed into deep trench for TSV which effectively suppresses Cu diffusion. The influences of Cu diffusion on device characteristics can also suppress by placing Cu-TSVs for power/GND lines apart from the active areas.

Annealing temperature dependence of mechanical stress induced by TSVs. The mechanical stresses increase with the increase in TSV diameter for both Cu-TSV and W-TSV. The compressive stress was observed in Si substrate between Cu-TSVs when the annealing temperature was lower than 200°C. This tensile stress changed to pure compressive stress not accompanied by the tensile stress when annealed at the temperature higher than 200 °C. Large compressive stress still remained even after annealing at 400 °C although the Cu extrusion (pop-up)

was observed [25]. In [24-44] TSV's with different dimensions are given. Due to the different dimensions a interpretation of the thermal-electrical and mechanical influences on the reliability is difficult, caused by the fact that all different cases have to be investigated. In table 3 the ITRS roadmap for TSV's is given [45].

Global Level	2011-2014	2015-2018
Minimum TSV diameter	4-8 μm	2-4μm
Minimum TSV pitch	8-16 μm	4-8 μm
Minimum TSV depth	20-50 μm	20-50 μm
Maximum TSV aspect ratio	5:1 – 10:1	10:1 – 20:1
Bonding overlay accuracy	1.0-1.5 μm	0.5-1.0 μm
Minimum contact pitch (thermocompression)	10 μm	5 μm
Minimum contact pitch (solder μbump)	20 μm	10 μm
Number of die per stack	2-5	2-8

Table 3: Dimensions of Interconnect Level 3D-SIC/3D-SOC ITRS roadmap [45].

TSV AND MECHANICAL STRESS

The need of TSV for advanced interconnects has considerably increased in different applications. TSVs packages have many advantages such as improving electrical performances; delivering higher density and higher performance. Nevertheless, TSVs are susceptible to failure due to thermal expansion mismatch between materials during temperature variation. Therefore, the study of the reliability of the through silicon via and of most critical areas for the emergence of failure remains a major concern. Research on TSV reliability and finite element modeling investigations are further needed. In a previous study, [20] three geometric parameters and five material parameters were used in the evaluation of stress and strain in the TSV. The geometrical parameters were the via diameter, the substrate thickness and the copper layer thickness. Materials factors were composed by three quantitative parameters related to copper properties (young modulus, yield stress and ultimate strength) and two qualitative parameters: via filling and TSV substrate. The via filling parameter was used to investigate the effect of the material in the via with the variation of stress and strain. Silicon substrate was compared to ceramic substrate to seek about the effect of substrate properties in the variation of stress and strain. Based on simulation performed in a local model of the TSV and a statistical tool, stress and strain results showed that the interface between copper and silicon is an indicator for a potential failure. Moreover, Young modulus of copper and the via filling mode are mainly influential parameters: using a carpeted via decreases the plastic strain and the stress in the TSV.

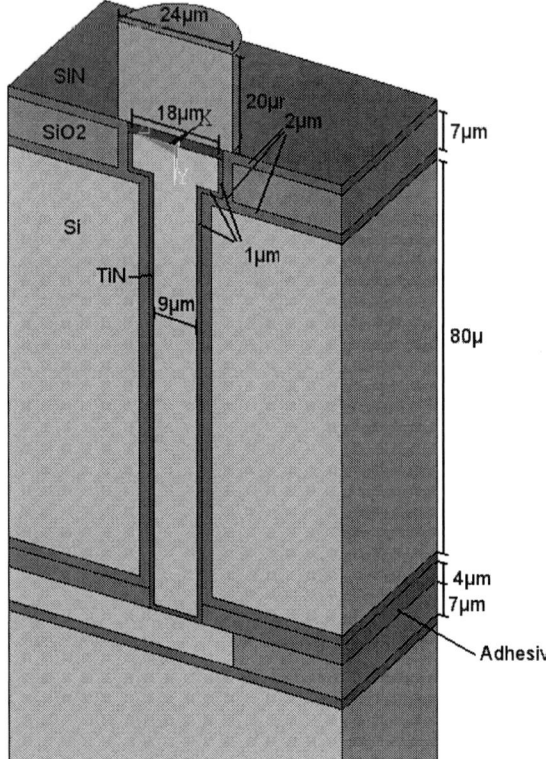

Figure 13: Model of the TSV.

MIGRATION IN A TSV WITH μ-BUMP

In Figure 13 the model of a TSV is shown. The diameter of the TSV was set to 9μm and the height was set to 80μm. The copper barrier in the TSV consists of SiN and TiN. The thickness of the interconnections is 7μm. The passivation on the Device is SiN.

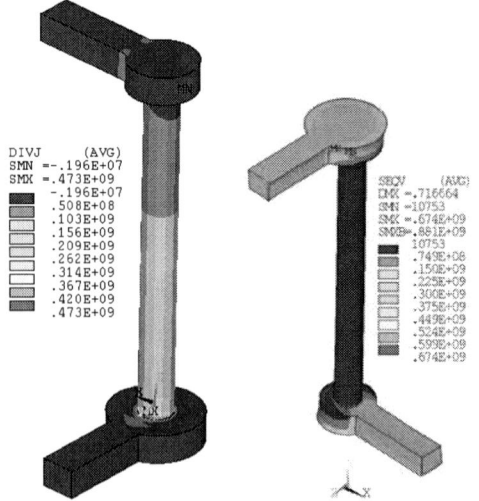

Figure 14: Mass flux divergence distribution upside down (left) and von Mises stress distribution (right).

A typical mechanical von Mises stress distribution is shown in figure 14 right. The maximum values occur in the copper metallization. The maximum mass flux divergences were found at the lower interconnect at the transition of the TSV barrier layers to the interconnect. The μ-bump shows in this case no influence on the electrical behaviour. The mass flux divergences in this case are three orders of magnitude lower compared to the bumps, but increasing applied current as well as enhanced local heating due to current crowding can lead to a reliability problem at this position.

CONCLUSION

With the help of simulations the weakest links as well as locations with high thermal-electrical and mechanical loads can be determined in 3-D applications. In this paper different examples like electro- and thermomigration effects as a failure mechanism in Package on Package PoP-BGA, CuSn pillars, μ-bumps, flip chip with power metal and TSV were presented. Furthermore the influence of the stress distribution on the possible local fracture appearances was shown in a (PoP).

The simulations help to avoid time consuming long term tests and give a hint for the possible failure location. Due to this work intensive micro sections for failure analyses can be placed more easily.

This paper proves that finite element simulation bring a big effort concerning reliability investigations in 3-D integration. In future the process induced stress should also be included in the simulations of bumps.

REFERENCES

1. Y. Tao, L. Ding, et.al: "Investigation of Thermomigration in Composite SnPb Solder Joints" ECTC 2011, pp 1190-1194
2. K. Weide-Zaage: "Exemplified calculation of stress migration in a 90nm node via structure", IEEE EuroSimE, 2010.
3. J. Ciptokusumo, K. Weide-Zaage, O. Aubel: "Mechanical Characterization of Copper based Metallizations with different Via-Bottom Geometries", Physical and Failure Analysis of Integrated Circuits (IPFA), 2010 17th IEEE International Symposium on the Juli 2010, p.1-5I. S, 1963, pp. 271–350.
4. Y. Bulur, R.J. Fishburn, et. al: Electromigration Study on the Interconnects of High Density Power Modules. CHIPS, Nürnberg 2012
5. L. Meinshausen, K. Weide-Zaage, H. Fremont: "Electro- and Thermomigration induced Failure Mechanisms in Package on Package, Micro. Reliability accepted for publication.
6. Y.-S. Lai, K.-M. Chen, et al., "Electromigration of Sn-37Pb and Sn-3Ag-1.5Cu/Sn-3Ag-0.5Cu composite flip-chip solder bumps with Ti/Ni(V)/Cu under bump metallurgy", Micro. Rel., Vol.47 (2007), pp. 1273.
7. Y. Liu, L. Liang, S. Irving et al.:"3D Modeling of electromigration combined with thermal-mechanical effect for IC device and package", Microelectronics Reliability, Vol.48 (2008), pp.818-824.
8. H. Wever, G. Frohberg, P. Adam: "Elektro- und Thermotransport in Metallen", 1973, Leipzig, Germany.
9. W. Feng, K. Weide-Zaage, F. Verdier: "Electrically driven matter transport effects in PoP interconnections", IEEE Mechanical & Multi-Physics Simulation, and Experiments in

Microelectronics and Microsystems EuroSimE, 2009.

10. L. Meinshausen, K. Weide-Zaage, H. Fremont: "Virtual Prototyping of PoP interconnections regarding electrically activated mechanisms", IEEE Mechanical & Multi-Physics Simulation, and Experiments in Microelectronics and Microsystems EuroSimE, 2010.

11. J. Ciptokusumo, K. Weide-Zaage, O. Aubel, Principles for Simulation of Barrier Cracking due to high stress', Proc. 11nd Int. Conf. Benf. Therm. Mech. Simu. Microelec, EuroSimE 2010.

12. A. Syed, K. Dhandapani, et.al: "Cu Pillar and µ-bump Electromigration Reliability and Comparison with High Pb, SnPb, and SnAg bumps, IEEE Electronic Components and Technology Conf. 2011, pp. 332-339.

13. ANSYS®, Inc. Software products, multiphysics, ANSYS, Inc. Southpointe, Canonsburg (PA), USA

14. P.G. Shewman: „Diffusion in Solids", Mac Graw-Hill Series in Materials Science and Engineering, 1963/1989.

15. Huffman, A.; Lueck, M.; et. al: "Effects of Assembly Process Parameters on the Structure and Thermal Stability of Sn-Capped Cu Bump Bonds", ECTC Conf. 2007, pp. 1589-1596.

16. Syed, A.; Dhandapani, K. et. al: "Cu Pillar and µ-bump Electromigration Reliability and Comparison with High Pb, SnPb, and SnAg bumps, IEEE Electronic Components and Technology Conf. 2011, pp. 332-339.

17. Labie, R.; Limaye, P.; et. al: „Reliability testing of Cu-Sn intermetallic micro-bump interconnections for 3D-device stacking", IEEE/Electronics-System-Integration-Conference (ESTC), Berlin, September 2010.

18. Meinshausen, L.; Weide-Zaage, K.; et. al: „Electro- and Thermomigration in Microbump Interconnects for 3D Integration", IEEE Electronic Components and Technology Conf., June 2011, pp. 1444-1451.

19. Z. Xuefeng, W. Yiwe, I. Jang-Hi: "Chip–Package Interaction and Reliability Improvement by Structure Optimization for Ultralow-k Interconnects in Flip-Chip Packages", IEEE Trans. Dev. Mat. Tech., Vol. 12, No. 2, 2012, pp. 462-469.

20. S. Barnat, H. Frémont, et. al.: "Design for reliability: Thermo-mechanical analyses of Stress in Through Silicon Via", IEEE - EuroSimE (2010)

21. Knickerbocker, J.U., et al, "Development of next generation system-on-package (SOP) technology based on silicon carriers with fine-pitch interconnection," IBM J. Res. Dev. 49 (4/5), 2005, pp. 725-754.

22. T. Frank, S. Moreau, et.al: "Electromigration Behavior of 3D-IC TSV Interconnects", IEEE ECTC 2012, pp.326-330.

23. T. Frank, et al., "Resistance increase due to electromigration induced depletion under TSV", IEEE International Reliability Physics Symposium (IRPS), 2011, pp. 347-352.

24. C. Kung, T.-T. Liao, et al. : "Parametric Analyses on Fatigue Reliability of 3D IC Packages with Built

Through Silicon Vias (TSVs)", Mechatronics and Automation (ICMA), 2012, pp.121-126.

25. M. Murugesan, H. Kino, et al.: "High Density 3D LSI Technology using W/Cu Hybrid TSVs," Electron Devices Meeting (IEDM), 2011, pp. 6.6.1 - 6.6.4

26. M. Jung, D.-Z. Pan, et al.: "Chip/Package Co-Analysis of Thermo Mechanical Stress and Reliability in TSV-based 3D ICs", Design Automation Conference (DAC), 2012, pp.317 - 326.

27. G. Plas, P. Limaye, et al.: "Design issues and considerations for low-cost 3D TSV IC technology" in Proc. IEEE 2010, pp. 148–149.

28. M. Jung, D.-Z. Pan, et al.: "TSV Stress-Aware Full-Chip Mechanical Reliability Analysis and Optimization for 3-D IC", Computer-Aided Design of Integrated Circuits and Systems, Vol.31 (2012), pp.1194-1207.

29. T. Tanaka, J. Bea, et al.: "3D LSI Technology and Reliability Issues", Electrical Design of Advanced Packaging and Systems Symposium (EDAPS), 2011, pp.1-4.

30. K. Lu, S.-K. Ryu, et al.: "Thermo mechanical Reliability of Through-Silicon Vias in 3D Interconnects", Reliability Physics Symposium (IRPS), 2011, pp. 3D.1.1 - 3D.1.7.

31. M. Koyanagi. : "3D Integration Technology and Reliability", Reliability Physics Symposium (IRPS), 2011, pp. 3F.1.1 - 3F.1.7.

32. T. Frank, C. Chappaz, et al.: "Resistance increase due to electro migration induced depletion under TSV", Reliability Physics Symposium (IRPS), 2011, pp. 3F.4.1 - 3F.4.6.

33. H. Kitada, N. Maeda, et al.: "Diffusion Resistance of Low Temperature Chemical Vapor Deposition Dielectrics for Multiple Through Silicon Vias on Bumpless Wafer-on-Wafer Technology", The Japan Society of Applied Physics, 2011, Vol.50 (2011), pp. 05ED02.34.

34. F. Carson, K. Ishibashi, et al.: "Development of super thin TSV PoP", CPMT Symposium Japan, 2010, pp. 1-4.

35. P. Ramm, M.J. Wolf, et al.: "Through Silicon Via Technology Processes and Reliability for Wafer Level 3D System Integration", ECTC 2008, pp. 841-846.

36. S. Yoon, K. Ishibashi, et al.: "Development of super thin TSV PoP ", CPMT Symposium Japan, 2011, pp 1-4.

37. K. Fujimoto, N. Maeda, et al.: "Development of Multi-Stack Process on Wafer-on-Wafer (WOW)", CPMT Symposium Japan, 2010, pp 1-4.

38. V. Sukharev, E. Zschech, et al. : "Multi-Scale Environment For Simulation And Materials Characterization In Stress Management For 3D IC TSV□Based Technologies Effect Of Stress On The Device Characteristics", American Institute of Physics (AIP), Vol.1378(2010), pp.21-49.

39. X. Xu, A. Karmarkar, et al.: "3D TCAD Modeling For Stress Management In Through Silicon Via (TSV) Stacks ", American Institute of Physics (AIP), Vol.1378(2010), pp.53-66.

40. K. Yeap, U. D. Hangenb, ct al.: "Nanoindentation Study Of Elastic Anisotropy Of Cu Single Crystals And Grains In TSVs" , American Institue of Physics(AIP),Vol.1378(2010), pp.121-128.

41. I.-D. Wolf, "Raman Spectroscopy Analysis Of Mechanical Stress Near Cu-TSVs", American Institue of Physics (AIP),Vol.1378(2010), pp.138-149.

42. S.-K. Ryu, K. Lu, et al.: "Stress-Induced Delamination Of Through Silicon Via Structures"'", American Institue of Physics (AIP),Vol.1378(2010), pp.153-167.

43. S. Niese, P. Karmarkar, et al.: "NanoXCT A High-Resolution Technique For TSV Characterization", American Institue of Physics (AIP),Vol.1378 (2010), pp.168-173.

44. M.-C. Hsieh, S.-T. .Wu, et al.: "Nonlinear thermal Stress Analysis and Design Guidelines for Through Silicon Vias (TSVs) in 3D IC integration", Microsystems, Packaging, Assembly and Circuits Technology Conference (IMPACT), 2011, pp.75-78.

45. itrs.net/Links/2011Winter/PublicPresentations.html

LOW-COST AND HIGH PERFORMANCE SILICON INTERPOSERS AND PACKAGES (LSIP) – A NEW GEORGIA TECH PRC INDUSTRY CONSORTIUM

Rao R. Tummala, Ph.D., Venkatesh Sundaram, Ph.D., Qiao Chen, Hao Lu, and Gokul Kumar
3D Systems Packaging Research Center
Georgia Institute of Technology
Atlanta, GA, USA
rao.tummala@ece.gatech.edu

ABSTRACT

The Low-Cost Silicon Interposer (LSIP) industry consortium at Georgia Tech is proposed to address limitations of current organic packages as well as wafer-based silicon interposers that are being produced in wafer fabs. Organic substrates are seen as approaching limits in wiring, I/Os, warpage and acceptable cost. Silicon Interposers, on the other hand, suffer from high cost and low electrical performance. This paper addresses both these limitations.

Key words: Wafer silicon interposer, panel silicon interposer, polysilicon, through-package-via, polymer liner, insertion loss, low cost silicon interposer, dry film dielectric

INTRODUCTION

The Low-Cost Silicon Interposer (LSIP) industry consortium at Georgia Tech is proposed for eight reasons: 1) to replace organic substrates as they approach limits in wiring, I/Os, warpage and acceptable cost; 2) to act as 2.5D interposer in the short-term and BGA package in the longer-term for interconnecting ICs and 3D ICs with ultra-low k dielectrics; 3) to act as 3D Interposer, an alternative to 3D ICs for achieving high bandwidth at low power between logic on one side and memory or other devices on the other side, enabled by ultra-short through-package-via interconnections (TPVs) at same or similar pitch as TSVs in the 3D ICs; 4) to further miniaturize package and sub-systems as a result of double-side mounting of both actives and passives enabled by TPVs; 5) to extend wafer-level fan-out technologies in I/Os, miniaturization and cost; 6) to provide a technology platform for MEMS and Sensors with ultimate reliability, performance and cost compared to plastic packages that absorb moisture and transfer stress to the die as a result of large CTE mismatch between die of 3ppm/C and plastic package of 17ppm/C; 7) to achieve ultra-low warpage in ultra-thin packages as a result of high modulus of silicon and balanced double-side structure; and 8) to go beyond interposers and develop them into surface-mountable BGA packages.

A two-year program is focused, however, on 2.5D interposer with 2-5μm lithographic ground rules. This presentation will review goals and describe accomplishments in low-cost silicon raw material, low-cost through via and metallization, and low cost RDL.

OBJECTIVES

The objectives of low-cost silicon interposer technology being pioneered by Georgia Tech's industry consortium are to: 1) provide I/Os at 20-50μm pitch that cannot be achieved in area array with organic packages, 2) lower the cost of wafer-based silicon interposers by 2-10X and improve electrical performance over traditional lossy silicon.

General Requirements of Interposers

Table 1 lists general requirements of interposers:

• Low Cost of Raw Substrate Material (Silicon)
  o Thin and large wafers or panels at low cost
  o None or minimal CMP cost

• Low Cost Processability of Through Vias and RDLs
  o Low cost through via hole formation and metallization
  o Low cost and high I/O multilayer redistribution layers

• Improved Properties
  o High electrical resistivity
  o Low electrical loss
  o High dimensional stability
  o High thermal conductivity
  o Appropriate CTE between IC, interposer or Package and board

Table 1. Primary Interposer Requirements

The critical characteristics include: availability in thin and large area raw substrates; processability to achieve high via and wiring density; good electrical and thermal properties such as CTE, thermal conductivity, electrical resistivity; and finally, relative overall cost that includes not only raw material cost but also the final processed cost with vias and wiring at 2-5μm technology, per square area. This comparison leads to low-cost wafer and large-area panel processing using non-traditional silicon as being the single most important factor in the final decision of interposer and package technology for low cost.

Table 2 compares and contrasts Georgia Tech's low-cost silicon interposer approach with the current wafer-based silicon interposers being pursued by industry.

Technology	Traditional Si Interposer		GT Low Cost Silicon Wafer and Panel Interposer
	BEOL	**WLP**	
Raw Material	CMOS Si		Polycrystal Si
Size	200-300mm wafer		300mm wafer to 700mm panel
Thickness	50-200μm with polish		100-200μm w/o polish
TPV Hole	DRIE		Laser
TPV Liner	Thin Oxide		Thick Polymer
TPV Metal	Barrier/Seed/Cu		Eless Seed/Cu
RDL	Single-side Process		Double-side Process
	Cu-Oxide	Cu-Pl	Dry Film Build-up
Dielectric	Thin SiO$_2$	Spin on Polymer	Dry Film Polymer
Via	Dual Dam	Photo Via	Laser
Cost • Wafer • Panel	High N/A	Medium N/A	2X 10X

Table 2. Comparison of Current Silicon Wafer-based Interposers with Georgia Tech's Polysilicon Interposer

Interposers, by definition, connect IC I/Os on the top side at 20-50μm pitch to organic packages on the bottom side at 100-150μm pitch for flip-chip assembly to organic BGA-like package substrates, which connect, at 300-500μm pitch for direct SMT-attach to board. Since the number of interposers coming from 200 or 300mm silicon wafers is low, particularly if the interposers are 50-60mm in size, serious cost concerns remain as the biggest barriers to silicon interposers. The second major concern with existing silicon interposer is the signal loss in vias and transmission lines as well as cross-talk between adjacent through vias at fine-pitch.

Georgia Tech's Packaging Research Center (PRC) proposes, therefore, a two-phase strategy as shown in Figures 1 and 2: 1) apply first, at largest wafer size by performing leading-edge R&D to demonstrate low loss silicon interposer with low cost TPV and RDL, and then commercialize using the improved and low cost materials and processes as described in Table 2, resulting in about 2X cost improvement; followed by 2) large panel processing of polycrystalline silicon for a broad set of applications.

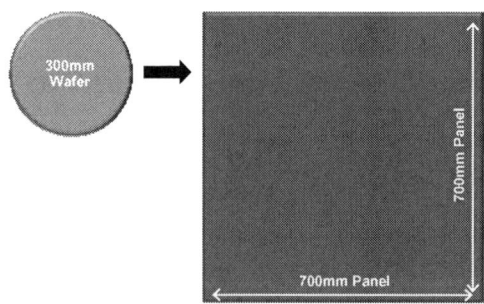

Figure 1. Georgia Tech PRC Strategy for Low-Cost Silicon Interposer – 2X Lower Cost at Wafer-Level First, Followed by Large Panel for potential 10X Cost Reduction

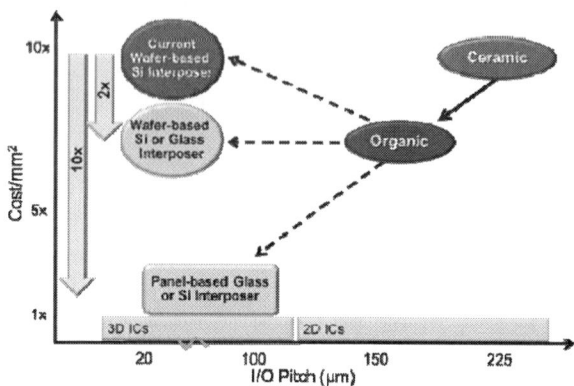

Figure 2. Georgia Tech PRC Low-Cost Strategy with Both Wafer and Panel Interposers

Such a strategy is expected to solve two major limitations of traditional silicon interposers: cost and electrical performance, thus leading to improved electrical, mechanical and thermal performances, as described below.

TECHNICAL BARRIERS
The barriers to the use of silicon are mainly its high signal loss and its high cost per mm^2, when processed with BEOL tools. Additional challenges include fine-pitch RDL with micron-size lines at high yield and low cost with minimum signal loss, and micro-bump assembly at fine-pitch.

The overall strategy is to address these barriers to enable commercialization of polysilicon interposers and packages. The Georgia Tech approach to low-cost silicon interposer is shown in Figure 3. It starts with polycrystalline silicon, which is a factor of 10 lower in cost than single-crystal silicon. The TPV hole is formed by laser at a cost lower than DRIE. The TPV liner is made of thick polymer by a low-cost process technology in contrast to thin SiO$_2$ liner. The RDLs are fabricated, not with liquid dielectrics, but with dry films. The initial program will focus on applying as many of the above low-cost technologies as possible at large wafer-level, which will then be applied at panel-level.

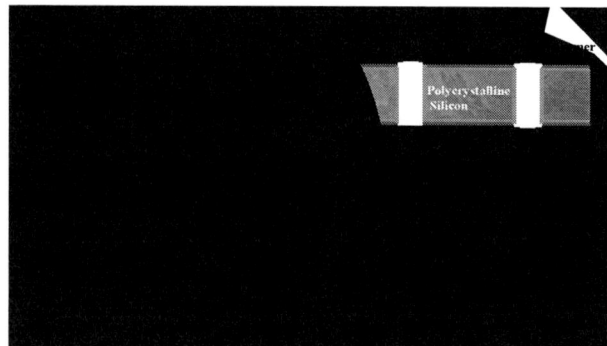

Figure 3. Georgia Tech's Non-traditional Silicon Interposer Materials and Process Flow

SILICON INTERPOSER TECHNOLOGIES

Low-Cost Silicon

The cost reduction focus starts with the use of upgraded metallurgical grade (UMG) polycrystalline silicon wafers, which are manufactured by directional solidification and wire-cut into 100-200µm finished thickness. This eliminates expensive steps including purification and ingot growth, as well as back-grind and CMP polish, required for CMOS-grade single crystal silicon wafers. The raw silicon wafer or panel used in this interposer research has been developed for photovoltaic applications, and can have varying electrical resistivity, ranging from 0.1 to 0.6 ohm-cm, based on the level of impurities in the silicon material. The polycrystalline silicon, not only is a factor of about 10X lower in cost than CMOS silicon, but is also scalable to large and thin 700mm x 700mm panels. Figure 4 shows a top view of a typical 156mm x 156mm wire-cut polycrystalline silicon panel at 200µm thickness. Novel thin silicon handling methods, for double-side processing, have been explored and demonstrated in this research [1].

Figure 4. Polycrystalline Silicon Panel at 200µm Thickness and 156mm x 156mm Panel Size

Via Formation

Bosch process is commonly used for via formation in traditional silicon interposers [2, 3]. However, such a process is not scalable to large-panel substrates. Therefore, low-cost and high throughput laser ablation process was developed in our study to form fine-pitch vias in polycrystalline silicon substrate. Three different laser technologies (UV, Excimer and Pico-second laser) were explored for via formation in polycrystalline silicon substrates. Table 3 compares the differences among the three laser technologies. The 355nm UV laser was able to achieve as small as 25µm vias at high throughput. While the excimer laser processing with a wavelength of 248nm was faster and resulted in small vias, this technology suffers from higher production cost. The 355nm pico-second lasers can further reduce the heat generated during the laser ablation process. Vias with 10-50µm diameter were formed by pico-second lasers. However, this method is currently limited by slow processing speed and serial via formation. Therefore the low cost 355nm UV laser was used as front-up method in our study. Figure 5 shows the top and bottom view of small and fine-pitch vias formed in polycrystalline silicon by UV laser.

Laser	Via Dimensions Achieved	Cost
UV (355nm)	25-150µm	Low
Eximer (248nm)	10-20µm	High
Pico-Second Laser (355nm)	10-50µm	High

Table 3. Via Formation in Polycrystalline Silicon Substrate by Three Laser Tools

Via diameter Via pitch	Top side (diameter)	Back side (diameter)
25um diameter 75um pitch	(25 um)	(14 um)
50um diameter 75um pitch	(50 um)	(37 um)

Figure 5. Via Formation in Polycrystalline Silicon by Low-Cost UV laser

Liner Formation

A SiO_2 layer with expensive barrier to prevent Cu diffusion is widely used as insulation liner in traditional silicon interposers [3]. Such a step always involves high-cost CVD or PVD processes. In addition, the traditional silicon interposer suffers from high signal loss due to the thin SiO_2 layer (0.1-0.2µm) that is typically and sparingly employed. The GT silicon interposer focuses on ultra-low-loss and thick polymer formation in fine-pitch vias to address the high signal loss in TPVs, while eliminating the diffusion barrier. Simulation results showed that the thick polymer helps to reduce the loss and resulting silicon interposer shows superior electrical performance than traditional silicon interposers, as shown in Figure 6. The technical approach for the liner formation is called double laser method, involving polymer filling of larger TPV, followed by laser ablation to form an "inner" via resulting in a via side wall liner of controlled thickness, as shown in Figure 7. The polymer was fully filled by a low-cost double-side vacuum lamination process. Such a lamination process can also form polymer layers on top and bottom sides of the silicon panel simultaneously, for insulation purposes. A low-cost UV laser process was then used to drill a smaller via in polymer. Alternative approaches to fabricate the liner, by conformal electrophoretic deposition and spray coating are also under study and will be presented in the future.

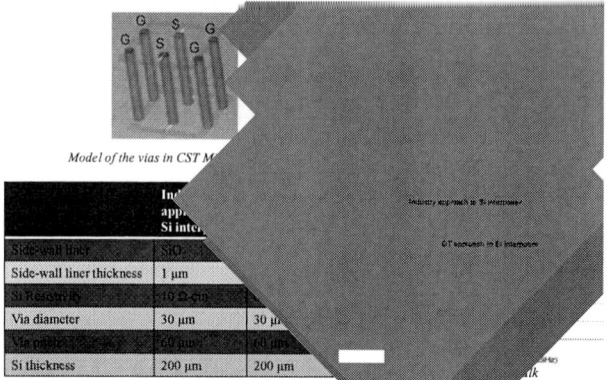

Figure 6. The GT Silicon Interposer Approach Shows Superior Electrical Performance Than Traditional Silicon Interposer

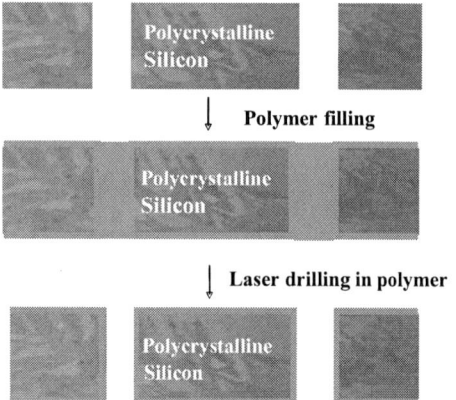

Figure 7. Double Laser Method to Form Thick Polymer Liner

Metalization

Comparing to the high-cost vapor deposition method for adhesion layer and electroplated Cu seed in traditional silicon interposers, the GT process involves a low-cost, all-wet metallization process, including electroless and electrolytic plating. The metallization process consisted of three steps: 1) Cu seed layer formation, 2) Cu electroplating and 3) pad formation. The polycrystalline silicon samples were first cleaned using plasma to remove any impurities on the surface. A fast, low-cost electroless plating process was then used to fabricate a 1μm thick Cu seed layer for further electroplating. The dry film was laminated on both surfaces of the sample followed by a lithography process. A void-free semi-additive Cu electroplating was then performed in the plating tank to fill the through vias. A final step, including Cu thinning, photoresist removal as well as seed layer etching, was conducted to form the Cu pad for connection. Figure 8 shows the cross-section of Cu-filled TPV with polymer liner.

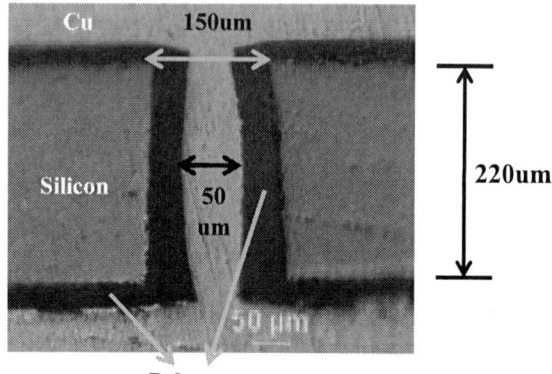

Figure 8. Cross-section of Polymer-lined and Metallized TPV

Ultra-fine line RDL wiring is essential for escape routing on silicon interposers, in as few layers as possible. Most of the published research on fine-pitch RDL fabrication is either expensive [4], or limited by relatively coarse wiring ground rules [5]. The GT target, in this project, is to realize fine-pitch RDL wiring on thin-panel silicon interposer at a lower cost than with wafer-based processes. In the 2.5D silicon interposer demonstration (Figure 9), 1000 interconnections between logic chips (10mm x 10mm) and memory chips were achieved with two metal layers of RDL. Bump pitch in these examples was 50μm, with 25μm landing pad diameter. Therefore, the wall-to-wall distance was 25μm. With 5μm wiring, three rows of I/Os can be routed in one metal layer (Figure 10).

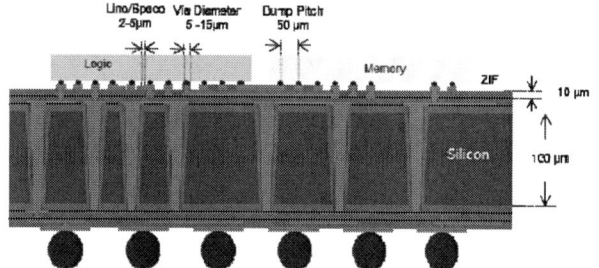

Figure 9. 2.5D Silicon Interposer Assembly

To achieve low-cost at ultra-fine-pitch RDL, the major challenges are transmission-line impedance match and the Cu-line formation processes. With 2μm line width and 4μm line height of microstrip transmission lines, the required dielectric thickness to achieve 50 ohm impedance is only 1.4μm. This is very hard to achieve. Thinner dielectric polymers and embedded lines are two ways to solve this problem. To achieve low-cost Cu wiring, semi-additive processes were applied. The semi-additive process involves Cu seed layer deposition, photolithography, and Cu-line metallization. In the GT approach, each step was controlled to achieve low cost. The GT approach uses electroless-plated Cu on ZIF polymer dry film dielectrics to form the RDLs.

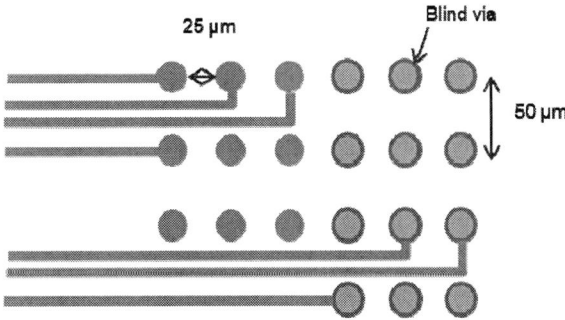

Figure 10. 2-Metal Layer Escape Routing Strategy

ELECTRICAL CHARACTERIZATION

Electrical design and characterization of TPVs and transmission lines were performed, as shown in Figure 11 to obtain S-parameter measurements. Four-metal layer test vehicles were designed and fabricated to form silicon interposers with coplanar wave guide (CPW) lines. The resistivity and thickness of the silicon used was 0.5 Ω-cm and 200μm, respectively, with a polymer-liner thickness of 40μm. The fabricated-CPW lines were 120μm wide. The gap between the signal and ground was 36.5μm. Frequency-domain simulations were carried out for TPVs using full wave EM solvers. The VNA measurement was performed with SOLT calibrations. Low signal loss is a key important parameter for longer RDL lines (4-8mm) that are routed between two dies in the 2.5D interposer. The 7mm signal lines fabricated were characterized to have less than -0.03dB of insertion loss, up to 10Ghz. This lower insertion loss is attributed to the thick polymer liner of very high resistivity. Thus, CPW lines in wafer or panel-silicon interposers are capable of handling high-speed digital signals with nominal distortion.

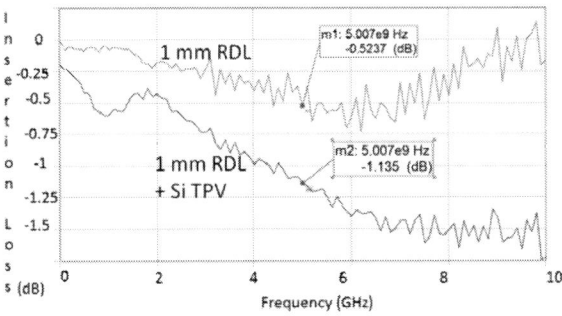

Figure 11. Insertion Loss of RDL Lines in GT's Silicon Interposers

SUMMARY

As organic packages reach their limits in I/Os, thermal and reliability performances, silicon interposers are being developed using BEOL processes in the wafer fabs. But such interposers have two main problems: high cost and low performance. The Georgia Tech approach addresses both these problems as described in this paper.

REFERENCES

[1] Dominique Sarti, Roland Einhaus, "Silicon feedstock for the multi-crystalline photovoltaic industry," *Solar Energy Materials and Solar Cells*, Vol. 72 Nos. 1–4, April 2002, pp.27-40

[2] Masahiro Sunohara, T. Tokunaga, T. Kurihara, M. Higashi, "Silicon Interposer with TSVs (Through Silicon Vias) and Fine Multilayer Wiring," *Proceedings Electronic Components and Technology Conference, 58th*, ECTC 2008, pp. 847-852

[3] K. Zoschke, J. Wolf, C. Lopper, I. Kuna, N. Jurgensen, V. Glaw, K. Samulewicz, J. Roder, M. Wilke, O. Wunsch, M. Klein, M.V. Suchodoletz, H. Oppermann, T. Braun, R. Wieland, O. Ehrmann, "TSV based Silicon Interposer Technology for Wafer Level Fabrication of 3D SiP Modules," *Proceedings Electronic Components and Technology Conference, 61st*, ECTC 2011, pp. 836-843

[4] Kirk Saban, "Xilinx Stacked Silicon Interconnect Technology Delivers Breakthrough FPGA Capacity, Bandwidth, and Power Efficiency," *WP380*, Vol. 1 No. 12, October 2011

[5] Katsura Hayashi, Kimihiro Yamanaka, Masahiro Fukui et al., "Advanced Surface Laminar Circuits Using Newly Developed Resins," *Components, Packaging and Manufacturing Technology, IEEE Transactions on*, Vol. 1 No. 12, 2011, pp. 1908-1915

BVA: SOLUTION FOR NEXT GENERATION VERY FINE-PITCH PACKAGE-ON-PACKAGE (PoP) APPLICATIONS

Vern Solberg and Ilyas Mohammed
Invensas Corporation
vernsolberg@att.net

ABSTRACT

Stacking heterogeneous semiconductor die (memory and logic) within the same package outline has compromised test efficiency and overall package assembly yield. Separating and packaging the semiconductor functions into sections for stacking, on the other hand, has proved to be more efficient. Although package-on-package methodology is currently considered a primary assembly solution for the logic and memory applications, the next generation of processors will have increased functionality, more I/O and more complex multiple die memory configurations. This factor has prompted developers to seek a packaging solution that is economical, physically robust and will enable a significant reduction in contact pitch.

This paper addresses the primary technological challenges for reducing contact pitch for package-on-package applications with current solder ball interface methodology. The authors follow with an introduction to the 'Bond-Via-Array' (BVA) package assembly process, a unique approach for forming smaller and finer pitch contacts on the organic based interposer substrate. The BVA contact forming process utilizes copper wire-bond technology that enables several contact profile variations and can furnish an array configured contact pitch as small as 200µm. This interconnect solution is very economical and lends itself to a wide variety of 3D packaging applications, including multiple-row, area array, fan-in and fan-out configurations.

Key words: Package-on-Package, PoP, Bond Via Array, BVA.

INTRODUCTION

Multiple die package technology has proved to be a practical solution for a number of high-performance, system-level applications. The initial multiple die products developed combined two or three semiconductor elements within a single package outline. Even though die elements are often supplied from several sources with differing certification methods, assembly process yields generally reached acceptable levels. These products were typically a vertically stacked set of semiconductors interfaced to a ceramic or organic BGA configured substrate using common wire-bond and encapsulation techniques.

As companies became comfortable with the die-stack package some pushed the technology further with significantly higher die counts. A key concern for these more complex package assemblies became the confidence level that, when completed, all elements would actually function properly. If all die were certified and 'tested good' then confidence remained high. If some of the die were designated only as 'probably good' based on wafer level yield data, then the die-stack process yield would be seriously compromised. Although some manufacturers achieved a level of success, many companies came to the conclusion that combining logic and memory within a single package posed significant risk.

PACKAGE-ON-PACKAGE IMPLEMENTATION

The packaging industry began furnishing stackable package sections that enabled the separate pre-testing of memory and logic before joining. The now familiar 'package-on-package' (PoP) process (**Figure 1**) has proved practical for a broad number of heterogeneous applications. Because each package section has been electrically tested before joining, the user can be assured that the multiple die products will meet all performance criteria.

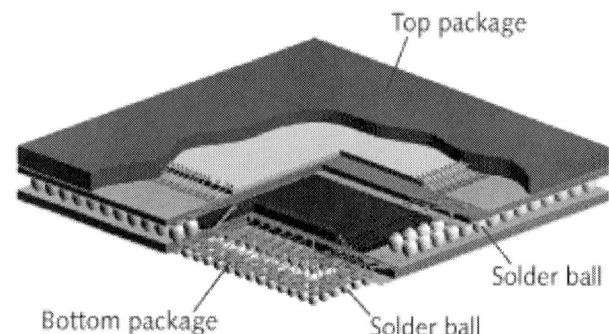

Figure 1. Commercial Package-on-Package Example.
(STATS ChipPAC, Inc.)

Although the PoP assembly methodology has become mainstream, a number of issues continue to trouble users. Many have observed that a great deal of improvement can be achieved in both performance and power control by closer coupling of primary signal paths and minimizing circuit interconnect length. As far as containing physical robustness and controlling package warp at high processing temperatures, the physical attributes (modulus, glass transition temperature and material compositions) for both substrate and mold compounds must be closely matched. And, in consideration for controlling PoP assembly cost, all elements of the package assembly process must continue to utilize the existing manufacturing infrastructure.

SOLUTION NEEDED FOR HIGHER DENSITY PoP

Interface density for PoP has improved to a degree using smaller diameter solder balls, solder filled laser drilled vias in the mold cap or by adopting higher density PCB interposers. The challenge now facing our industry is how we address the growing number of multiple die package assembly concerns:

- How to reduce the overall package outline
- How to accommodate greater I/O semiconductors
- How to enhance package performance
- How to maintain lower power operation
- How to minimize package warp during SMT assembly
- How to maximize existing infrastructure capability

To overcome the limiting aspects of the current FBGA PoP assembly methods an alternative substrate interconnect solution has evolved for both memory and logic sections. The base section of the package adopts a unique molded bond via array (BVA™) process. This is a Cu wire-bond based package stacking interconnect technology developed to enable a substantial reduction in interface contact pitch between the lower and upper PoP sections. The main features of the concept are that the wire-bond contact is only 50μm diameter and is encased in a mold compound that encapsulates the lower package semiconductor. The wire tips remain exposed after molding to extend from the top surface of the lower package section to matching contact locations on the bottom surface of the upper package section (**Figure 2**).

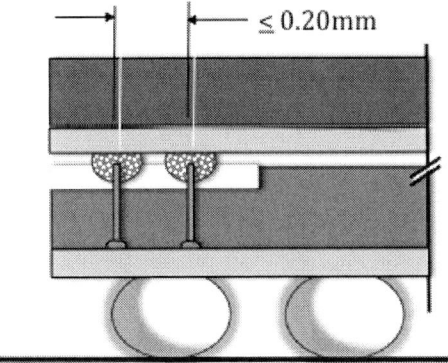

Figure 2. Molded BVA Cu wire contact protrusion.

The reduced contact pitch easily accommodates a higher number of interconnects between the lower and upper package sections. The molded wire process (developed at the Invensas Laboratories in San Jose California) furnishes a very robust, warp resistant stacked package innovation that enables significantly greater I/O capability while utilizing the existing wire-bond infrastructure. If needed, the BVA assembly process can provide a contact pitch as small as 100μm, a significantly closer pitch than possible with solder ball configured package technology.

Copper wire bonding was selected as the method for forming the PoP interconnects. Copper provides advantages over gold in cost, stiffness and electrical conductivity. Further more, the length of the bonded Cu wire can be extended to very precise elevations accommodate upper and lower package interface. This interconnect technology lends itself to a wide variety of 3D packaging, including multiple-rows and area array, fan-in and fan-out configurations.

MOLDED Cu WIRE PROCESS DEVELOPMENT

The lower package substrate layout for the BVA package-on-package test vehicle was based on the current JEDEC standard for PoP packaging having 0.5mm pitch solder balls for PCB mounting. The outline dimensions established for both lower and upper sections of the PoP test vehicle was 14mm x 14mm. Several substrates were configured in a panel format to better accommodate package assembly processing. The 'Phase 1' test vehicle substrate design enabled the Cu wire contacts to be bonded onto the top surface of the lower section in a two-row pattern with 0.30mm pitch. This pattern provided a uniform perimeter array with more than 400 contact locations designed to mate to an identical contact pattern provided on the bottom surface of the upper section.

The first stage of the BVA package development was to terminate the wire length at a prescribed height and maintain an acceptable positional tolerance. Various approaches to bonding and forming the copper wire on the bottom package substrate were evaluated. Wire formation with Electronic Flame-Off (EFO) balls was studied, as was wire formation without EFO balls. A free-ended wire without EFO ball formation was ultimately selected because it enabled a finer contact pitch in both X and Y directions (**Figure 3**).

Figure 3. Free ended Cu wire contact prior to molding.

The next step was the development of a method for exposing a solder-compatible wire tip above the mold cap. The following represents some of the methods explored during the mold process development:

- Fully molding the wires and then exposing the loops, balls or wire tips by means of surface grinding, laser ablation, sand blasting or wet blasting the entire top surface of the molded package.

- Coating the wire tip with a water-soluble protective coating before molding and then rinsing the coating off to expose the wire.
- Laser ablation to form a semi-spherical cavity around every wire ball or tip.

The ultimate wire-mold solution employed a film assisted molding process, a mature technology commonly found in many packaging assembly operations. The process adapts a mold chase design with mold cavities only slightly deeper than the formed Cu wires. When the mold is clamped to the substrate, the Cu wires are pushed through the mold film.

A TCE matched molding compound is injected into the mold cavity and cured. When the mold tool is opened the mold film is pulled away to expose the protruding wire tips. The photograph in **Figure 4** clearly shows the exposed wire tips completely free of any mold compound residue.

Figure 4. High magnification photo of Cu wire contact tips exposed after molding and mold film release.

The wire protrusion maintained an average value of 110μm with a standard deviation of 9μm as shown in **Figure 5**.

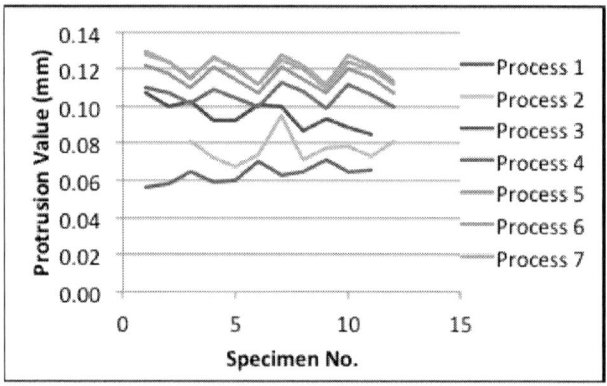

Figure 5. Comparing BVA wire tip protrusion variation

The final stage of the lower package assembly placed and reflow soldered the array of 300micron diameter solder balls on the bottom surface. These contacts are furnished to accommodate the eventual mounting of the package to the host PCB assembly.

A process was developed in parallel for mounting and reflowing the solder balls onto lands provided on the bottom surface of the top package. The solder ball diameter selected

for this program was ~150μm diameter. A commercial high-speed ball placement system dip-coated the each solder ball into a tacky flux solution and distributed them onto their respective land features. The solder ball to substrate joining process utilized a conventional hot air/gas reflow system. Following reflow processing the assemblies were cleaned, singulated and made ready for electrical test. Samples were selected at this phase for shear testing to compare relative strength of the solder ball to substrate interface and examine any failure mode potential.

BVA PACKAGE-on-PACKAGE ASSEMBLY
With both upper and lower package sections complete they were made ready for the package joining process. A wide range of soldering experiments were performed to evaluate the process repeatability and reliability performance of reflow soldering the upper section solder balls onto the exposed Cu wire tips on the lower package section. The example shown in **Figure 6** illustrates the BVA upper and lower package wire contact-to-solder ball interface.

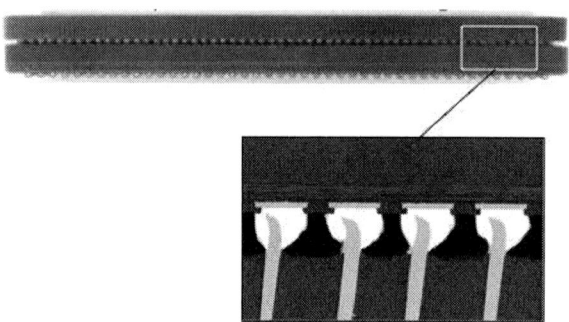

Figure 6. BVA molded Cu wire contact-to-solder ball interface.

BVA enables the shortest possible routing path between the over-molded semiconductor die on the lower package and the semiconductor(s) encased in the upper package.

SYSTEM PACKAGE APPLICATIONS
Developing a multiple die, system level package with related but dissimilar functions within a single package outline enables greater PCB surface utilization and the potential for providing enhanced electrical performance. There are a number of candidate semiconductor functions that could be integrated into a common PoP package configuration.

A likely configuration for BVA package stacking would begin with the processor die element packaged onto the base substrate. This processor die will likely have significantly more I/O than any other closely related semiconductor functions. The processor may be mounted face-up for wire-bond assembly or (after wafer level redistribution and bumping) mounted facedown for direct solder attachment. The lower packaged processor is most often packaged individually because it will require a very specific electrical test methodology. The mating package assembly attached on top of the processor component will most likely furnish the memory functions. Packaging memory using the die stacking process is commonly utilized for the upper section because the wafer level fabrication for memory, when mature, is a high yield process.

ENVIRONMENTAL TEST for BVA

A full suite of reliability tests has been completed for the BVA package-on-package configuration and the results are summarized in **Table 1**.

Table 1:
BVA PoP reliability testing results

Test	Standard	Test condition	Sample size	Evaluation	Result
Moisture sensitivity Level 3	IPC/JEDEC-J-STD-020C	125°C for 24hrs; 30°C/60% RH for 192 hrs, 3X Pb-free reflow	22 logic and 22 memory packages	C-SAM inspections at T0 and post-MSL3	Pass
High temp. storage	JESD22-A103D-condition B	150°C, 1000 hours	22 PoP off-board	E-test after 168, 500 and 1000 hours	Pass
Unbiased autoclave	JESD22-A102D-condition D	121°C/100%RH/2atm for 168 hours	22 PoP off-board	E-test after 96 and 168 hours	Pass
Drop test	JESD22-B111	>30 drops, 1500 G, 0.5 msec of half sine pulse	20 PoP on board with underfill	In-situ monitoring	Pass (128 drops)
Temp. cycling (board level)	JESD22-A104D Condition G	MSL3 and then -40°C to 125°C, 1000 cycles	45 PoP on board with underfill	In-situ E-test up to 1000 cycles	Pass

Moisture sensitivity level (MSL) testing was performed on individual memory and logic packages before joining. The high temperature storage and unbiased autoclave testing was conducted after joining the top memory package and bottom logic package. Prior to testing joined package configuration, a polymer underfill reinforcement was applied between package sections. The board level thermal cycling and drop tests were performed with the pre-joined memory and logic BVA test vehicles soldered onto a FR-4 test board. Before conducting these two tests an additional polymer underfill was dispensed between the lower package section and the FR-4 test board surface. Underfill is a common implementation for wireless phones and other mobile devices. All the tests passed without any failures. The drop test was extended up to 128 drops before stopping the test and no failures were detected.

To study the effects of copper-tin diffusion between the protruding copper bond wire and the solder ball, a number of BVA test packages were subjected to accelerated thermal testing (3x solder reflow cycles followed by 230 hours at 175 °C). This test allowed the Invensas engineers to evaluate the affect of intermetallic formation during board level assembly processing and high temperature storage testing. To inhibit diffusion and intermetallic formation, the copper bond wire used for the lower BVA package section were furnished with a thin palladium alloy coating. Prior to joining the package sections a number of samples were cleaned using a wet-etch process while others were cleaned using a more aggressive wet-blast method. The examples

shown in **Figure 7** shows that wet-etch method (example a and b) has not damaged the wire tip palladium coating and has served as an effective barrier against intermetallic growth. The wet-blast method (example c and d), however, has damaged the palladium coating leading to extensive intermetallic growth between the copper wire and solder.

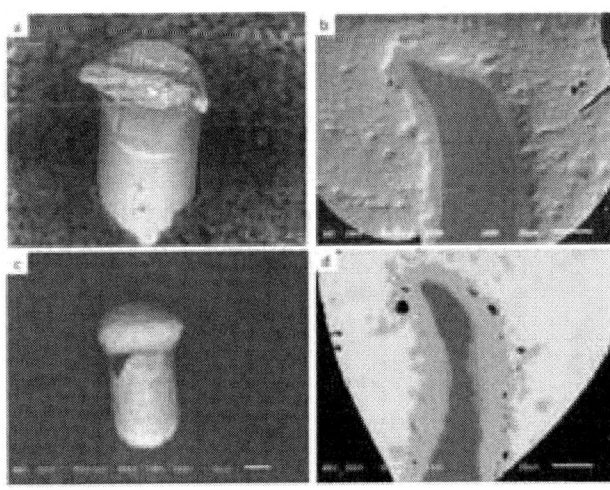

Figure 7. Comparing intermetalic growth of the the wet-etch cleaned Pd coated Cu wire and the wet-blasted cleaned Pd coated wire.

SUMMARY AND CONCLUSIONS

The BVA package technology has the potential for providing more than 400 interconnects within 2 rows on a 14mm x 14mm package body size. This unique molded wire contact process enables the shortest possible memory to processor interface. The process enables several profile variations and can furnish an array configured contact pitch at or below 200μm. The benefits are immediately seen. This interconnect solution is very economical and lends itself to a wide variety of 3D PoP packaging, including multiple-rows and area array, fan-in and fan-out, flat or step mold profile.

Although the BVA process may be a game changer in regard to semiconductor package performance there is no disruptive impact to the existing package assembly infrastructure. Although the open-end BVA wire forming process is quite unique the processes and equipment required for BVA package level assembly and stacking are quite common, In regard to board level (SMT) assembly, most commercial EMS suppliers currently processing FBGA will have the necessary capability for assembly and reflow solder attachment of the BVA package assemblies.

REFERENCES

1) Damberg, P.,Mohhammed, I.,Co, R., "Fine Pitch Copper PoP for Mobile Applications", Proceedings, 59th Electronic Components and Technology Conference (ECTC), San Diego CA 2012.

2) R. Crisp[1], W. Zohni[1], B. Haba[2], G. Pelissier[3], V. Bui[3] ([1]Invensas Corp., San Jose, Ca. USA, [2]Tessera

Intellectual Property Corp., San Jose, Ca, USA, [3]Dell Inc., Taipei, Taiwan) '*A multi-die DRAM package for solder-down memory in UltraBook and Tablet PC applications*', proceedings of IEEE ICEP 2012.

3) R. Crisp, B. Gervasi, W. Zohni, B. Haba, '*Cost-minimized Double Die DRAM Packaging for Ultra-High Performance DDR3 and DDR4 Multi-Rank Server DIMMs*', proceedings of ESQED 2012.

4) S. McElrea, '*3D Packaging Solution Providing DDR & LPDDR Co-Support for UltraBook and Next Generation Servers*', proceedings, IMAPS 8[th]International Conference and Exhibition on Device Packaging, 2012.

5) Kim, Jinseong, et al, "Application of Through Mold Via (TMV) as PoP Base Package", Proc. 58th Electronic Components and Technology Conference (ECTC), Orlando FL 2011, pp. 1089-1092.

6) Solberg, Vern, "Basic PCB Level Assembly Process Methodology for 3D Package-on-Package", technical proceedings of IPC APEX-Expo 2010.

7) JEDEC PUBLICATION 95 DESIGN GUIDE 4.5, '*Fine-pitch, Square Ball Grid Array Package (FBGA)*'.

Invensas Corporation licenses the use of its technology innovations worldwide. The company's headquarters and development laboratories are located in San Jose, California (www.invensas.com).

JETTING FINE LINES FOR HIGH VISCOSITY FLUIDS ONTO 2D AND 3D ELECTRONIC PACKAGES

Horatio Quinones
Nordson ASYMTEK
Carlsbad, CA, USA
horatio.quinones@nordson.com

ABSTRACT

For many years technologies such as inkjet printing and fine screen printing have been successful in placing dots and lines on mostly flat topographies. However these technologies have required either low viscosity fluid inks with several properties tailored to meet the inkjet hardware operation window, and in the case of the former method, where higher viscosity fluids and pastes are used, a flat and unobstructed topography of the surface is required. We are proposing in this study an alternate method where a hybrid jetting mode is used to dispense small dots and fine lines in which the above restraints are overcome, i.e., the low viscosity fluid requirement and the flatness of surface upon which we place the fluids. The implementation and application process for 2D and 3D packages is described, where nano hybrid jet dispensing is practiced resulting in high conductive lines of 40μm to 160μm in width on 80μm to 200μm pitch. Detailed descriptions will be shared about how traditional issues in jetting are overcome during this work.

Key words: conductive lines, jetting, keep-out-zone, jet streaming, 3D dispense.

INTRODUCTION

The ever increasing demand for smaller dispense doses including fine lines, small dots, edge definition improvement and more accurate dispensing requires examination of available products and technologies. Potential solutions include piezo-driven valves, positive displacement jets, nano-dose dispense valves, multi-nozzle inkjets, and other non-contact valves capable of dispensing high viscosity and doped fluids. Although printing electronics may offer a process solution to some family of fluids and applications, this inkjet technology at present is limited to fluids of single-digit centipoises viscosities and other material properties i.e., thixotropy, magneto-electrical, that may make such fluids incompatible for the electronics industry in other multiple applications. There are some limitations for stenciling or screen printing high viscosity conductive pastes in the form of small dots or thin lines for numerous package formats including multiple components layouts, and accessibility of the printing head. Three dimensional electronic packages, i.e., stacked die, also present major challenges to such processes. An alternative to the traditional processes can be printing or jetting technologies which are non-contact fluid dispensing. Electrical conductive lines are a feasible solution for multiple applications provided that their electrical and physical properties are within well-defined boundaries and ranges. This requirement often requires thin lines and homogeneous lines. In another application where fluid is to be kept from wetting or flowing into Keep-Out-Zones (KOZ), a thin hydrophobic dam can be dispensed with a non-contact technology. Fluid viscosities can vary from a few mPa-s to 5.0E5 mPa-s, therefore dispensing systems capable of operating in those ranges is required.

EXPERIMENTAL WORK

System in package (SiP) and other highly populated boards often require fluid dispensing on well defined areas only, thus avoiding KOZ. Geometries involved, material wetability, osmotic and diffusion of the fluid onto the solid areas violate such zones, and hence a possible solution can be a fluid dam.

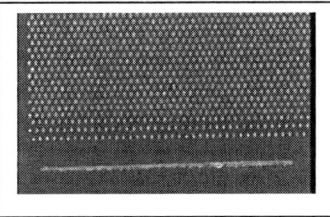

Figure 1. Ink line jetted as a dam to prevent underfill fluid from reaching KOZ.

Such dam can be manufactured as part of the board topology or can also be built by a thin fluid line dispense. A thin line was dispensed using an inkjet print head with a nozzle orifice of 25μm; the line of about 65μm was placed within the allowed zone and near the side of the die where the underfill fluid was jetted. The ink used for this purpose was underfill phobic. Figure 1 depicts the jetted line and part of the footprint of the bump lands (die has not been attached). The property of the fluid as well as the geometry of the dam is expected to stop the underfill from passing over the dam. Figure 2 depicts the efficiency of the dam as underfill ceases to flow. Another application is the writing of conductive lines on substrates, resistors and inductors can be built by simply printing or dispensing conductive lines.

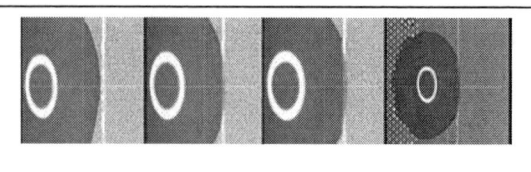

Figure 2. Ink jetted line dispense on die carrier as to stop underfill from reaching KOZ.

A non-contact or jetting process can overcome stencil printing requirements for surface topography and be able to draw conductive lines on 3D topographies and potentially connect multiple 3D layers. Traditionally wire bonding is the connection of choice; we propose here jetted conductive lines to electrically coupled multiple layers of a 3D stacked die. Figure 3 depicts a typical pyramid structure of eight layers.

Figure 3. A multi-level 3D with a pyramid-like geometry, jetted conductive lines at 200µm pitch provide electrical interconnection among the eight layers.

The challenge to meet desired electrical properties, i.e., electrical resistance/conductance, depends on the material properties[1] and the geometry of the line. Highly conductive materials tend to have high viscosity from the fact that the fluid matrix is doped with suspended conductive particles; this rheology state makes dispensing with the non-contact jetting process more difficult. A valve capable of droplet formation in small doses is used[2] ,a hybrid dispense mode where sudden stopping of the fluid stream to form droplets and a fast retrieving motion of the valve from the substrate surface results in a successful dispense. Figure 4 shows conductive lines dispensed in this mode on a 3D package with eight layers, four layers are connected at a time in two opposite sides of the stack, the line pitch is 200µm and the lines are 160µm wide. The cross section of the part shows the fluid migration into the layers, which wets the die pads resulting in a robust connection; line geometry is essential to obtain proper electrical and mechanical attributes. Given the size of the lines and rather complicated geometries and surface wetability, a consistent dispense gap is of most importance to obtain line repeatability.

The jet valve can also be used in a hybrid mode where the pulsing of the piston is used to stop a fluid stream rather than continuously generate droplets at higher frequency of actuations.

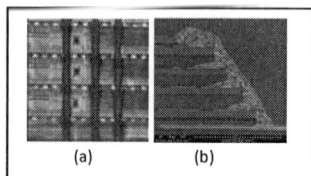

Figure 4. (a) Jetted conductive lines on 3D package. (b) Cross section of an interconnection line.

This "streaming mode of the jet" can generate thin lines at very high rates. The starting and ending of these lines tend to be wider than the bulk of the line as seen in figure 5, a "dogbone effect" for the streaming mode as well as the dot mode. A dynamic dispense control i.e., a dispense rate correlated to the head motion can mitigate this issue. For the present work some modifications to the software i.e., addition of z-motion gave acceptable results. In figure 5b one can notice that for the dot mode the end of the line has no dogbone issue.

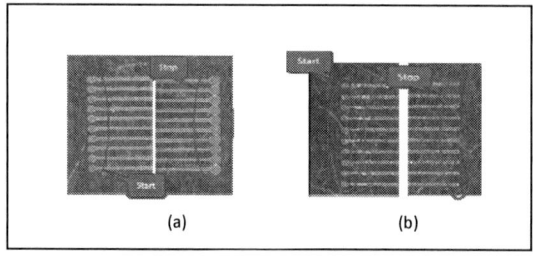

Figure 5. The end of line issues, dogbone effect for: (a) jet in line streaming mode and (b) jet dot mode.

However the beginning has a wider geometry, this is due to the fact that the nozzle orifice has a meniscus from the previous dispensed that get deposited during the jetted first droplet. Figure 6 shows the improvement upon dynamic dispense algorithm implementation and by minimizing the shutoff spike in the fluid delivery. The shutoff spike was minimized by altering hardware geometry like fluid path ducts and by piston driver dynamic deceleration during the jet actuation and just prior to piston to seat impact.

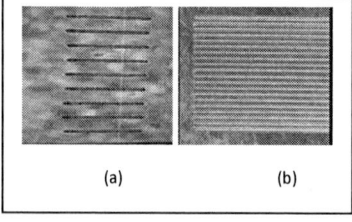

Figure 6. Jetted conductive lines with minimum "dogbone effect for (a) jet streaming and (b) jet dotting.

In some applications, however, the dogbone effect may be advantageous since at those end locations a connection to a pad is needed. Obtaining small conductive lines for long periods of dispensing presented a large clogging problem in the jet ducts. The particle doping needed to make these fluids good electrical conductors have the tendency to segregate upon impact and form a solid powder that eventually causes obstruction of the dispenser ducts; traditionally the smallest diameter of the fluid path is the orifice of the nozzle, and hence it is the location most prone to clogging. For the case of the jet however, the solidification of the particles occurs in the area of high pressure such as where the piston impacts the seat surface and the compacted powder eventually clogs the seat. Figure 7 shows a cross section of the duct and the location of most likelihood for clogging.

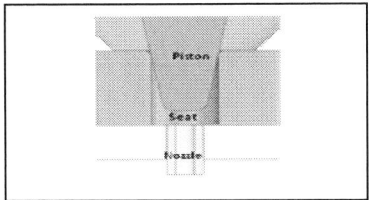

Figure 7. Cross section of the jet valve near the exiting orifice, the seat is the area where material clogs.

This issue becomes apparent after a few hundred actuations of the jet; it has been a common issue for jetting this kind of doped fluid as it is the case for most solder pastes. The valve-on-time was shortened for the pneumatic jet taking advantage of the fact that the impact force to prevent leakage is low and the velocity of the piston does not need to be very high (<1 m/s) since these materials stream well under high pressure differential (>2.5 bars.) Volumetric dispense accuracy and consistency of the line mass density is another issue when dispensing small doses. The jet when used in the streaming mode has a very precise timing in the shutoff (positive shutoff) and opening making it ideal for this type of application. Figure 8 depicts the dispensing accuracy of the jetting process.

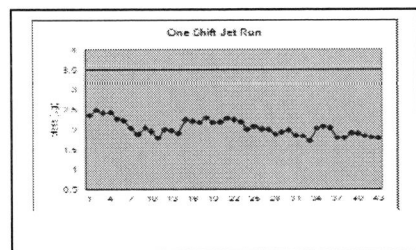

Figure 8. Plot of mass dispense during a simulated shift time frame, accuracy of about +- 0.2μg.

To obtain line widths of less than 160 μm the droplets need to be between 15 to 2.5μg in mass. Lines of <70μm width were dispensed on flat surfaces as depicted in Figure 9 on a 100μm pitch. Similar geometries were accomplished on 3D stacked packages where a vertical device (side of the die) was to be electrically connected to a pad on the substrate to it.

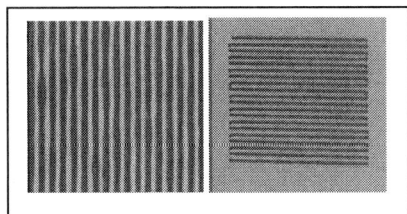

Figure 9. Conductive lines jetted on a flat surface, line width is <65μm.

A similar application process was exercised for solder paste. Thin lines (<230μm width) were jetted. The traditional clogging problem of these materials when jetted was not present in this process. More than 70,000 droplets were jetted without clogging issues.

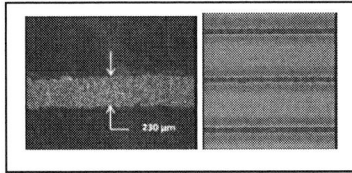

Figure 10. Solder paste line jetted on a flat surface. Line width is 230μm.

Type 500 solder paste was used for this study (a Nordson EFD[3] product). Dispensing was performed during several days without clogging problems with a high rate of consistency.

PROCESS IMPLEMENTATION
The process described above has been implemented in a manufacturing environment. These jetted conductive lines are replacing traditional wire bonding and at the same time providing interlayer interconnection without through silicon vias (TSV.) Figure 11 shows the 3D line jetting process for 125μm jetted lines of conductive silver paste on 3D stacked die with a total of 16 layers, of which eight are connected with the jetted conductive lines on the opposite side of the package; the pitch is 200μm. The ends of the lines exhibit a somewhat wider thickness; this feature was dispensed by design to enhance the conductivity on the connected pads. It was important to not generate debris or satellites that could create electrical connection among lines. The stack was built on steps in a pyramid manner. In another design, the 3D stack was built without the steps with the connected side normal to the die stack carrier.

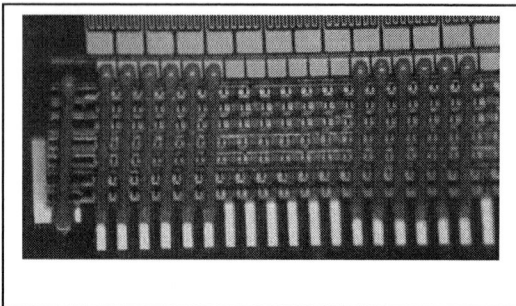

Figure 11. Jetted lines making electrical connection between multiple layers of a 3D package.

Figure 12 depicts the perfectly vertical package and the conductive lines as well.

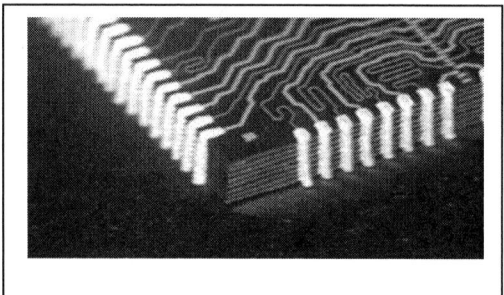

Figure 12. Sixten-layer package with a total height of about 0.8mm.

CONCLUSIONS

Conductive lines with widths of less than 200μm down to 45μm were jetted successfully. These lines were dispensed on flat surfaces as well as in a 3D space. Highly doped fluid with conductive particles was successfully jetted.

RECOMMENDATIONS

Continuous hammering of the material by the motion of the needle needs to be attenuated: Softer elastic modulus of the spring driving the needle motion. Implement needle tip shape change, from spherical to conical to minimize applied pressure on the material. Use a larger seat inner diameter to further minimize likelihood of clogging. Fluid pressure increase to drive (extrude) the material from jet nozzle orifice, i.e., pressure differential jetting. To decrease the shot size and line width, one needs to implement shortening of the valve-on-time, reduce nozzle inner diameter orifice and dispense gap, assure high Z-motion resolution. For lines, apply dog bone geometry elimination software with dynamic line dispense control and minimize the momentum transfer mode from the jetting process.

ACKNOWLEDGEMENTS

Special thanks to: Mr. Tom Chang for the experimental and application work during the entirety of the project; ORMET Circuit Inc. and finally thanks to Ms. R. Smith-Foster for her help in preparing the manuscript.

REFERENCES

[1] Ormet Circuit Inc.: http://www.ormetcircuits.com.

[2] Nordson ASYMTEK: http://www.nordson.com/en-us/ divisions/asymtek/products/JetsPumpsValves/DispenseJet-Series/Pages/DJ-9000-DispenseJet-Jet-Technology.aspx.

[3] Nordson EFD: http://www.nordson.com/en-us/divisions /efd/.

AEROSOL JET® PRINTING OF CONDUCTIVE EPOXY FOR 3D PACKAGING

Michael J. Renn, Ph.D., and Kurt K. Christenson, Ph.D.
Optomec, Inc.
St. Paul, MN, USA
mrenn@optomec.com and kchristenson@optomec.com

Donald Giroux
Resin Designs, LLC
Woburn, MA, USA
dongiroux@resindesigns.com

Daniel Blazej, Ph.D.
Assembly Answers, LLC
Dracut, MA, USA
daniel.blazej@assemblyanswers.com

ABSTRACT

Traditional dispensing technologies are typically limited to feature sizes above 100 μm and have limited 3D capability due to near- or direct-surface contact with the substrate. Conductive adhesives for these systems typically contain large metallic flakes which lead to inconsistent conductivity in small features such as dots. In this paper we combine a new, nanoparticle conductive adhesive with Aerosol Jet® dispensing and demonstrate non-contact printing over 3D surfaces and into recesses such as through silicon vias (TSV). The conductive adhesive is a high-solids, nanoparticle system with resistivity as low at 5 μΩ-cm at 150 °C cure temperatures. The adhesive has superior thermal conductivity, adhesion, and elasticity, which are desirable for fabricating heterogeneous connections, such as between IC chips and circuit boards. The material can be dispensed with dot sizes down to 25 μm, which enables high density die attach. Similarly, TSV filling has been demonstrated with via diameters down to 50 μm and depths of 300 μm. The high-solids formulation reduces the material shrinkage during curing resulting in a dense conductive plug.

Key words: Aerosol Jet, conductive adhesive, die attach, through silicon via (TSV), 3D packaging.

INTRODUCTION

Various dispense technologies are commercially available, including both contact and non-contact approaches.[1-3] Among the contact approaches are needle and valve dispense as well as screen printing. Non-contact approaches include jetting and ultrasonic deposition. These technologies can be used to dispense a wide variety of materials including conductors, insulators and adhesives, and it is also possible to use them to fabricate multilayer circuitry by sequentially depositing conductors and insulators. However, because the dispensers require either contact or near-contact with the dispensing surface, the printed circuitry is generally limited to planar geometries. Furthermore the types of inks used in these dispensers typically limit feature size. For example, many of the jetting approaches require low viscosity inks.[4] These inks tend to spread on the substrate, which limits the aspect ratio that can be achieved. A related limitation is that needles and valves have increased clogging frequency when the orifice size gets smaller.

Most of these technologies have been applied to dispensing the conductive adhesives used for attaching chips to Printed Circuit Boards (PCB). As chip packages continue to shrink there is a continuous need to reduce the size and volume of the adhesive dots. For example, the pad size on an 0201 metric SMT package is 0.2 x 0.1 mm, which implies that a dot size of approximately 0.05 mm is needed for attaching the device to a PCB. This dot size is on the low end of what conventional technologies can achieve.

Aerosol Jet technology has been available for several years and has mainly been used for dispensing of nanosilver inks for photovoltaics, printed circuits, and printed antennae.[5] One major advantage of this technology is that fine feature printing, down to 10 μm feature size, can be achieved with a surprisingly large standoff height (up to 10 mm). This capability allows for printing in recessed regions such as vias, over step edges such as SMT devices, and over non-planar substrates such as populated boards.[6] In contrast to other jetting techniques, the ink droplets in the Aerosol Jet system do not actually interact with the surfaces of the nozzle. This feature results in greatly reduced frequency of clogging and less sensitivity to specific ink rheologies, such as viscosity, surface tension and solids loading.

One further consideration is that even with accurate dispensing of small volumes, inhomogeneities in the material itself become a factor at small dot volumes. The conductive phase of typical die attach epoxy consists of large particles of silver (>5 um). As the dot volume shrinks, the number of silver particles in the dot becomes small enough that any fluctuations in the number of particles leads to statistical variability in the conductivity. So even if the hardware was dispensing perfectly uniform volumes, fluctuations in the amount of silver particles would lead to fluctuation in electrical connection and fluctuation in the contact resistance between the epoxy and contact pad.

In this paper we describe efforts to improve both the dispensability of small volume dots and correspondingly improve the consistency and conductivity of the dots. To that end, Optomec, Resin Designs and Assembly Answers have teamed up to develop a new nanosilver epoxy that can be Aerosol Jet printed to very fine features. The nanosilver epoxy is designed for high electrical and thermal conductivity, excellent adhesion and shear strength, and low temperature curing. This paper will discuss results of the print properties of the material including dot size, volume, and consistency. The electrical properties under various sintering conditions, including laser sintering, will be described as well as measurements of bond strength. Initial results of chip attach, printed circuits, and via fill will be described.

PRINTING TECHNIQUES
Figure 1 shows a conceptual layout of the Aerosol Jet printing apparatus. Liquid inks are first atomized to create a dense aerosol cloud of 1-5 μm diameter droplets. A pneumatic nebulizer is typically used when the liquid viscosity is in the range of 20-1000 cP and an ultrasonic atomizer is used for low viscosity fluids (1-20 cP). A wide range of liquid inks can be atomized in the system such as pure solvents, solutions, dispersions, and mixtures of particles and resins. However, the particle diameters are limited to below 1 μm for the pneumatic nebulizer and below 0.2 μm for the ultrasonic atomizer. Flake and nanotube particulates can have larger lateral dimensions and still be atomizable.

Once created, a carrier gas transfers the aerosol to a printing head. This carrier gas is typically dry nitrogen, but other gases and vapors can be introduced into the gas stream as needed. For example, air sensitive materials may benefit from adding a fraction of forming gas to the carrier. Within the print head, a second, co-flowing sheath gas focuses the droplets to a 10-100 μm diameter jet. The droplets exit the nozzle at velocities up to 100 m/s. At these velocities the droplets have sufficient momentum to travel ballistically and impact on a target substrate. The droplets do not slow significantly over a distance of 5 mm. Even larger working distances can be achieved with large diameter jets that can extend the standoff up to 10 mm from the nozzle. The print head and substrate motion are controlled by computer for

high resolution patterning and alignment. The flow of droplets is gated with a high-speed mechanical shutter.

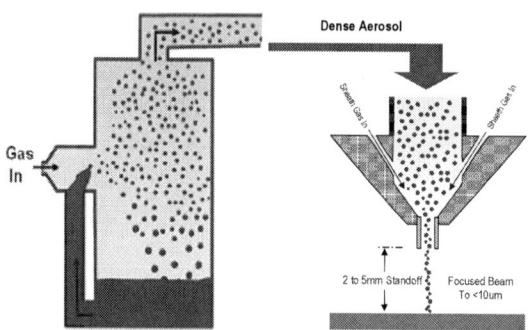

Figure 1. Schematic diagram of an Aerosol Jet-based printing system.

Aerosol Jet printing differs substantially from other dispense techniques. First, the aerosolization and aerodynamic focusing steps are unique. Once the droplets are created in the atomizer, they do not contact any other surfaces until impaction on the target. This makes the system relatively clog free, robust, and material friendly. Second, since the droplets travel ballistically over several mm from the nozzle, non-contact, fine feature deposition can be achieved at a large standoff. For example, the aerosol can be jetted into via, over 3D steps, and across complex topography. Finally, since the droplet size is on the order of 10 femtoliters, very small volume features can be dispensed with excellent uniformity. For example a 1 mil (25 um) cube of material will consist of more than 1000 individual femtoliter droplets. The printed features can be scaled by printing more or fewer droplets at a given location. The overall printing rate is also scalable by adjusting the droplet atomization rate and carrier gas flow rate. The print rates can typically be scaled from 10 nl/min to 10 μl/min but with tradeoffs on the feature size, i.e. larger print rates correlate to larger printed features.

Print Module and System
The Aerosol Jet print engine is shown in Figure 2. In addition to the dispensing head (6) described above, various other components are incorporated to allow for alignment, process visualization, and *in situ* material curing. The vision system consists of alignment camera (1) and a print process viewing camera (5). The alignment camera is used for substrate alignment (positioning) prior to execution of a printing toolpath. Up to three fiducial points can be recognized by the camera and software to allow translational, rotational, and skew corrections. A teach-and-learn utility can be used to program specific fiducial points into the software to automate the fiducial locating process. The offset distance between the camera and print head is calibrated so that once the start point is indentified by the

108

camera, the print head can be translated to that location and the toolpath executed.

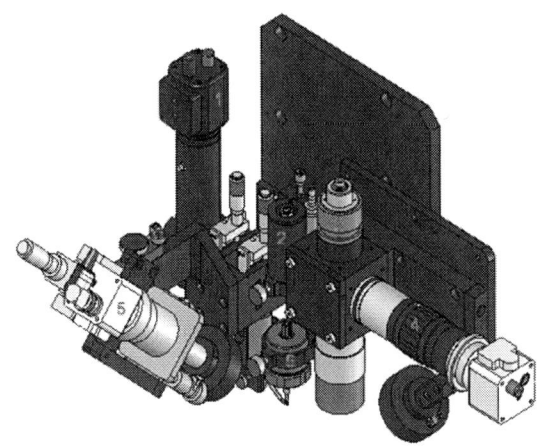

Figure 2. Print module for the Aerosol Jet print system showing (1) Alignment camera, (2) Tube heater, (3) Laser module with (4) Laser alignment camera, (5) Process viewing camera and (6) Print head. The atomizer is not shown.

The process camera is used to view the printing in real time. The target region on the substrate is displayed live on the computer monitor. The process camera is useful for monitoring toolpath progress as well as for developing new processes. For example, when filling TSVs, the ink fill level can be observed when the via fill approaches the top surface of the die.

A tube heater (2) is incorporated in order to modify the aerosol droplet viscosity while in flight to the print head. The tube heater raises the surface temperature of the transfer tube connecting the atomizer and print head and consequently causes partially drying of the aerosol. Volatile solvents in the ink can be partially evaporated which causes the droplets to become more concentrated and viscous. The more viscous droplets can be printed with less spreading compared to the low viscosity starting material. Consequently, high aspect ratio features are achieved.

While most materials printed on an Aerosol Jet system are cured in an oven after printing, an optional laser module (3) can be used to post-process materials *in situ* without removing the product from the machine. The laser consists of either a fiber-coupled diode laser at 832 nm or a fiber-coupled diode pumped solid state Nd:YAG laser operating at 532 nm. The laser power is up to one Watt continuous wave (CW). The laser beam is focused onto the substrate at an offset from the print head and so the laser processing can be performed subsequent to the printing. The printed material can be laser processed by translating the laser to the start point of the print pattern and re-executing the toolpath motion. Also included in the laser beam path is a beam splitter, which is used to deflect the image from the laser process zone into a camera (4). This vision system allows the laser process to be viewed in real time. When laser curing silver epoxy, a distinct color change is observed with the silver epoxy transitioning from brownish-black to bright silver. One significant advantage of the laser curing is that it allows immediate *in situ* curing of printed materials without the need to remove the substrates for oven processing. Among other things, this allows more efficient printing of multiple material patterns as it eliminates the need to remove and externally process substrates between print layers.

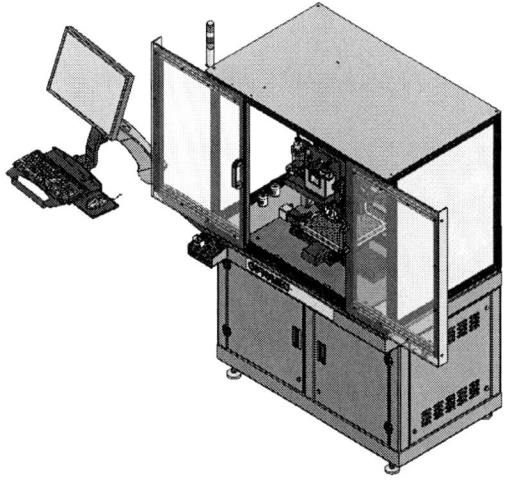

Figure 3. CAD rendering of the AJ300 print system.

A CAD rendering of a complete Aerosol Jet printing system is shown in Figure 3. The basic system includes the print engine from Figure 2, a granite base, heated vacuum chuck, linear X-Y stages and controller, and an environmental enclosure. This system provides users with the ability to produce components directly from CAD drawings of the desired circuits. The enclosed system keeps the work surface clean and any solvent fumes are extracted through an exhaust port. The window incorporates a laser safe glass to provide a Class I protection. Industry standard, power, emergency machine override, and reset switches are located in a breakout box immediately outside the enclosure, along with a safety light stack indicating the status of the tool.

MATERIALS AND PRINTING QUALITY
The Resin Designs E8074 silver epoxy formulation consists of a homogeneous mixture of nanoparticle silver and proprietary epoxy resin. The hardener is designed with a threshold reaction temperature of approximately 100 °C and full curing of the resin occurs at 150 °C in 60 minutes. The

working life of the epoxy is greater than one week at room temperature and the storage life is six months at -10 °C.

E8074 silver epoxy is supplied at a viscosity of 100-200 cP with a carrier solvent of γ-butyrolactone to facilitate nebulization. During printing, the target substrate is heated to 60 °C to evaporate the added solvent. Since the threshold reaction temperature is set at 100 °C, the printed silver epoxy remains in an uncured state until the post-processing step.

Figure 4 shows a 3D optical profile of an array of small volume, silver epoxy dots. These data were measured with a Cyber Technologies CT100 optical profilometer. Each dot in the array was dispensed by opening the shutter for 100 ms with the print head centered over the dot location. Subsequent dots are printed by shuttering the material flow and stepping at relatively high velocity to the next position. The 2D array is fabricated at a rate of 5 dots/sec. For this example, the dot diameter at the base is 120 um and the full width at half maximum is 60 um. The average height is 6.8 um, the peak height is 10.3 um, and the volume is 40,000 um^3 (40 pL). The placement accuracy is better than 1 um in both x- and y- directions. It is possible to print dots as small as 1 mil (25 μm) diameter, but currently the volumetric variation is significantly higher than with larger dots.

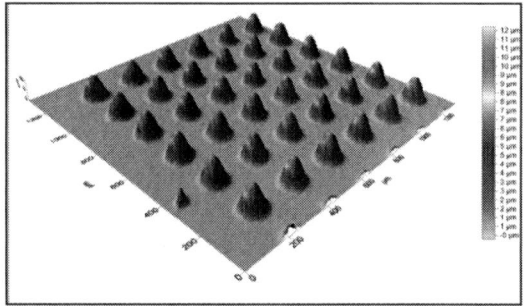

Figure 4. 3D profile of bump pattern array.

Printed traces of variable lengths are used to test the linearity of resistance with line length. Figure 5 shows an array of lines printed on an alumina plate that are terminated with large contact pads. The sample was cured at 150 °C for 60 minutes. The two point resistance measurements below show a linear increase in resistance with line length. The linearity suggests that the material has uniform resistivity and that the traces have a consistent cross section.

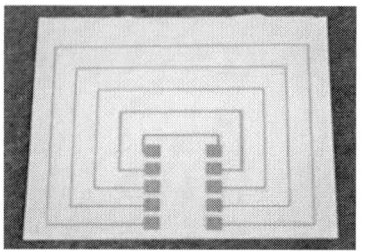

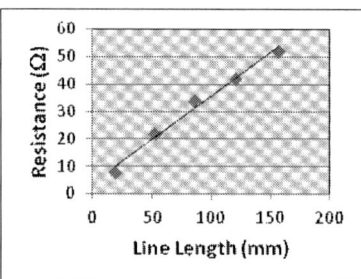

Figure 5. Test coupon for determining dc resistivity of printed lines (upper). Resistance as a function of line length when cured at 150 °C for 1 hr (lower).

Oven Curing

Figure 6 shows the material resistivity of printed traces when cured for variable periods at 150 °C. The traces are profiled with a Dektak stylus profilometer to determine the geometric cross section. Four-point resistance measurements along with geometric cross section data are used to derive the material resistivity through the relation: R=ρL/A where R is the 4-point resistance, L is the trace length, A is the geometric cross section, and ρ is the material resistivity. As shown in Figure 6, the resistivity when cured at 150 °C in 1 hour is 1.6 μΩ-m or 100 times the resistivity of bulk silver. Although the epoxy resin is designed to cure at 150 °C in 60 minutes, we find that the resistivity continues to drop with extended heat treatments. With a 2 hour cure the resistivity decreases another factor of 10. With extended curing the resistivity reaches an asymptotic value of 3 μΩ-cm (approximately 2x bulk silver). This value is substantially lower than traditional silver epoxy and is more comparable to sintered nanosilver inks.[7]

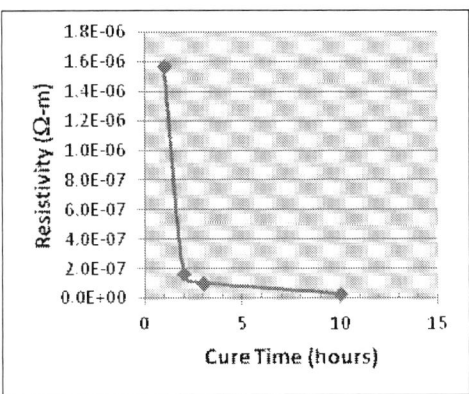

Figure 6: Resistivity as a function of cure time at 150 °C. Resistivity continues to decrease over several hours approaching an asymptotic value 2x of bulk silver. SEM analysis shows that after an initial cure the silver flakes sinter and form an interconnected metallic network.

Figure 7 shows cross section SEM micrographs of uncured and cured silver epoxy. The uncured micrograph shows distinct, overlapping silver flakes. However, the micrograph of cured material clearly shows that the flakes are merging to form a connected metallic network. Consequently the increased conductivity observed with longer cure times may be a result of continued merging and densification of the silver flake. The high silver content in the epoxy then suggests that the primary conduction occurs through the metal network rather than by tunneling from particle to particle.

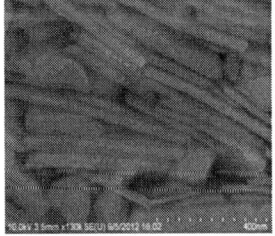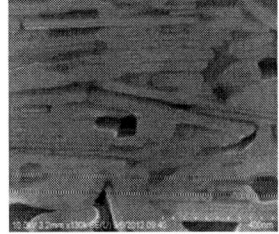

Figure 7. Cross section SEM of uncured silver epoxy (left) and silver epoxy cured for 2 hrs at 150 °C (right). The individual plates merge over extended cure times.

Laser Curing
Since extended curing schedules may not be acceptable for some applications, we have also examined laser assisted curing. Figure 8 show a schematic of the laser curing technique used in the Optomec system as well as resistivity data. As mentioned above, the focused CW laser beam acts as a point source of heat to rapidly cure material under the impinging spot. In this case, a 200 mW, 832 nm laser is focused to a spot size of 20 µm to give a peak intensity of 2×10^5 W/cm^2. The laser spot is scanned at a rate of 10 mm/s. Based on color changes observed while scanning the laser over a silver epoxy film, the heat affected zone is estimated to extend to 50 µm. In other words, even though

the laser is focused to 20 µm, the lateral heat spread is sufficient to process a 50 µm spot. Printed features larger than 50 µm can be cured by scanning the laser in a raster pattern with a 50 µm pitch. A larger pitch and larger spot size would be possible with a more powerful laser.

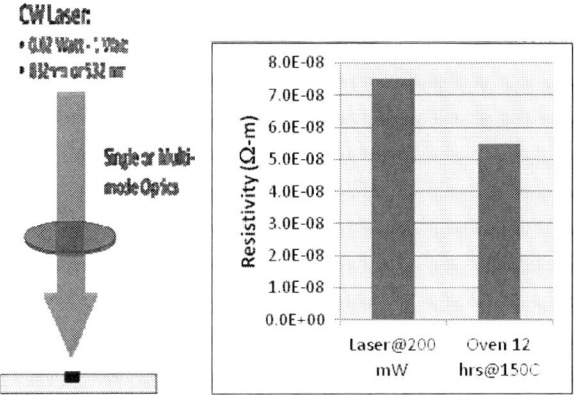

Figure 8. Schematic diagram of laser sintering configuration (left). Resistivity of laser sintered silver epoxy compared to oven cured epoxy (right). While similar resistivity values are obtained, the laser process greatly reduces the overall cure time and can be performed without removing the substrate from the system for oven processing.

The resistivity data in Figure 8 compares the resistivity of laser cured silver epoxy to the resistivity of epoxy that was cured in an oven at 150 °C for 12 hrs. The values are similar to within 30%. The advantage of the laser cure process is that high conductivity can be achieved within minutes of printing the ink, whereas the oven curing requires hours for the lowest values. The laser curing can also be performed without removing the substrate from the system, so it can reduce the number of steps required for printing multiple layers of dissimilar materials. On the other hand, the laser requires a line of sight to the substrate, which is not always possible, especially if the epoxy is used to bond between a chip and PCB.

Adhesive Properties
One of the primary applications for dispensing nanosilver epoxy is for attaching chips to PCBs. Figure 9 shows images of 15 mil (375 µm) Kovar tabs attached to gold coated alumina. The epoxy pads are printed with a 20 mil (500 µm) width and 1 mil (25 µm) thickness. When the tabs are placed, the epoxy clearly wets the vertical edge and forms a fillet. Die shear measurements indicate that the highest shear strength is obtained when the epoxy wets the sidewall in this way. The measured die shear strength is 2000 PSI which is comparable to standard die attach epoxies and meets the MIL-STD-883, method 2019 minimum requirement of 882 PSI. Cohesive failure is seen in all cases. The glass transition temperature is approximately 100°C.

Figure 9. 15 mil (375 µm) Kovar tab bonded to gold coated alumina substrate using silver epoxy.

Applications

Multiple bond pads can be printed at high density for attaching multiple I/O chips. Figure 10 shows the attachment of 0402 SMT resistors to ceramic substrate and QFN devices to glass plates. As described above, the substrates are heated to 60 °C when jetting the silver epoxy pads. This temperature is sufficient to evaporate residual solvents in the epoxy, but low enough that the epoxy does not cure. The printed epoxy remains uncured until subjected to elevated temperatures of 100 °C or higher. Consequently, it is possible to print a large number of bond pads sequentially and then attached the chips later. The chips shown in Figure 10 were manually placed with assistance of an optical microscope. The QFN sample also shows that the chips are physically connected to printed conductor lines.

 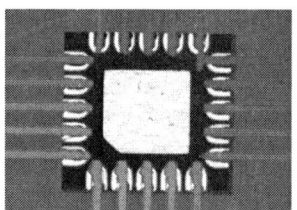

Figure 10. Images of attached SMT devices using printed silver epoxy. The 0402 resistor (left) is attached to ceramic using 8 mil (200 µm) pads with an 8 mil (200 µm) gap between pads. The QFN chip (right) is attached to printed silver lines on glass substrate. Dots of epoxy have been printed on the ends of the lines and then the chip was placed manually with the assistance of a microscope.

Compared to conventional nanoparticle silver inks, the silver epoxy exhibits far less shrinkage during curing. After solvent evaporation, the measured volumetric shrinkage during curing is less than 2.5%. Consequently, the silver epoxy may be more effective at producing solid conductive fills in confined geometries. One application where low shrinkage is required is in producing conductive via plugs in Through Silicon Via (TSV) die. Figure 11 shows an SEM cross section of a 50 µm diameter by 300 µm tall TSV, which contains jetted and cured silver epoxy. The fill process consists of several steps. The silver epoxy can be jetted into an open via, but there is substantial leakage from the bottom side. Consequently, the amount of time required to fill the open via becomes variable, depending on the amount of leakage. In this work, an adhesive tape is applied to the back side of the die to convert from an open via to a blind geometry. The next step is to fill the via completely by dispensing liquid, silver epoxy down the center of the via. The epoxy flows to the bottom of the via and fills from the bottom upward. When full, the liquid epoxy is flush with the top surface. The platen is heated to 60 °C to evaporate the solvent and in this drying process the top level of the epoxy recesses into the via. If needed, the filling step is repeated at least once more to return the liquid level to the top surface. Repeated partial fills can bring the dried epoxy level flush to the surface, but at the expense of additional process time. After filling and allowing the solvents to evaporate, the samples are fully cured at 150 °C. Since the dried epoxy has low shrinkage, highly dense metal plugs are achieved.

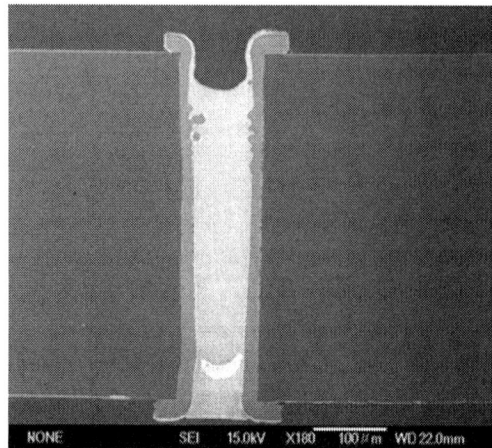

Figure 11. SEM cross section of silicon via filled with silver epoxy. The cured silver epoxy appears bright white and a sidewall coating of plated copper is also evident.

FUTURE WORK

Future development work is planned for measuring thermal conductivity of the silver epoxy. High thermal conductivity is important for attaching power devices and heat generating chips. Given the high metal loading, the silver epoxy is expected to have an excellent thermal conductivity. Additional work is needed to determine contact resistance between the silver epoxy and various chip and board metallizations. Finally, a full characterization die shear strength as a function of bond line thickness is under way.

CONCLUSIONS

A new nanoflake silver epoxy has been developed for small feature, die attach applications. Using the Aerosol Jet tool, uniform dot sizes as small as 1 mil (25 µm) can be dispensed. The epoxy is curable at 150 °C and at that temperature the die shear strength exceeds MIL-STD-883, method 2019. The electrical resistivity is 100 times greater than bulk silver with a 1 hour cure, but can go down to 2 times greater with extended curing. Laser assisted curing results in similar resistance values but with dramatically shorter processing time. The Aerosol Jet dispensing system is capable of non-contact printing at high standoff heights. Consequently, the silver epoxy can be printed into recesses and over steps. This capability should enable various TSV and chip-on-chip configurations and support high density I/O packages.

REFERENCES

[1] Jetting and Valve dispense technologies are described: www.nordson.com

[2] A description of syringe dispense technogies can be found at: www.newport.com

[3] A description of ultrasonic aerosol deposition can be found at: www.sonotech.com

[4] F. Bibelriether et al., Print the Printed Circuit Board - Inkjet Printing of Electronic Devices, Proc. NIP26, pg. 715. (2010).

[5] Aerosol Jet technology information available at: http://www.optomec.com

[6] M. O'Reilly and J. Leal, "Jetting Your Way to Fine-Pitch 3-D Interconnects," Chip-Scale Review, 14, 18, (2010).

[7] Examples of nanosilver ink products can be found at: www.cabot-corp.com

DIELECTRICS FOR EMBEDDING ACTIVE AND PASSIVE COMPONENTS

J. Kress, R. Park, A. Bruderer, and N. Galster
Atotech Deutschland GmbH
Basle, Switzerland
juergen.kress@atotech.com

SH Cho
Dongyang Mirae University
Seoul, Korea
coolcsh@dongyang.ac.kr

ABSTRACT

Embedding of actives and passives is a quickly growing field which is being investigated by many companies and institutes. Reasons for this increasing interest are to reduce the complexity of the packages, achievement of higher degree of miniaturization, shorter electrical connections, and a reduction of layer count.

Among other technical challenges, warpage is a major concern. The inherent different thermomechanical properties of the different materials involved cause internal stresses. Those stresses show up as warpage which makes handling during production more difficult, reduces the overall yield, and imparts reliability.

Presented in this paper is a composite type material which can be used as dielectric for the embedding of actives and passives. It combines the advantages of the mechanical stability of prepregs and the good encapsulation properties and ease of handling of resin coated copper foils (RCC). Also described is a concept for simulating the warpage of packages using such composite build-up materials taking into account different resin properties.

Key words: Embedding material, actives, passives, simulation, warpage

INTRODUCTION

In the area of chip embedding there are currently two main technologies which find increasing interest. Packages which are based on a molded wafer infrastructure (Fan out wafer level packages, FOWLP), a technology which is mainly driven by the large packaging houses which already have much of the required infrastructure or which have the necessary capital for the needed investments [1].

The other principal technology uses the know-how and the equipment which is used in printed circuit board (PCB) manufacturing. This panel based approach is of main interest for PCB companies which want to extend their product portfolio and see a chance to enter the market of chip packaging.

Advantages of FOWLP are that there is no need for an organic substrate like a copper clad laminate (CCL), the use of existing supply chains because the existing packaging players can implement this technology, and huge investments which are already made in the last two years. Technical challenges are currently the restriction to 200 mm / 300 mm wafer size and the need to shift to panel formats to reduce costs, the need to develop "3D" capability, current package size limitations of 8 mm x 8 mm, and die shift during shrinkage of the mold compound during curing [1].

The technology of die embedding using a PCB infrastructure has the intrinsic advantage to use panel size, the possibility to connect both sides of a panel ("3D" intrinsically given), and the fact that relatively thick copper tracks can be made which opens the way to embed high power components like insulated gate bipolar transistors (IGBT) or metal oxide semiconductor field effect transistors (MOSFET). Issues with this approach are the complex supply chain, a lower rerouting density, currently low manufacturing yields, warpage due to CTE mismatch, and laborious process steps like pre-machining (cavity formation) of prepregs [1].

The two main techniques in the die embedding arena are "chip first" and "chip last" [2,3,4,5]. Whereas "chip first" has the highest potential for miniaturization, formation of thin packages, and good thermal properties, it suffers from very high cost of yield loss. For final testing the substrate manufacturer requires detailed information regarding the test program and the IC, which is sensitive information and which the semiconductor IC company does normally not want to disclose.

The "chip last" approach has the advantages that only known good dies (KGD) are actually used and assembled after most of the PCB processes are completed and that all kind of cooling systems and interconnect technologies can be applied directly on top of the die. On the other hand the thickness cannot be reduced so much, it is less suitable to system in package (SiP) applications and a complex process for cavity formation in the prefabricated PCB is needed.

Both approaches have their merits and drawbacks and will find their niches in the market place. Among others there is interest for high power applications [6], for electromobility [7], for high frequency applications [8], embedding of passive devices [9], embedding for flexible and medical devices [10].

TECHNICAL CHALLENGES WITH DIE EMBEDDING

Embedding technology provides a lot of advantages but also several risks and challenges that have to be overcome. One of the major concerns is on process yield related to the fact that the value of PCBs with integrated components is increased by orders of magnitude and so is the costs of yield loss.

Warpage of boards after lamination/curing and in application are a particular challenge as a variety of different materials with significantly different thermo mechanical properties is packed jointly into one build-up. This affects both, yield and reliability.

Some early work regarding chip embedding using the "chip first" approach used RCC for the embedding of the dies [11]. Resin coated foils are readily available, handling is easy, and there is no need for cavity formation in the resin, i.e. the resin can directly encapsulate any given structure. The disadvantage of this methodology is the limitation of resin thickness of the RCC and therefore a limitation in the thickness of the dies which can be embedded. Warpage is a concern for unsymmetrical designs because common RCC materials have a rather high coefficient of thermal expansion (CTE). Another technical limitation of RCC for embedding, apart from its high CTE, is the limited resin thickness of RCC materials and therefore also a limitation regarding the thickness of the dies which can be embedded.

Prepregs are an alternative to RCC. Prepregs are commercially available with a variety of resin systems and thicknesses. In a first step cavities are formed by laser. The pre-machined prepregs are then stacked on top of each other and the chips are placed into the openings. The whole stack is then vacuum pressed and cured, this is illustrated in Figure 1. Warpage is better controlled in this approach due to the lower CTE of the prepregs. One problem can be protrusion of glass fibers into the cavity enclosing the chip. A huge disadvantage is the need for the cavity formation process which is an additional costly process step.

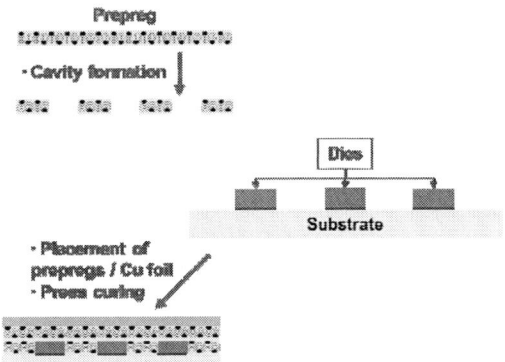

Fig. 1: Embedding approach using pre-machined prepregs

Figure 2 shows an example of a power demonstrator which was part of the HERMES project [11]. This shows that in reality a number of prepregs have to be pre-machined, depending on the thickness of the structure.

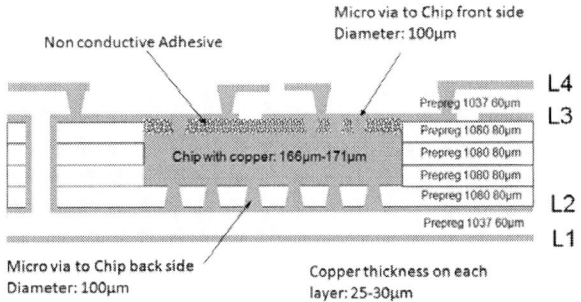

Fig. 2: Example of a power demonstrator which needs numerous pre-machined prepregs (courtesy of Infineon/HERMES consortium)

COMPOSITE MATERIAL FOR EMBEDDING MANUFACTURING

Atotech has been active in the development of RCCs for the last years. A proprietary solvent free technology for the manufacture of dielectrics for the high end PCB and packaging industry was developed. Here the focus is on the manufacturing of resin coated copper foils (RCC), reinforced resin coated copper foils (RRCC), and coated polymer foils. Those semi-finished goods are used at the customer site for the formation of build-up layers, the production of coreless structures, and for embedding. The overall process is entirely solvent free and is to some degree similar to powder coatings which are used in other industries [12].

The unique procedure which was developed over the last years and which is used to manufacture such products is illustrated in Figure 3.
From solid raw materials (resins, hardeners, flame retardants etc.) a powder is produced by melt extrusion and subsequent milling [13].

The powder is scattered on the substrate, for example a copper foil, in a continuous roll to roll process. Shortly after the powder deposition the powder is molten in an oven and forms a closed film which sticks to the substrate surface. At the same time the oven conditions can be set in order to achieve a defined b-staging of the resin.

A glass fabric is being laminated in this resin layer of this RCC in a subsequent roll to roll production step. Since the stress on the glass fabric is negligible, there is no risk of damaging or destroying the fabric and therefore also ultrathin glass fabrics can be deployed. The degree of b-staging is defined by the lamination parameters.
A second resin layer is then deposited and molten on top of the reinforced layer of the RRCC. This final product, a composite type material, can be applied to embed actives and passives.

The absence of solvent is an important aspect of the overall process. Solvent could swell the underlying resin of the reinforced layer. This could impart the position of the glass fabric which was already defined in the previous lamination step. Solvent which penetrated this resin layer would be difficult to be fully removed by evaporation and might partially remain in the resin which would then pose a huge risk in terms of reliability after being applied on a PCB. Furthermore due to the low solid content and low viscosity of solvent based lacquers the maximum thickness which can be achieved is always inherentlylimited. Such limitations are absent with a 100 % solid system like powder coatings.

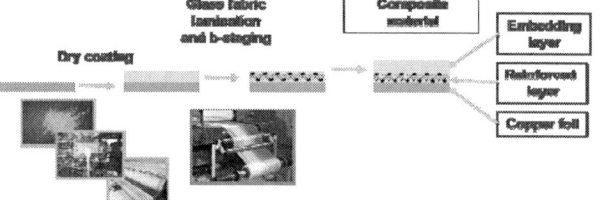

Fig. 3: Sequential process to manufacture composite embedding material: Dry coating of the first layer, lamination of glass reinforcement, dry coating of embedding resin layer.

THICKNESS DISTRIBUTION
The suggested composite material can be used to embed active and passive structures by vacuum pressing or lamination. In contrast to prepreg embedding there is no need for a cavity formation step. The sheets of the composite material are laid up on the structure to be embedded and pressed and cured in one single step.

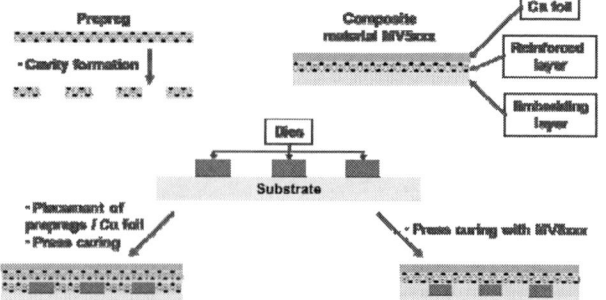

Fig. 4: Embedding using pre-machined prepregs and composite type material

Two representative examples are shown in Figure 5. All the area of the lead frame and around the die is filled with the resin of the embedding layer. The degree of b-staging allows the optimization of the rheology of the resin which guarantees void free encapsulation of the entire structure under the given processing conditions. The glass fabric reinforcement is situated above the chip. The reinforcement defines the distance and the uniformity between the chip surface and the copper foil, while the glass cloth is kept in a defined position.

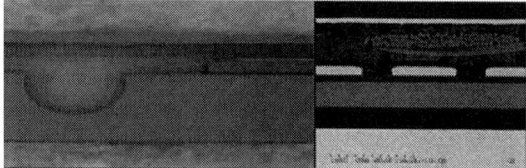

Fig. 5: Left: Embedding of chips on a Cu lead frame; void free resin encapsulates the whole structure; Right: well defined glass fabric position on top of the die

A test design with dummy structures was created in order to evaluate the lamination performance, the thickness distribution, to optimize the press profile, to compare different resin formulations, and to determine the influence of the rheology of the resin. This design is shown in Figure 6. The notation (A1, A2 etc.) indicates the position where measurements are taken, e. g. the thickness by cross section.

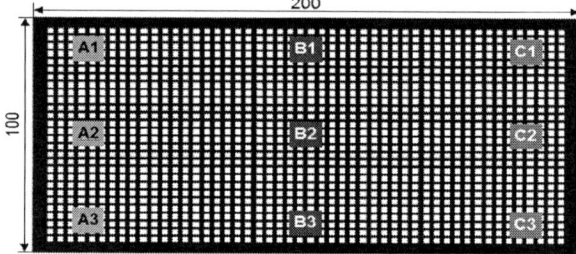

Fig 6: Test layout used for the evaluation of composite material performance

A typical result of the thickness distribution is shown in Figure 7. In this case the reinforced layer was made using a glass type 1027 and the thickness above the structure is very uniform. This good thickness distribution will facilitate the laser drilling of microvias which will connect the chip to the next layer.

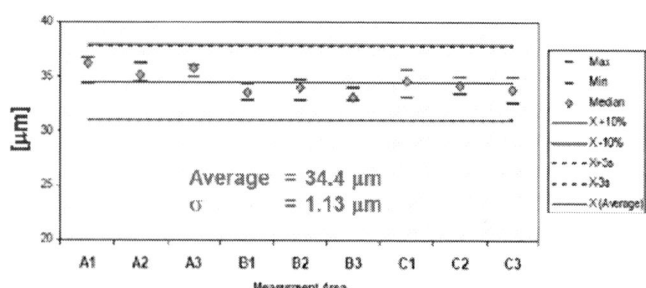

Fig. 7: Thickness measured above the dummies by cross section

116

STRUCTURES WHICH CAN BE EMBEDDED

The composite materials can be used to embed a variety of different layouts and different chip thicknesses. According to Yole report on "FOWLP & Embedded Die Packages" from 2012 the thicknesses of chips to be embedded by 2013/2014 will be in the range of 30 - 350 μm [1].

The dielectric resin thickness which is required to encapsulate a given design can be estimated consulting the chart shown in Figure 8. It shows a plot of the free area for a given layout versus the thickness of the active/passive which is to be encapsulated. The different colored zones indicate different thicknesses of the embedding layer of the composite material. The thickness of the dies and the free area on the layout are the two factors which determine the amount of resin which is needed. For example in order to embed a structure with chips having a thickness of 150 μm and the design having a free area of 70 %, the thickness of the embedding layer should be in the range of 140 μm.

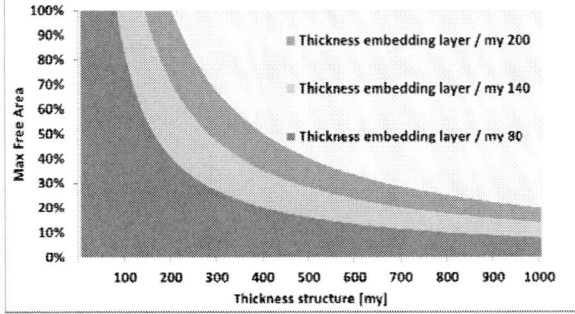

Fig. 8: Free area for a given die thickness and the required thickness of the embedding layer

Another example is illustrated in Figure 9. Two cases are shown: A die with a thickness of 150 μm and multi layer ceramic capacitors (MLCC) of similar height surrounding it. If the design has a free area of 38 %, a thickness of the embedding layer of 80 μm would be sufficient for encapsulation. The second example shows a die with a height of 600 μm, again surrounded by MLCCs having a lower thickness than the dies. In this case the required embedding resin thickness is in the range of 220 μm. This last example also shows the potential to embed structures in which parts of different size and height are present at the same panel, for example actives and passives or different kind of dies.

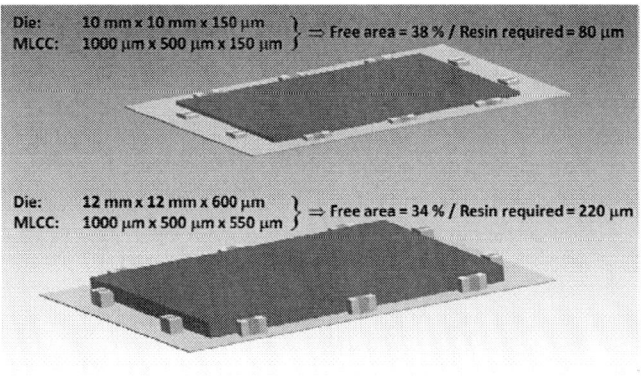

Fig. 9: Example of two design layouts with different form factors and the required resin thickness needed

The capabilities of the composite type material will reach its limits when it comes to the embedding of layouts which have a high free volume which has to be filled with resin. This is either the case with very thick structures and/or a low density of components per area unit. The limiting factor for the maximum resin thickness is not the powder coating step itself, but rather the adhesion of the b-staged embedding layer on the reinforced layer during manufacturing. The b-staged resin can flake off in the reel to reel process if the layer becomes too thick.

The actual current available composite materials in terms of thickness ranges are indicated in Figure 10.

The carrier copper foil can be ultrathin low profile copper or standard profile copper. The standard profile foils can be used for subtractive structuring, the ultrathin low profile foils are intended for modified semi additive plating (MSAP) fine line structuring.

Regarding the reinforced layer as thin as 25 μm can be realized (1015 glass), thicker glass fabrics like 1027, 1037 can also be used. Even thicker glass fabrics can be employed which could be of interest for embedding of thick copper structures for high power applications. The embedding layer thickness which can be currently produced ranges from 20 μm - 200 μm.

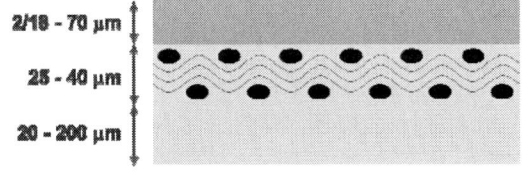

Fig 10: Thicknesses which are currently accessible for composite type embedding material

WARPAGE CONSIDERATIONS AND SIMULATION

In addition to the already mentioned advantages of such composite materials, there emerge some interesting possibilities. This is because the reinforced layer and the embedding layer can in principle be tuned independently according to technical requirements. For example the ubiquitous problem of warpage might be tackled in a novel way. In order to be able to better understand which factors have the largest influence on warpage, a more detailed study has been started.

The reinforced layer and the embedding layer can have quite different physical properties, e. g. in terms of CTE, modulus, and glass transition temperature (Tg). The hypothesis to be tested is that warpage for a given design is minimized under the following conditions: The reinforced layer having a high Young's modulus and a low CTE, and the embedding layer having at least a rather low modulus, ideally in combination with a low CTE. This would allow the embedding resin layer to take up whatever stress might be formed during the different process steps such as vacuum pressing, curing, cooling, soldering etc. This is illustrated in Figure 11.

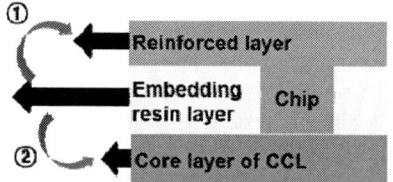

Fig. 11: Compensation of the CTE mismatch between the core material and the reinforced layer by the embedding resin having a low modulus

Warpage after pressing and curing and the evaluation of the significance of the applied finite element method (FEM) itself were to be evaluated first. Therefore the warpage predicted by the FEM has to be compared with the actual warpage determined on physical samples.

The mechanical analysis (MA) or the technology of thermal mechanically coupled analysis (TMCA) was applied to cured materials in the field of PCB technology in the past [14]. However, in this study we wanted to use FEM to calculate the warpage taking into account the changing physical properties of the resin during the curing cycle. The physical properties which were determined for each resin type are summarized in Figure 12.

The test layout is shown in Figure 12. It has an array of chip dummies placed on a low CTE copper clad laminate having a thickness of 40 µm. The dimension of the dummies is 2.5 mm x 2.5 mm x 0.06 mm, the distance between the dies being 1.2 mm.

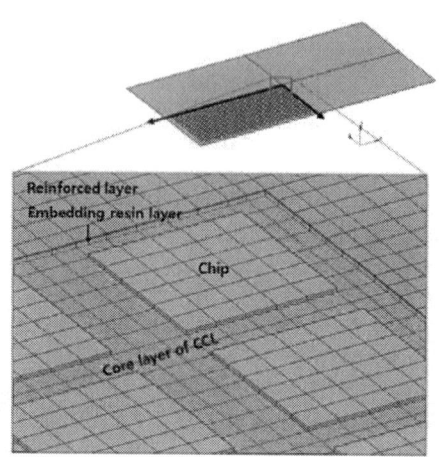

Fig. 12: Layout of the test design

In order to test the utility of the simulation, physical test boards will be built using the same resins the properties of which were used in the simulation. The three different resins have different moduli, CTE, cure shrinkage, and Tg. The glass fabric used in the reinforced layer was in all cases 1027 glass type (layer thickness 30 µm). The thickness of the embedding layer was 60 µm in all cases. The material properties of the embedding resin layer were measured for each resin formulation over the range of the entire press cycle which is summarized in Figure 13.

Reinforced layer with 1027 glass	
Modulus / [GPa]	Measured on cured sample by DMA (tensile mode)
CTE / [ppm/K]	Measured on cured sample by TMA
Embedding resin layer	
Heat conductivity / [W/mK]	Measured on the cured sample
Modulus / [GPa]	Determined by rheometer over the entire curing cycle starting with uncured resin
Shrinkage / [%]	Determined by volume change over the entire curing cycle starting with uncured resin

Fig. 13: Measured physical properties which were determined of three different resin

After establishing the principal capability of this approach, further simulation work will be carried out in order to estimate the warpage after other process steps like soldering and the long term reliability of a given composition of the composite material.

SUMMARY

A solvent free coating process allows for the production of composite materials with relatively thick resin layers which can be used for embedding of structures of various height and density. The application process is as easy as the use of an RCC and uses the infrastructure which is already available in PCB shops. Inherently there is no risk of glass fabric protrusion. The thickness distribution after press curing is excellent which will allow for reliable laser drilling of the microvias and therefore better yield of the overall production sequence. The modular approach of two distinct layers within the composite material allows in principle the adaptation of each layer in order to reduce technical issues like warpage. Theoretical and practical work to prove this concept are under way.

Future improvements will include resin compositions having higher thermal conductivity, the extension of the current maximum thickness of the embedding layer, and low profile substrate foils which would allow for finer pattern beyond the current MSAP capability.

CONCLUSIONS

New material types can help to overcome some of the challenges of embedding technologies. We presented a type of composite resin material consisting of two distinct dielectric layers which are attached to a carrier foil. In a solvent free manufacturing, ultrathin glass fabrics can be easily employed and relatively thick resin layers for embedding can be created. Such a material combines the advantages of a prepreg (mechanical stability) with the advantages of an RCC (encapsulation properties). The thickness distribution above the dies is excellent which is important for reliable laser drilling processes.

A study was initiated in which the warpage is to be predicted for a defined chip layout and a defined set of material properties. The results of this investigation might allow to minimize warpage for a given layout by adjusting the material properties of the individual layers of the composite.

REFERENCES

[1] J. Baron, L. Cadix, FOWLP & embedded die packages", Yole Development report 2012.

[2] A. Ostmann, A. Neumann, P. Sommer, H. Reichl, "Buried components in printed circuit boards", Advancing Microelectronics, May/June 2005, pp. 13-18.

[3] M. Brizoux, A. Grivon, W.C. Maia Filho, E. Monier-Vinard, Thales, AT&S, "Industrial PCB development using embedded passive & active discrete chips focused on process and DFR", www.pcb007.com, May 2010

[4] N. Kumbhat, F. Liu, V. Sundaram, G. Meyer-Berg, R. Tummala, Georgia Tech, Infineon, "Low cost, chip-last embedded ICs in thin organic cores", ECTC, June 2011.

[5] V. Sundaram, Georgia Tech, "Chip last embedded actives and passives in thin organic package for 1-110 GHz multiband applications", ECTC 2010, pp. 758.

[6] L. Boettcher, S. Karaskiewicz, D. Manessis, A. Ostmann, IZM, "Development of embedded power electronics modules for automotive applications", SMTA, February 2012.

[7] T. Hofmann, S. Gottschling, Continental, "Integration of electronic components into PCB for electromobility application", SMTA, February 2012.

[8] F. Liu, V. Sundaram, S. Min, V. Sridharan, H. Chan, N. Kumbhat, WW. Lee, R. Tummala, D. Baars, S. Kennedy, S. Paul, Georgia Tech, Rogers, "Chip last

embedded actives and passives in thin organic package for 1 – 110 GHz multi band applications", ECTC, 2010.

[9] T. Löher, J. Marques, M. Haubenreisser, A. Ostmann, N. Bauer, TU Berlin, IZM, Murata, "Embedding and reliability of discrete capacitors into build up layers of printed circuit boards", SMTA, February 2012.

[10] G. Kunkel, T. Debski, J. Link, A.E. Petersen, W. Christiaens, J. Vanfleteren, Hightec MC AG, Oticon A/S, IMEC, "Ultra-flexible and ultra-thin embedded medical devices on large area panels", ESTC 2010, Sep 2010.

[11] L. Boettcher, D. Manessis, A. Ostmann, H. Reichel, TU Berlin, IZM, "Realization of system in package modules by embedding of chips", IMAPS Device Packaging, 2008

[12] Pieter Gillis De Lange, Powder coatings chemistry and technology, 2nd ed., Vincentz Network, 2004

[13] N. Galster, R. Park, A. Bruderer, "Dielectrics for the embedding of active and passive devices", Atotech, SMTA, February 2012.

[14] SH. Cho, S. Cho, J.Y. Lee, Microelectronics Reliability, 2008, pp. 300.

A NANO SILVER REPLACEMENT FOR HIGH LEAD SOLDERS IN SEMICONDUCTOR JUNCTIONS

Keith Sweatman[1], Tetsuro Nishimura[1], and Teruo Komatsu[2]
[1]Nihon Superior Co., Ltd.
[2]Applied Nanoparticle Laboratory Co., Ltd.
Osaka, Japan
k.sweatman@nihonsuperior.co.jp

ABSTRACT

While it is now widely accepted that most electronic assembly can be reliably effected with lead-free solders, a practicable alternative to the high-lead high-melting-point solders has not been available. That reality has been acknowledged by the interim exemption from the requirements of the EU RoHS Directive granted for solders with 85% or more of lead. With no direct replacement yet found by conventional alloying of elements permitted by the RoHS Directive the search for a replacement for these high-lead solders has extended to alternative joining materials. One approach has been to take advantage of the reactivity of nano particles of silver to make a product that while ultimately having a melting point at or near the silver melting point of 961.8°C can combine to form reliable connections at temperatures much lower than that. The challenge in this approach is that the very reactivity that makes the formation of a joint possible at a relatively low temperature means that the nano silver tends to be unstable. In this paper the authors report the development of a unique nano silver material that is manufactured and stabilized in an alcohol environment to produce a material that can be used to make reliable joints between a wide range of the substrates commonly used in electronics in process conditions similar to those used with high-lead solders. This material can be used to make joints to ferrous materials (e.g. stainless steel) as well as non-ferrous materials such as copper and nickel. And most importantly for component manufacture this new material bonds strongly to semiconductor materials such as silicon. Where even longer life in thermal cycling is required the silver structure can be reinforced by the addition of other materials in the form of particles of the appropriate size. The paper will include details of mechanical and reliability testing of joints made with these materials under a range of temperature, pressure and atmosphere conditions.

Key words: Nano Silver, die attach

INTRODUCTION

The main application for high- Pb solders that resulted in the granting of an exemption from the requirement of the EU RoHS Directive has been in semiconductor packaging. The relatively high melting point of these solders (around 300°C) means that joints made with them in earlier stages of component assembly are not disturbed by subsequent stages of assembly with lower melting solders (step soldering) or by the high peak temperature sometimes required in final reflow soldering.

For power semiconductors the die-attach material has to be able to maintain a reliable joint at the relatively high operating temperature of these devices while at the same time providing a high thermal conductivity path for dissipation of the heat generated during their operation.

In this application the trend to wide bandgap semiconductor devices such as SiC power diodes and transistors means that there is an advantage if the die-attach material can sustain a higher operating temperature than can the high-Pb solders. With SiC devices a substantial increase in power density can be achieved because of its higher thermal conductivity, higher breakdown voltage and higher saturated carrier velocity. The larger bandgap of SiC can allow higher junction temperatures without compromising performance. [1]

The conventional metallurgical alternatives to the high-lead solders such as the gold-tin eutectic have a processing temperature that is higher than the semiconductor can tolerate without degradation, And another alternative, silver-filled epoxy cannot always provide the thermal conductivity that is required to keep the die at a temperature at which it can operate at maximum efficiency [1] [2].

The relatively high surface activity of nano particles means that, for example a 2.4nm Silver particle would be expected to have a melting point of 350°C [3], much less than the 961.8°C melting point of bulk silver (Figure 1). The outer layers of such a particle would have a mobility similar to that of the molten state at even lower temperatures so that they will bond to each other or to other compatible materials by wetting and interdiffusion at temperatures well below those required for conventional sintering of conventional Silver powder.

Although the application of external pressure during the sintering process does increase the area of contact of the particles it is not essential in sintering nano silver particles. Even at temperatures less than those used in reflowing the high-lead solders the capillary forces generated by the mobile atoms at the surface of the silver are sufficient to ensure the wetting of adjacent particles with which they are in contact.   Since silver is so much

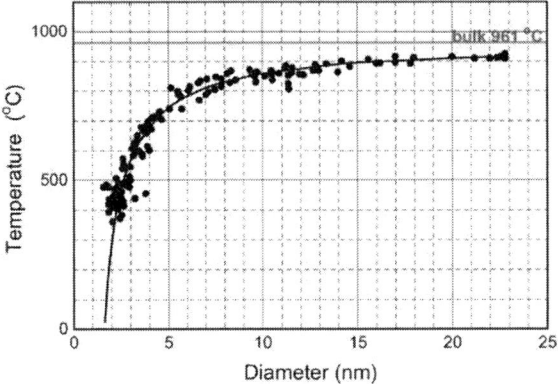

Figure 1. Silver melting point as a function of particle size [4]

stronger than solder the full density of silver is not required to achieve the strength required in this application.   In fact the lower modulus of the porous structure is an advantage in reducing the stress generated in the chip during thermal cycling because of the differences in the coefficient of thermal expansion

When properly formulated, pastes based on nano particles of silver can be processed to form reliable joints at temperatures that fall within a range similar to that required for the high-Pb solders and even lower.

NANO SILVER
There are significant challenges to be faced in delivering the highly reactive nano silver particles to the joint area and then facilitating their bonding to form a structurally strong joint that delivers the required levels of electrical and thermal conductivity.

Manufacturing the nano silver particles in itself is a major challenge but if they are to remain as nano particles with the properties required for them to sinter at the lowest possible temperature they have to be stabilized until they are in place in the joint gap ready for sintering.   That is achieved by passivating or capping the particle during the manufacturing process with a coating that bonds to the nano silver surface while presenting a resistant external surface.

Chemicals such as thiols, amines and carboxylates have been identified as effective capping materials [2] but these are bound strongly to the silver and require a high temperature for their removal, to some extent negating the chief advantage of the nano particles they are protecting. Some capping materials require sintering to occur in air so that they can be removed by oxidation and that can compromise other parts of the component that

need the protection of a nitrogen atmosphere at sintering temperatures.

Another disadvantage of some of these capping agents is residues of sulphur and nitrogen compounds that can interfere with the performance of the sintered silver and contribute to corrosion problems in service.

There is thus a strong motivation to identify a capping material that would not suffer from these disadvantages and make possible wider application of nano-silver bonding as a reliable die-attach material.

For practical application in commercial mass production the nano-silver bonding materials must also have the physical properties required by the application process, which might be screen printing, dispensing or dipping. This is achieved by dispersing the capped nano- silver particle in a suitable vehicle.   The rheology can also be optimised by the inclusion in the mix of sub-micron particles of silver or other metals such as Cu and by adjustment of the particle size distribution

While offering the possibility of relatively low process temperatures and high temperature reliability, silver suffers from the disadvantage of a relatively high elastic modulus so that potentially damaging levels of stress can be developed in the die by the strains arising from CTE differences during the thermal cycling that can occur in service.   Another objective in optimizing nano silver bonding materials is, therefore, the control of the microstructure to minimise the elastic modulus while retaining a high level of thermal conductivity.

ALCOXIDE-PASSIVATED NANO SILVER
In an attempt to address the problems encountered with established nano silver capping materials the possibility of using alcohols was explored.   These can attach to the silver through the formation of alcoxides with silver atoms on the surface of the nano particle. The advantage of these chemicals in this application is that the oxygen-silver bond, while strong enough to stabilize the nano particle during manufacturing processes and subsequent storage and handling, is weak enough that it can be broken at a relatively low temperature to expose the active surface of the nano particle so that it can bond to adjacent particles. This opens the way to nano-silver materials that can effect bonding with thermal profiles comparable with those used for reflow soldering.

For example, alcoxide-capped nano silver particles were produced by reacting silver carbonate (Ag_2CO_3) with n-dodecanol ($CH_3(CH_2)_{11}OH$) or n-decanol ($CH_3(CH_2)_9OH$) under a nitrogen atmosphere.   In this reaction the hydrogen atom (H) in the hydroxyl group at the end of n-dodecanol or n-decanol molecule (Figure 2 (a)) is replaced by a silver atom on the surface of the nano particle as it forms by reduction of the silver carbonate (Figure 2(b)) [5]

The result, a nano silver particle, protected by the n-dodecanoxide molecules, is illustrated schematically in Figure 3.

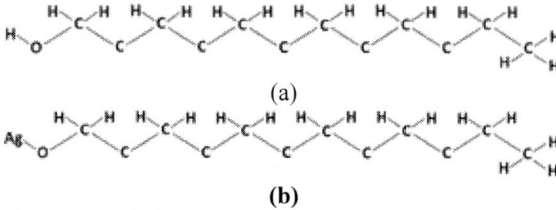

(a)

(b)

Figure 2. n-dodecanol (a) and Ag n-dodecanoxide (b)

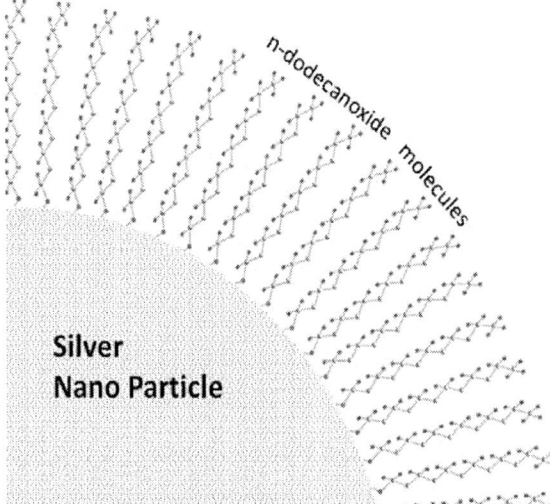

Figure 3. Schematic representation of alcoxide-passivated nano silver particle.

The resulting mixture was cooled and the stabilized nano-silver particles filtered out washed in ethanol, dried and dispersed in hexane. The nature of the material so produced was confirmed by placing some of this dispersion on a carbon film substrate for examination in a transmission electron microscope (TEM).

The particles were observed to be roughly spherical with a diameter of <5nm (Figure 4). The lattice planes of these crystalline particles is apparent in the TEM image and the estimated 0.23nm spacing between planes is consistent with the spacing of the [111] plane of silver's close packed cubic crystal structure.

The arrangement of the nano-silver particles on the carbon film in an approximately hexagonal pattern (Figure 5) suggests that in that area they are close packed so that their separation distance is determined by the thickness of the protective alcoxide coating. The estimated gap between the nano-silver cores of these particles of 2.5nm is close to the expected length of two n-dodecanoxide molecules aligned end to end, which is consistent with the passivation model illustrated schematically in Figure 3.

Similar reactions can be effected with shorter alcohols but as the number of carbon atoms in the alcohol falls the size of the nano particle produced increases (Figure 6).

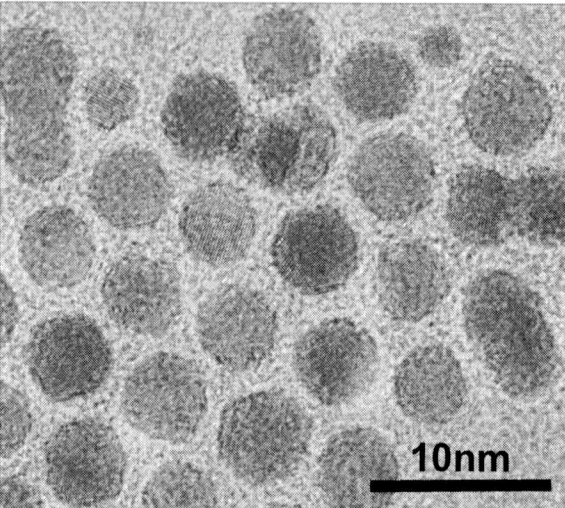

Figure 4. Dodecanoxide-passivated nano-silver particles recovered from reaction with Ag2CO3.

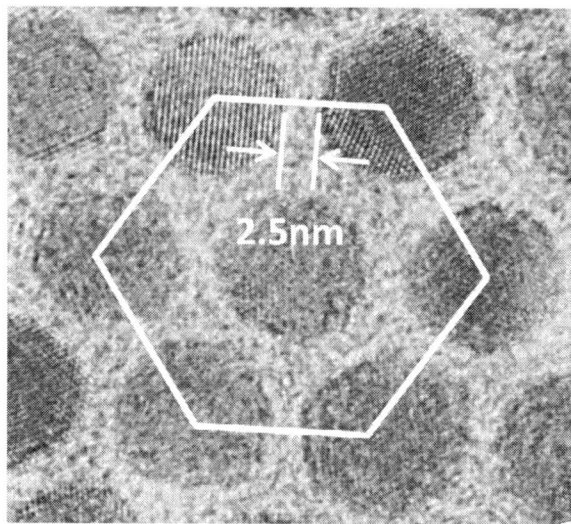

Figure 5. Close-packed arrangement of nano-silver particles with spacing approximately twice the length of the passivating n-dodecanol molecule

SINTERING NANO SILVER

Nano-silver particles were mixed with 0.4μm silver powder in an isobonyl cyclohexanol (IBCH) vehicle to achieve a total metal content of 87% with the dodecanoxide-passivated nano-silver. This mixture was subjected to thermogravimetric and differential thermal analysis in air and nitrogen atmospheres with the temperature being raised at the rate of 10°C/minute.

The stages in the sintering process are apparent in Figure 7. The first stage is evaporation of the carrier and removal of the passivation layer. In air there is a highly exothermic reaction as the passivating alcohol is oxidized but in a nitrogen atmosphere the alcohol simple detaches with little heat evolution. With the passivation layer removed the active surface of the nano silver particles is exposed and the sintering proceeds.

The temperature at which sintering occurs varies with the number of carbon atoms in the passivating alcoxide molecule and it can be seen in Figure 8 that sintering can be effected at temperatures under 120°C nano silver passivated with alcoxides with 2-6 carbon atoms. There is, however, a trade off in that the passivation by these shorter alcoxide molecules is not as stable as that of the longer molecules.

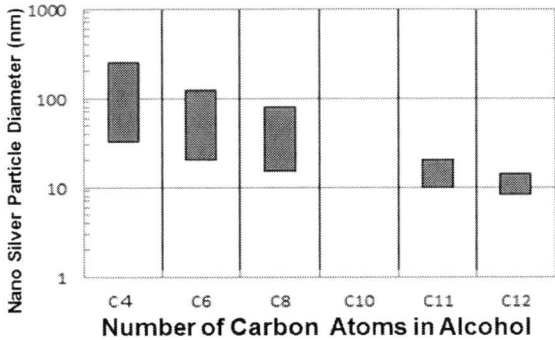

Figure 6. Relationship between the size of the silver nano particle and the alcohol used in its manufacture.

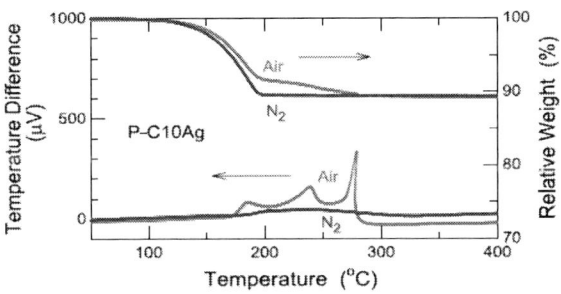

Figure 7. TGA-DTA plots for decanoxide-passivated nano-silver and 5µm silver mixtures dispersed in IBCH.

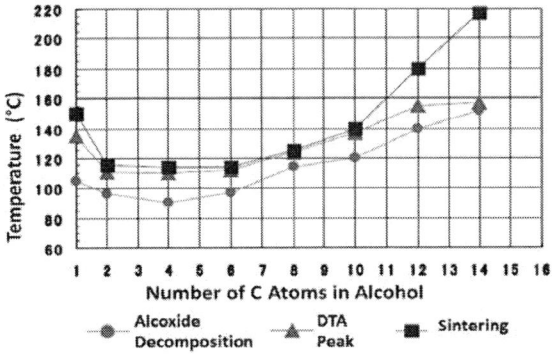

Figure 8. Sintering behaviour as a function of alcoxide passivator.

OPTIMIZING THE NANO SILVER PASTE

While it is possible to form a joint by sintering pure nano-particles there are cost and performance advantages to be achieved by mixing nano particles with conventional particles to which the nano particles can also bond at low temperatures. This arrangement is illustrated schematically in Figure 9.

The idea is to use a range of particle sizes that when bonded will fill the volume to the level required to achieve the required mechanical, electrical and thermal properties. The sub-micron voids in the sintered paste (Figure10) mean that the bond is more compliant than if it were solid silver so that the stress imposed on the die during thermal excursions due to CTE differences can be at least partly accommodated. However, the electrical and thermal conductivity of the sintered bond are still adequate.

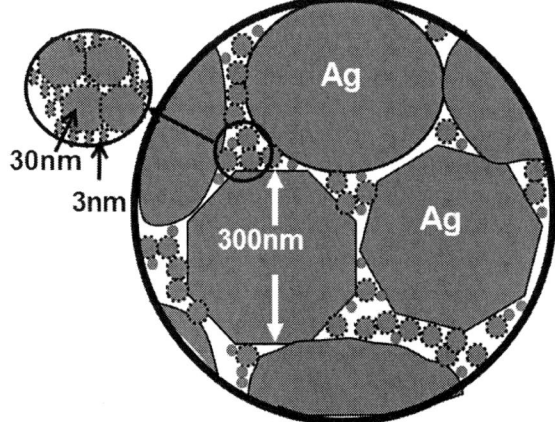

Figure 9. Schematic representation of a typical mix of nano silver particles with larger silver particles

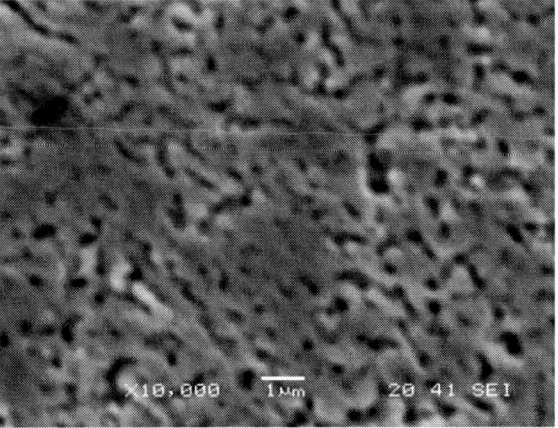

Figure 10. Submicron voids in sintered alcoxide-passivated nano silver

PROPERTIES OF SINTERED NANO SILVER

The electrical conductivity of the sintered silver was assessed by printing a film of paste onto glass and measuring its resistance using the four terminal method. The data summarized in Figure 11 indicates that resistance varies in the range 2-15μΩcm, which is generally greater than the 1.6μΩ resistivity of bulk silver but at least as good as the resistivity of reflowed Pb-5Sn solder paste and better than the resistivity of silver/epoxy paste.

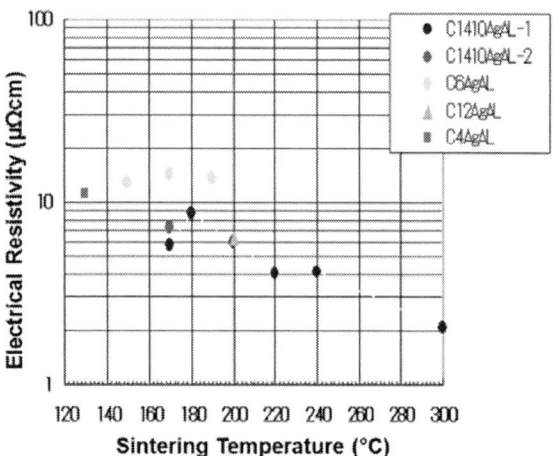

Figure 11. Electrical resistivity of nano silver paste as a function of sintering temperature. Sintering temperature varies with the number of carbon atoms in the passivating alcoxide.

AN EVALUATION

As a test of the practical applicability of pastes based on this nano silver technology a silicon diode package such as that represented in schematically in Figure 12 was assembled. The finish on the 2.3mm x 2.3mm x 0.25mm diode chip was gold and the substrate copper.

In the nitrogen atmosphere the assembly was heated to 300°C, a temperature similar to that which would be required to reflow the high-lead solder paste normally used in the manufacture of this product. In air the temperature was raised to 350°C to allow for burning off of the passivating material. The total time of the sintering process was 25 minutes in the nitrogen atmosphere and 30 minutes in air. No pressure was applied to the joint during the sintering process. After the joint was effected the components were packaging in resin according to normal production practice.

The stability of the electrical and thermal resistance of the joint was assessed by exposing the package to thermal cycling -55°C to 150°C. The results are plotted in Figure 13.

The electrical resistance was taken to be indicated by the forward voltage V_F at a constant current of 3A. Thermal resistance was estimated by measuring the difference in V_F before and after Joule heating with 100ms pulse of 10A.

In the as-manufactured condition diodes assembled with pastes based on nano silver passivated with both dodecanoxide and decanoxide had electrical and thermal resistances similar to those of that assembled with high-lead solder paste. The electrical resistance of those assembled with nano silver in air suffered rapid increase in electrical resistance. The increase in resistance of the diodes assembled with nano silver paste and sintered in nitrogen was similar to that of the diodes assembled with high-lead solder paste.

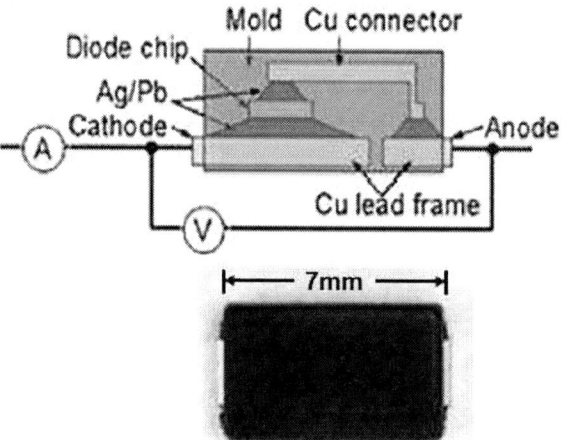

Figure 12. Nano-silver evaluation package

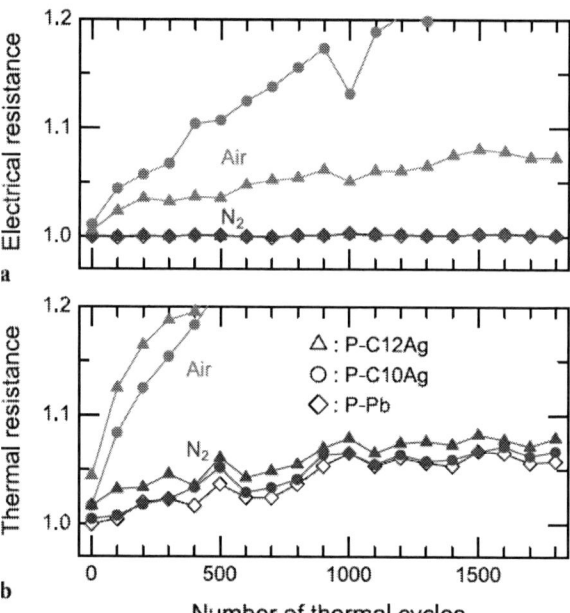

Figure 13. Electrical and thermal resistance of silicon diode package sintered in air and nitrogen as a function of the number of thermal cycles of -55°C-150°C. Results are normalized to that of the diode assembled with high-lead solder paste as manufactured. In the product code C12 and C10 indicate the number of carbon atoms in the passivating alcohol

While the thermal resistance of the diode assembled with high-lead solder paste or nano silver pasted sintered in nitrogen remained stable, that of the diodes assembled with nano silver paste passivated with dodecanol and deconal pastes sintered in air suffered significant loss of conductivity. Further investigation is required but it is presumed that the lower stability of joints sintered in air is a consequence of oxidation of the silver. An advantage of the alcoxide passivation is that oxygen is not required in its removal during the sintering process so that the process can be carried out in a nitrogen atmosphere. Simple breakdown of the alcoxide bond is sufficient to allow the sintering process to proceed.

Metallographic examination of a cross-section through the joint to the copper substrate effected by the nano silver paste indicates that it differs from that formed by the high-lead solder paste in that there is no intermetallic layer at the interface. The silver is bonded directly the copper substrate (Figure 14).

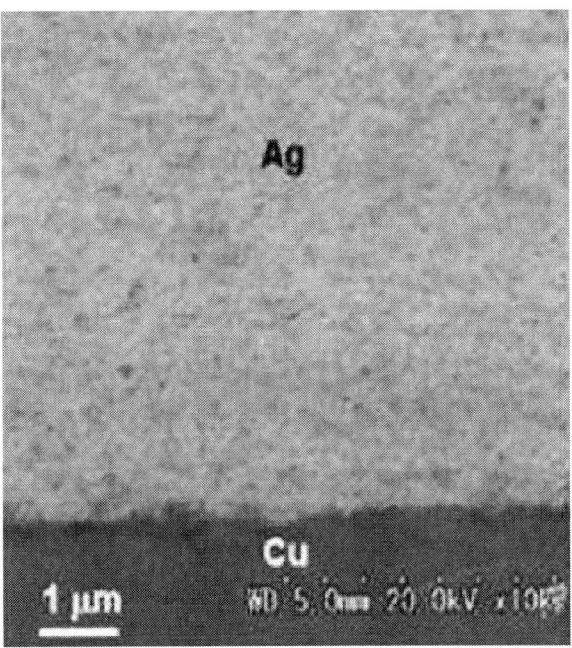

Figure 14. The bond line between the copper substrate and the sintered nano silver paste.

MECHANICAL PROPERTIES
Shear testing indicated the greater strength of joints made to copper with the the sintered alcoxide-passivated paste compared with those made with reflowed Pb-5Sn solder paste (Figure 15).

Although the shear strength of joints was greater when 1MPa of pressure was applied to the shear test piece during sintering the strength of joints made with the alcoxide-passivated nano silver paste exceeded that of a reflowed Pb-5Sn joint (Figure 17) even when the paste was sintered without applied pressure.

As explained earlier the sintering temperature can be adjusted by selecting the number of carbon atoms in the alcohol used. The results of shear testing joints made between copper surfaces with nano silver material fired at low, medium and high temperatures indicates that although those sintered at high temperature are stronger, even those sintered at lower temperatures are at least comparable in strength to joints made with reflowed Pb-5Sn solder.

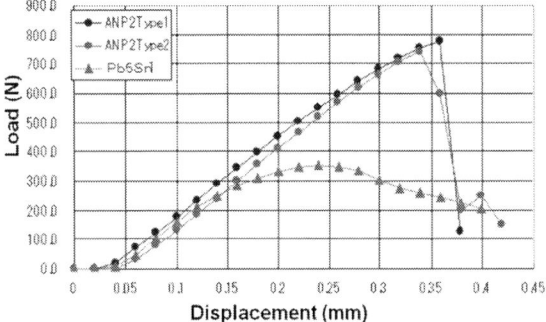

Figure 15. Shear testing load-displacement plots for sintered alcoxide-passivated nano silver joints to copper compared with reflowed Pb-5Sn joints.

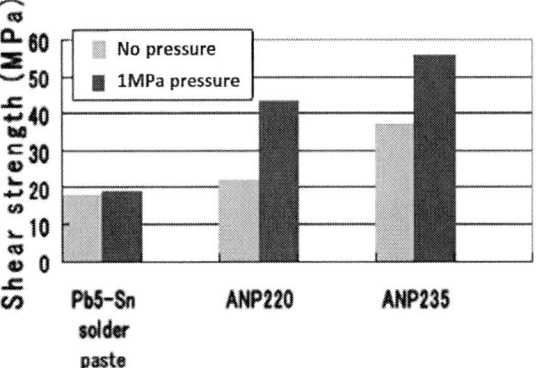

Figure 16. Shear strength of joint made to copper with nano silver solder paste with and without 1 MPa pressure compared with that of Pb-5Sn solder paste.

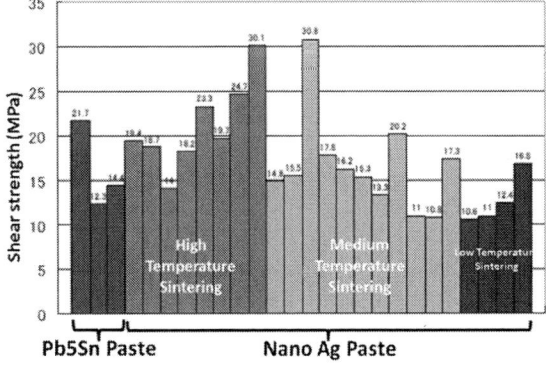

Figure 17. Shear strength of silver pastes sintered without pressure compared with reflowed Pb-5Sn solder paste.

APPLICATIONS OF NANO SILVER PASTE

In addition to die attach applications such as that described earlier in this paper alcoxide-passivated nano silver pastes can used to create printed wiring on substrates such as polyimide (Figure 18) and to be effective in wire bonding (Figure 19)

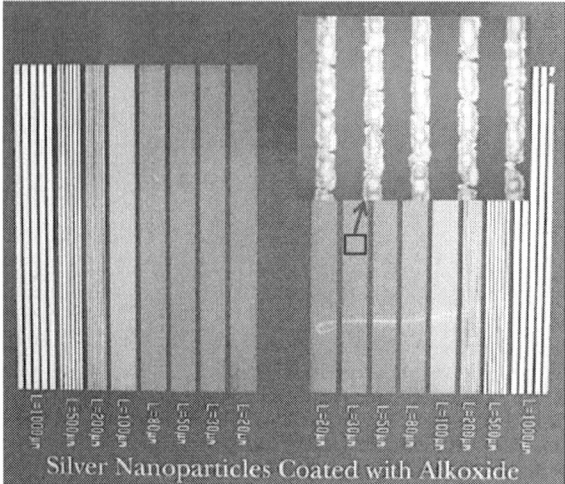

Figure 18. 20μm to 1000μm traces screen printed on polyamide with alcoxide-passivated nano silver paste sintered at 150C.   Inset is 30μm traces

The alcoxide-passivated nano silver paste has been proven to have high bond strength to silver, gold, platinum, copper, iron and 304 stainless steel surfaces.   The tenacious oxide films on nickel and aluminium interfere with the formation of the metal to metal bond by the highly active surface of the nano particles.

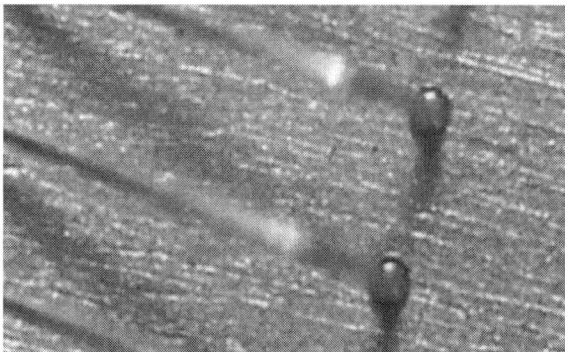

Figure 19.   Wire bonding with alcoxide-passivated nano silver paste.

CONCLUSIONS

Sufficient evidence has been accumulated to prove that nano silver manufactured in an alcohol environment and passivated with alcoxides formed by reaction with the silver atoms on the surface of the particles can provide the basis for joining materials that can be sintered at temperature comparable with those used in the reflow of high-lead solder pastes and lower.   The strength and electrical and thermal conductivity of the bond so formed is at least comparable with that formed by the reflow of high-lead solder

While the nano silver so formed can be used alone it has also been shown to be possible to reduce the cost of the material and enhance mechanical properties by mixing the nano particles with sub-micron particles of silver and copper.

The temperature required for sintering the nano silver paste varies with the length of the carbon chain of the alcohol used for its passivation with sintering temperatures under 120°C possible with nano silver passivated with alcohols with 2 to 6 carbon atoms although there is trade off in regard to stability.   Selecting a nano silver paste formulation it is therefore a matter of choosing the balance between sintering temperature and storage and handling stability that best fits the application.

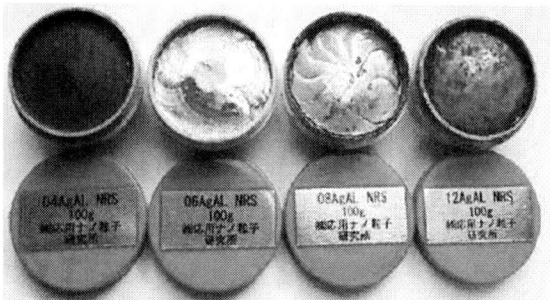

Figure 20. Examples of pastes based on alcoxide-stabilized nano-silver.

ACKNOWLEDGEMENTS

This paper is based on work carried out and reported by Dr Teruo Komatsu and his colleagues at Osaka City University.   Continuing work on alcoxide-passivated nano silver technologies is now being supported by Nihon Superior Co., Ltd

REFERENCES

1. J.N. Calata, T.G.Lei and G-Q Lu, Emphasis, November 2006
2. Kim S. Siow, Journal of Alloys and Compounds, 514(2012) 6-19.
3. W.D. Kingery, J. Appl. Phys. **30**, 301 (1959)
4. Ph. Buffat and J-P Borel, Phys. Rev. A13 (1976) 2287
5. M. Maruyama, R. Matsubayashi, H. Iwakuro, S. Isoda, T. Komatsu, Appl Phys A (2008) 93: 467–470

TECHNICAL COMMUNICATIONS: STRATEGIES FOR SUCCESS

Chrys Shea
Shea Engineering Services
Burlington, NJ, USA
chrys@sheaengineering.com

ABSTRACT

This paper is an accompaniment to a keynote presentation at the 2013 Pan Pacific Microelectronics Symposium. It discusses various technical communication vehicles, typical audience, exposure and interest levels, and how to adapt information to best fit the delivery method. It also identifies specific opportunities to maximize the ROI of technical communications and offers simple tips for improving writing quality.

INTRODUCTION

In technology-based organizations, everyone needs to communicate technical topics effectively, regardless of job function or level. People need to share information with coworkers, suppliers and customers. Customer communications are particularly important to companies that want to demonstrate their leadership in the marketplace. Technical publications can draw attention to an organization's superior technologies or value propositions, strengthen the brand, and raise market awareness of new products.

There are many ways to communicate with customers about new products or technological advancements, and employing the right ones can be instrumental in a product's early success. Strategic communication efforts select the best media to optimize exposure to potential customers, and then deliver information at a level with which the audience connects comfortably.

Successful technical communications reach the right people, deliver messages that are easily understood, maximize the utility of the information, and use the simplest language possible.

SELECTING THE RIGHT MEDIA

Media selection is based on understanding the target market. Is it comprised of advanced technology researchers or production engineers? Operations managers or process developers? The key is in knowing the audience and where they typically look for information. The best media choice for a technical message is the one that provides the best exposure to the potential customer base. Figure 1 shows a variety of technical communication vehicles, each designed to reach different portions of the technical community.

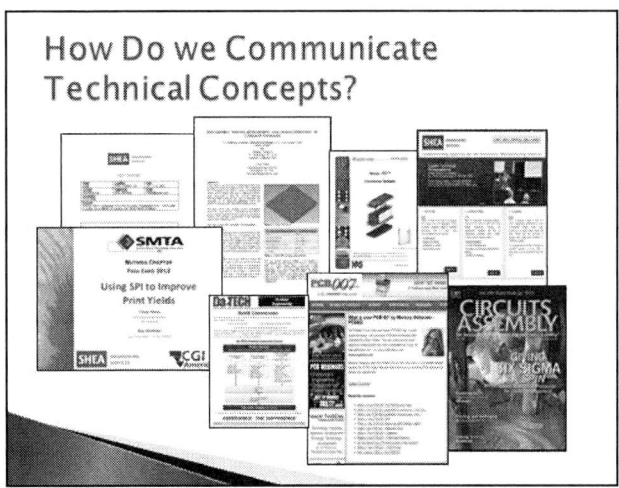

Figure 1. Technical communications vehicles

Technical Papers at conferences reach higher level technologists and technical managers who attend conference presentations or review conference proceedings. The exposure of a technical, or "white" paper, is narrower than other media, but the audience always has a strong interest in the topic. In supporting sales or marketing efforts, white papers generate high credibility levels for the organization, demonstrate technology leadership, and provide good sales collateral.

Magazine Articles extend to a wider audience than white papers. Magazines are browsed by technical personnel at all levels from operator to executive, and in a variety of different disciplines like engineering, purchasing or quality. The readers' interest levels in any specific topic vary considerably, so while the message becomes available to many people, not all will read it.

Magazine articles often offer the highest ROI of any market-focused publication because they can reach so many potential customers, have excellent electronic accessibility and are more user-friendly than most white papers.

Presentations and Webinars provide broad exposure to an audience with above-average interest in the topics. Technicians, engineers and first-level managers typically attend on-line or in-person presentations to learn more about their primary job function or broaden their knowledge of related fields. Attendees often bring pre-determined questions for the presenter based on their experience, and can be expected to formulate new questions based on the presentation's content. Presentations and webinars offer organizations excellent opportunities to showcase their capabilities and help technologists develop personal connections with colleagues.

Applications Notes are targeted to - and accessed by - product users or potential users. The exposure of product-specific information is limited to the technical personnel, but with extremely high interest in the content of the documents. Current users will access applications notes for help or information; potential users will consider their quality and usefulness as an indicator of customer service capability when making purchasing decisions.

Blogs also provide excellent avenues to showcase talent, personalize technical connections and discuss specific products, but present more risk factors than traditional communication conduits. Successful blogs represent companies professionally, with fresh installments on a regular basis, and build a fan base that helps propagate their message through peer-to-peer networks. However, blogs that contain too many unsubstantiated opinions, exhibit excessive commercialism or have long lapses between entries can reflect poorly on the sponsoring organization and turn away more readers than they draw.

ADAPTING STYLES FOR VARIOUS MEDIA AND AUDIENCES

Information can be conveyed in a variety of formats. Optimizing the delivery style for the medium and the audience improves the effectiveness of the message. Figure 2 lists some of the different communication vehicles that use varying styles.

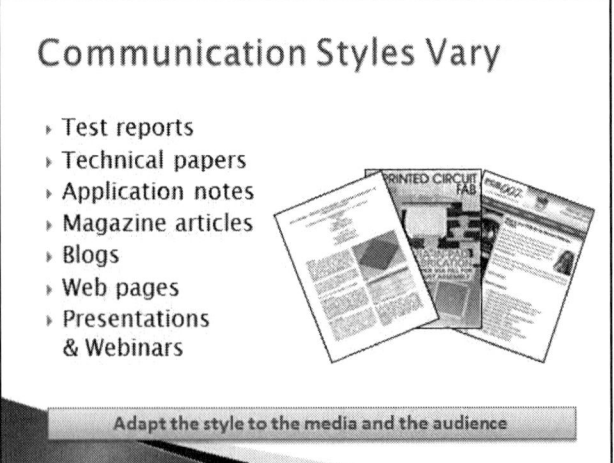

Figure 2. Communications tools with varying styles

Test reports are most often used internally or with development partners. They are the record of an experiment or test, and should include as much information as possible because they are likely to be reviewed by researchers in the future. Documentation should include test equipment models, parameters, vehicles, participants, materials and lot numbers, dates and times. Most of this information can be captured photographically. Reports should contain as much raw data as practical, and the names and locations of files containing large amounts of data should also be included.

Organizing and retrieving information can be facilitated by using presentation formats (e.g Power point) instead of traditional written formats (e.g. Word). Pictures and numbers tell stories better than words, are usually easier for technical professionals to produce and are less affected by language barriers. Emphasis should be placed on report content, not format.

The technologists who review test reports are far more interested in the details of the test than the format of the delivery. They will be using the information as background for future research or as the basis for customer-focused technical communications on the product.

Technical Papers formalize test information for publication at conferences and in journals. They should focus on experimental methods and findings, and present as much relevant data as possible. Results should be discussed and interpreted in the context of the application.

White paper language is the most formal of all publications types: avoid the use of slang, personal pronouns or attempts at humor in technical papers, as they are published by respected institutions. Proper formatting is important to comply with publication guidelines and project a professional image. Technologists reviewing white papers will most likely be seeking data or test methods, and are familiar with accepted reporting formats and industry-specific jargon.

Magazine Articles need to quickly capture the attention of the potential reader. They should be short (less than 1500 words), focus on results, and demonstrate the data with good graphics. The graphical element is critical to capturing the attention of a potential reader browsing a magazine; time spent on improving graphics quality is a wise investment. Another key element to pique the reader's interest is a compelling introduction – three sentences that describe a problem, its implications, and the payoff of a new solution.

Raising the article's visibility with potential readers who do not browse the magazine is also important. In the age of electronic media, an article published in any trade journal is now instantly available via the internet. Key words should be embedded within the article to optimize search engine hits. Social media, networking groups, blogs and even email signatures can provide links to new articles when they are initially published; hardcopy reprints are available for handouts. If the article is based on a previously published technical paper, the original study should be cited. Links should be provided for interested readers to learn more about the product or contact the primary author.

Web Pages, like magazine articles, should be compelling, providing the payoff in the headline or introduction, and should also embed key words to optimize search engine performance. Current guidelines for key word placement include usage at the top, middle and end of the copy, at a rate of 4-5% of the word count. The graphical layout of the page should not be too visually intimidating or busy and key information should be visible without scrolling. Numerous calls to action – options like "contact us," "request a quote," or "download a brochure/tech paper/presentation" should be present on each page.

Web page visitors are technically-oriented individuals investigating new products or solutions, and a company's web page is often that potential customer's first interaction with the provider. It needs to make a great first impression, and encourage the visitor to click to another company page instead of the "Back" button.

PowerPoint Presentations are the most frequently propagated form of information. They are often forwarded among parties and reviewed without the benefit of a presenter. Therefore, points should be very clear in the copy, leaving little open to interpretation. Summaries of complex slides or important takeaways should be highlighted with boxes or different fonts to make the key information as prominent as practical.

Graphics should be used wisely. Ideally, at least every other slide should provide some visual interest, but only relevant graphics should be incorporated. Graphics are meant to communicate, not decorate. Relevant graphics would include depictions of processes, equipment, usage or applications associated directly with the product; irrelevant graphics would include clip art and random subject-related images. Images that are not original should cite the source, even if they are copied from the internet.

Different technical levels of presentations should be prepared based on the expertise of the audience, and are discussed in the next section.

Live presentations are personal representations of an organization, and should be executed with the highest levels of professionalism. Time constraints should be respected. Most presentations average one minute per slide; they should be sized accordingly prior to the presentation. Extra slides should be hidden or removed, not quickly passed by mid-presentation. In-person presenters should dress professionally, review and understand the slides, face the audience, and make eye contact with audience members.

On-line presenters should prepare their presentation and check their internet connection, audio and video feeds prior to the web meeting. Headsets typically provide clearer sound than speakerphones; different camera positions, subject backgrounds and room lighting can dramatically affect image quality, and should be determined in advance.

To maximize the ROI of Webinars, they should be recorded and made available online for on-demand viewing.

Applications notes should focus on proper usage as determined by the product's specification or results of its development testing. Language should be formal and to the point; format should be well-organized and not distract from content. Readers are usually very familiar with the technology and want quick and easy access to information.

Blogs or Newsletters are the most casual form of technical communications, and offer authors a great deal of latitude in content and style. Language is more casual, often including slang, abbreviations and personal pronouns. Humor is typically rewarded in the form of increased readership and fan mail. Bloggers can offer opinions or insights without the typical requirements of substantiating data or citation of references.

Blogs offer new levels of freedom, but bear new levels of risk. To be effective, they need to stay fresh, demanding time from the blogger on a regular basis. Controversial or polarizing statements can draw heavy traffic and raise visibility, but can also draw heavy fire. A blog's audience could be anyone in the world – including a potential customer or boss.

MAXIMIZING ROI
There are several ways to maximize the ROI of technical marketing communications campaigns. Low cost pieces that offer high potential exposure like magazine articles or webinars naturally provide good ROI at the individual level. At a higher level, returns can be maximized across the entire campaign by leveraging existing data, using the same information across multiple communication platforms, and unifying the appearance of the communications tools wherever possible.

Typical product development tests produce reams of good information that is often underutilized. To capitalize on performance, benchmark or beta testing, review the research results from the user's perspective, which may be different from the developer's. Enlist a skilled user to perform an in-depth review, if necessary. Publish test results from the user's view point, highlighting the features and benefits that apply directly to them.

Producing technical documents from laboratory data requires an investment to formally write up the work and develop the graphics. But once these baseline items are established, they can be used over and over again in all publication platforms. A conference paper and presentation pose the biggest initial challenge, but yield everything required to quickly and easily produce articles, webinars, web pages, blog entries and other sales tools. Figure 3 illustrates a typical progression of communication tools that result from a research study.

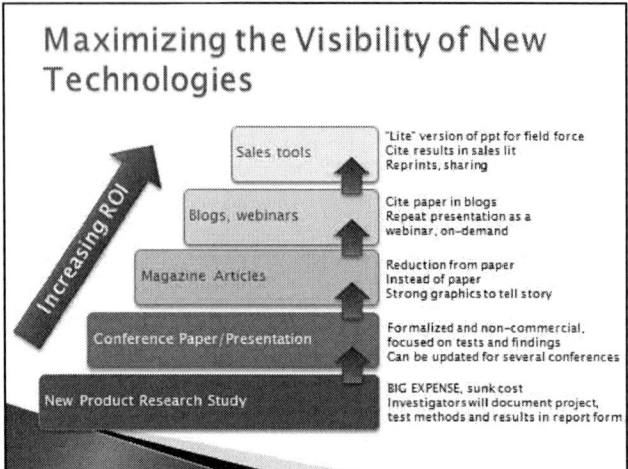

Figure 3. Technical publications evolved from research studies

The choice between a white paper and a magazine article depends on target market and the level of technological advancement. In some cases, skipping the formal paper and publishing articles are the best choice. If a technical paper is published, however, it should be cut down to produce at least one magazine article following its initial publication.

Conference presentations usually contain detailed information of interest to audiences highly skilled in the subject matter, but may be too complex for the typical customer to appreciate or the sales associate to confidently deliver. Conference presentations often contain a great deal of higher level information like statistical analyses and in-depth discussions. They communicate concepts among experts, in the parlance of experts. To communicate similar concepts among professionals of typical proficiency, the key information should be structured appropriately, introducing any new terminology and focusing on the impact of the results. The simpler versions of the information are more likely to garner the attention of average customers than the complicated versions designed for the experts. If presentations are based on conference publications, they should always reference original papers and presentations (when available) to offer more details for interested individuals.

Well-written web page copy is often repurposed for brochures, advertisements or other sales materials. Because web pages offer so little space but must be very compelling, their messages are usually as concise and as powerful as possible. Good web copy not only improves customer perceptions over the internet, but also pays dividends for the general sales and marketing efforts.

Unifying the appearance of data, charts and graphics improves interchangeability of graphical elements among different documents and platforms, and helps identify the brand. Microsoft Office easily enables the use of corporate identity colors with its Theme Colors function. Setting the colors takes less than five minutes, and once the theme is set, Excel charts, Power point graphics and Word illustrations all automatically apply the company colors to every creation. Using consistent colors and fonts for chart areas, data points, trend lines, labels, titles and other elements across all publications helps develop a style that is easily recognized and associated with a company or brand. In situations where publication guidelines preclude the use company of logos, templates or trade names, organizations can rely on their use of color and style to differentiate themselves from their competitors.

TIPS FOR EFFECTIVE COMMUNICATIONS
Technology workforces are very diverse in background, education and location. The effectiveness of any technical communication can be improved by simplifying its language, reducing its word count and integrating graphics.

Simplify language:
- Avoid plurals and articles (a, an, the), as they do not exist in some languages and do not translate well
- Use only present verb tense and active verb voice
- Substitute small words for big ones
- Write short sentences or cut long ones into two

Reduce word count:
- Eliminate useless modifiers like quite, very, rather
- Edit out superfluous phrases like "in other words," "that is to say that," "of course," "generally speaking"
- Target prepositional phrases for condensing
- Bulletize lists

Use graphics:
- An annotated picture is more effective than a written description
- Pictures and graphs translate better than words
- Visual images capture attention

A final tip for authors: *model your writing style based on what you like to read, not how you speak.* The information will naturally flow in a more concise, easy-to-read format.

CONCLUSIONS

Technical communications about new products or technologies can raise market awareness, demonstrate technology leadership, create new avenues for connecting with customers, and strengthen a brand.

Creation of technical communications can present numerous challenges in terms of selecting the right communication vehicle and delivery style, and in ensuring clarity of the message. Understanding the target market is the first step in formulating a communications plan and serves as the cornerstone on which everything is built. Ask three questions:

- Who is the target audience?
- Where do they typically search for information?
- Are they interested methods, data or results?

The answers provide the keys to publishing the information in the right place and delivering it at the right level.

Strategically planning technical communications and integrating them as part of new product introduction marketing initiatives will improve the overall effectiveness and efficiency of the marcomm campaign, and speed the return on investment in the new product.

TOOLS AND TECHNIQUES FOR MATERIAL ASSESSMENT IN ADVANCED TECHNOLOGIES

Martin Anselm, Ph.D., and Wayne Jones
Universal Instruments Corporation, Advanced Process Laboratory
Conklin, NY, USA
anselm@uic.com, jonesw@uic.com

ABSTRACT

As complexity in advanced manufacturing increases, especially for consumer electronics, the need to characterize the materials and processes used in electronic assembly also increases. OEM and EMS companies look to perform characterizations as early as possible in the process to be able to limit quality related issues and improve both assembly yields and ultimate device reliability. Many analytical methods are available to us on the market that each has their own risks and benefits. This paper will help identify some of these key limitations in the methods used for characterizing and evaluating solders, circuit board materials and surface finishes available in the market today.

BACKGROUND

The real cost of failures in manufacturing is significant but is one that is not accounted for during up front calculations. Line-down situations, product recalls, engineering time spent on customer interactions and failure analysis can quickly add up to millions of dollars depending on the product. It is critical that all resources are optimized in order to effectively determine root cause in the shortest possible timeframe. Unfortunately the industry is moving away from a skilled labor force that can accurately assess failures and determine root cause. Often, too much time is spent tracing false positives and incorrect assumptions leading to ineffective corrective actions and "Band-Aid" solutions. In an industry that values "5S" practices, fishbone diagrams and "5 whys" we have lost our ability to employ intuition and experience. Lean manufacturing practices can be very beneficial for failure analysis since often Lean manufacturing practices are associated with tracking lot and date codes of materials used during production, which can be linked to failures. Having this data can be critical in determining root cause and assessing the extent of a failures effect on a population of fielded products.

To that end we must therefore assess failures using techniques that will be able to isolate material and process variations. Whether it be manufacturing process, material quality issues, product design, excessive stresses (in factory or in field), or an inherent weakness in a material selected for the product (e.g. lead-free solder alloy susceptibility to failure). Most companies do not have the resources to employ a staff of engineers and purchase software to conduct physics of failure (PoF) analysis techniques. Also product modeling techniques may only highlight an area of high stress in an idealized condition. An experienced failure analyst needs to take into account the outliers of a manufacturing process or design in order to properly determine and consequently implement a successful corrective action plan.

This paper will begin by isolating some key questions that can be asked of the supplier, manufacturing engineers, supplier quality engineers, and reliability engineering teams. Once these critical questions have been answered, only then can we assess what analytical techniques should be employed. From this high level perspective limitations and opportunities in low and high cost analytical techniques will be discussed.

INTRODUCTION

This paper is written in a logical format that follows the procedure that an engineer should take in performing a material (product or process) assessment. Initially one must understand the scope and nature of the defect or failure. This is followed by material inspection and finally root cause or corrective action strategies. In this paper, the focus is not only on a discussion of optical and Scanning Electron Microscopy (SEM) procedures for material inspection. There are many more techniques available to the engineer. The optical techniques discussed in this paper can be performed with little resources that could be very helpful in determining root cause (if performed correctly). With that in mind, this paper includes some possible risks with performing these techniques that should be kept in mind. SEM has been included in this paper since it is a common first resource when selecting more sophisticated analytical techniques. This paper presents a common error in SEM analysis of solder joint cross-section inspection.

ASSESSING DAMAGE

It is critical that the extent of a failure is assessed, whether the product is a million unit cell phone or a $10,000 military circuit board assembly where less than 10 are being manufactured. The difficulty in determining the extent of the failure is the same; the success of a product is typically defined by high yield and high reliability. The engineer responsible for determining root cause for a failure must segregate the failure into categories and determine how many opportunities there are for further failures. These categories will often determine if the failure is being caused in house or by a supplier, subcontractor, or user. Questions must be asked that will determine if the failure is die level (0th), die attach level (1st), component attach to PCB (2nd), or final assembly (3rd). Areas that can fall between these levels are often material specific such as circuit board failures or post component attach process defects (cleaning, coating, test). Often the failures can fall into the following catigories;

A. Material Quality

Material quality can fall into many categories however more often we consider paste, board, component, adhesives, coatings, cleaners, etc. Each of these materials has their limitations and complications. For example, circuit board manufacturing is a complex process utilizing mechanical (drill), thermo-mechanical (press/cure) and chemical (plating/etching/stripping) processes. Each process has its own unique limitations and characteristic failures.

Q: What are the date or lot codes of the failed devices/boards/paste?

Q: What solder alloy was used for the SMT process? (Sn/Pb paste with lead-free component?)

Q: What component broker was used? Are they on our Approved Vendor Lists?

Q: How thick is the solder mask?

Q: What plating is being used on the component?

Q: What surface finish is defined on the board drawing? What thickness requirements for the surface finish are outlined on the PCB drawing?

B. Assembly Process

Assembly processes vary widely for electronic devices in our industry. Each process has operational windows that will produce high yield and reliable product. In order to assess the possibility of failure in each we must first understand the stresses that the product may face during assembly.

Q: What processes are being used for this product (print, placement, inspection, reflow, cleaning, dispense, final assembly, test, etc.)?

Q: How was profile development performed for this specific product?

Q: Were printing materials changed? New stencil?

Q: Is full I/O inspection being performed on placement machine?

Q: What torque specification is used for tooling whole locations when mounting product to chassis? What order are screws placed? How are boards supported?

Q: How are boards handled following assembly?

Q: How is the multi-up panel singulated?

Q: Is paste being under or over printed for a particular design? (1-2 mil reduction?)

C. Design

Design can affect many aspects of yield and reliability of a product. Simply following component manufacture recommendations for land patterns and stencil apertures may not be sufficient to overcome some unique product requirement. Proper design must be taken into account for managing reliability and determining root cause of failures.

Q: How close are fragile capacitors or associated passives to edge of PCB?

Q: What are the aspect ratios of the stencil?

Q: How close are critical components to tooling holes?

Q: Has the PCB manufacturer made modifications to PCB design from drawing?

Q: Is the failed part in a location of high stress? Has it been moved as compared to previous revision of the product?

Q: Is conformal coating being used on this product? What material has been selected?

D. Reliability

Functional testing, ICT, drop, vibration, ESS, HALT, HAST, are methods used to determine susceptibility of failure in manufacturing and in the field, however correlating them to true field reliability is difficult if not impossible for most reliability engineers. In order to interpret the failure modes identified by common failure analysis practices we must understand all the mechanical and thermo-mechanical stress conditions a product was subjected to, prior to the failure occurring. Often reliability issues are not associated with a single root cause. Therefore it is common in today's research to see topics in assembly pre-stress. It has been shown that thermal or mechanical pre-stress can dramatically affect the reliability of components [1,2]. Root causes for these accelerated failure conditions are not fully understood.

Q: What manufacturing and final assembly stresses is this product subjected to?

Q: What is the end use condition of this product?

Q: Is ICT fixturing designed properly (functioning as expected)?

Q: How were profiles developed and product fixtured in thermal and mechanical testing equipment?

Q: Were components removed immediately following failure or allowed to be tested far beyond their failure point in accelerated life testing?

Q: Can a particular I/O be identified as the failure location? A component? A circuit?

Q: Is the failure a short or an open?

Q: What environmental temperatures, corrosive media, or humidity was the product subjected to prior to failure?

OPTICAL ANALYTICAL TECHNIQUES

Once a failure has been identified, prior to root cause determination, the first objective should be identifying failure mode. Failure mode must be established using techniques that do not subject the product to further stresses and risk of damage. Several non-destructive and destructive techniques are considered low cost and can be very effective in assessing root cause. However if handled or interpreted incorrectly can be costly.

The simplest example of low cost analysis is optical microscopy. The IPC-A-610, E-2010 standard section 1.9 recommends limiting magnification for inspection purposes to 1.5x-40x depending on the size of the land pattern [3]. In cases of cleanliness or conformal coating inspection the maximum suggested magnification defined in IPC-A-610 is 4x [3]. Often this is done to limit the uneducated user of identifying anomalies that may not affect the overall performance of the product. Therefore it is best to have comparative samples from passing lots of product. These baseline samples often can segregate typical conditions from non-characteristic conditions. Having baselines of good products can also allow for higher magnification inspection of design and quality while reducing the risk of misinterpretation.

Optical microscopy is inherently non-destructive and can be used to identify failures in any of the categories listed in the previous section. Lighting techniques should be diversified in order to highlight defects. Low angle lighting, co-axial lighting, spot lighting and ring lighting can all be used at low magnification in order to illuminate surfaces. Lighting can have dramatic effects on illustrating contamination or fracture conditions that may normally be invisible. Often tin whiskering (Figure 1) and other surface conditions (figures 2-4) will only be visible when adjusting lighting techniques.

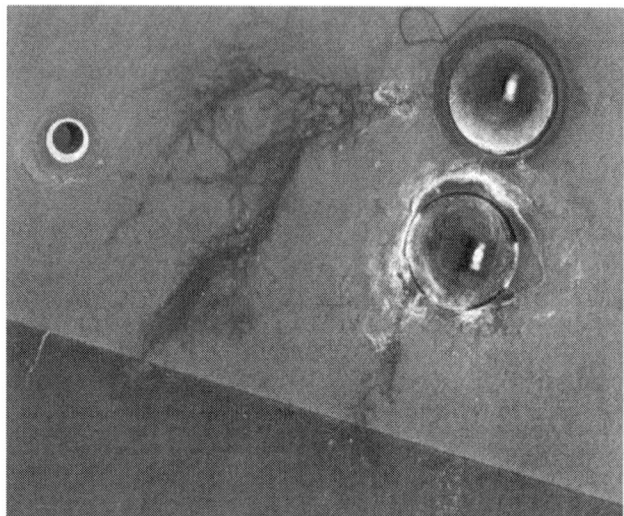

Figure 2, Optical microscopy of dendrites

Figure 3, PoP Head in Pillow failure

Figure 4, Cracking in conductive adhesive illuminated with low angle lighting

Figure 1, Low angle lighting to identify whiskering

Optical microscopy at higher magnification can be useful for assessing lead free solder joints. There is a lack of contrast in lead-free solder joints since they are 95% or more tin (Sn). In order to differentiate between alloys and precipitate structures in lead-free it is often useful to employ dark field or cross-polarized lighting techniques. A schematic of a polarizing microscope can be seen in figure 5 along with examples of various lighting techniques in figure 6.

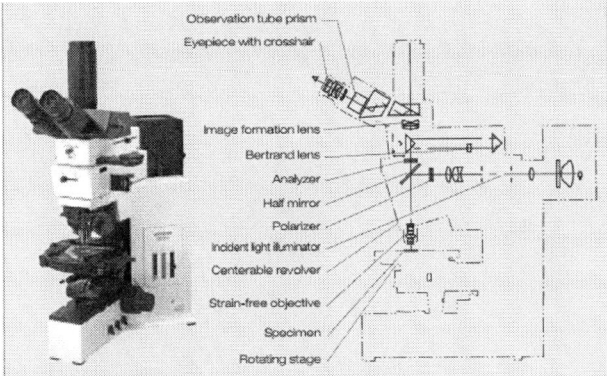

Figure 5, External view and construction of an incident light polarizing microscope [4]

It should be stated in order to get the contrast produced by images in figure 6c cross-sectioning techniques must be optimized and perfected to eliminate not only scratches but the damage caused by the grinding and polishing steps to soft Sn-based solder. The details for preparing a sample for polarized light inspection are not covered in this report.

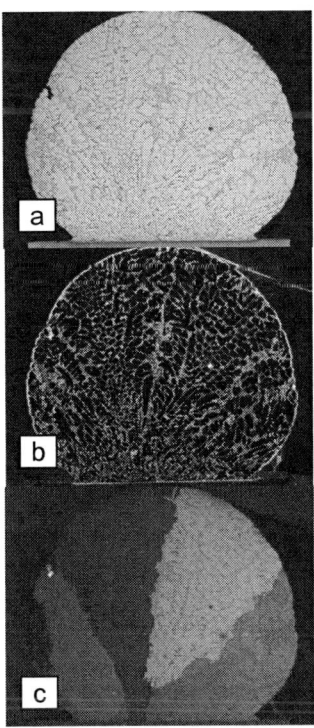

Figure 6, Identical SAC305 solder joint cross-section observed in a) bright field b) dark field and c) cross-polarization

In addition to solder joint condition laminate failures may also be difficult to view using standard lighting techniques. Failures described as pad craters, where the top layer copper is separated from the circuit board due to thermal or mechanical stresses can be difficult to identify. Often the investigator optimizes lighting for inspection of the solder joint. This leaves the laminate material dark and underexposed. In order to properly image the laminate the solder joint requires over exposure as shown in figure 7.

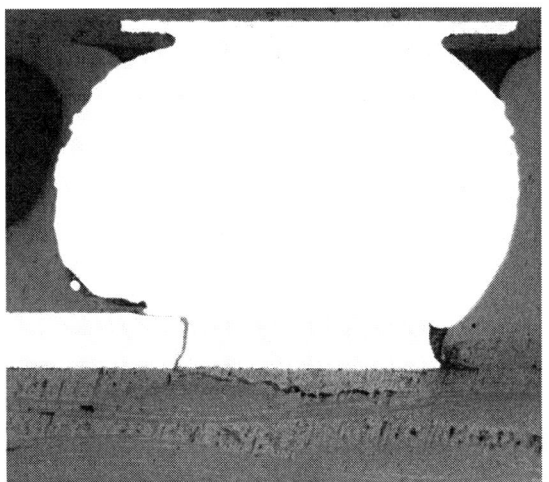

Figure 7, Pad cratering, overexposed bright field image

Poor cross-sectioning techniques can make evaluation of solder joint conditions difficult. Examples of poor sections can be seen in figure 8. Improperly polished sections, where scratches and debris from the initial grinding operations occur, should be avoided. Improper visual interpretation of these "Laboratory artifacts" can produce false positives with respect to fractures or separations, conductive particulates or foreign materials, intermetallic anomalies, and laminate or dielectric defects.

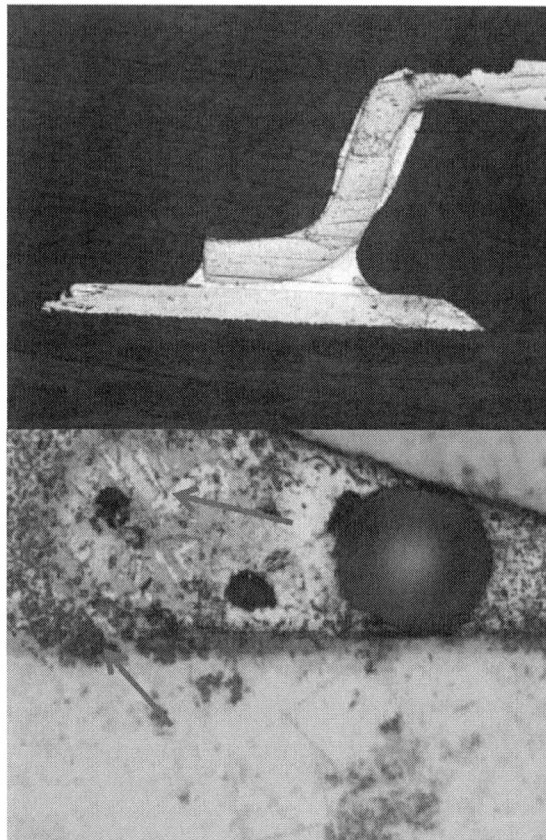

Figure 8, Poor cross-sectioning results in difficult to interepret samples. In the above image large scratches are observed. In the below image polishing residues cover the surface.

Dye penetration can also be an effective low cost analysis technique; in particular for failures were a specific target I/O has not been identified. Multiple devices can be tested and multiple failure modes can be identified. Graphic representations can be developed with locations of failures, percentages of fracturing and types of failure modes (e.g. component side pad crater, component side IMC, bulk solder, PCB pad IMC, PCB pad crater).

A short description of the method is listed below with the dye used for the test. At highest risk for processing error is the curing (step 6) of the dye. One must ensure that all the dye is dried prior to removal of the components. Otherwise liquid dye could migrate onto surfaces causing false interpretation. Figure 9 also shows the result of the test on pad cratering failure modes and IMC failure modes.

Dye Penetration Procedure
1. Carefully cut the region of interest from the assembly by using a low stress technique. A water cooled diamond band saw is often an effective extraction method. Ensure at least ½ inch spacing

exists between the edge of the coupon and the component being tested.
2. Clean the assembly with isopropyl alcohol (IPA) or an appropriate flux remover using an ultrasonic bath, and dry. This step should also clean most cutting debris from step 1.
3. The assembly is immersed in red dye (Dykem steel layout fluid #80496) to stain all exterior and fracture surfaces.

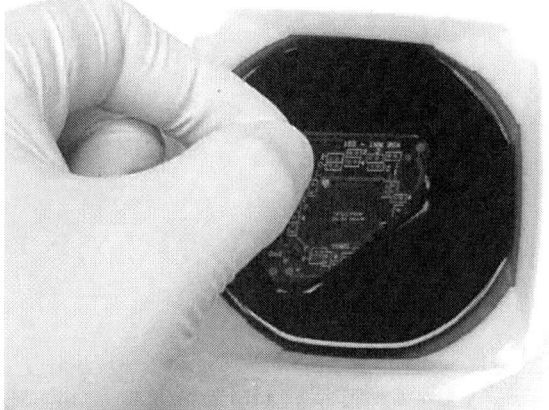

4. While submerged in the dye bath is placed in a vacuum of 9 in Hg for 1 minute to eliminate air from under the device. Ultrasonic baths are also useful during this step. When using an ultrasonic bath the circuit board should be placed vertically in the dye. The liquid is allowed to penetrate for 1 hour.
5. Excess dye is removed. Dye removal can be optimized by placing the coupon vertically and placing a paper towel at the bottom edge of the device to wick dye from under the component.
6. The component is dried 30-60 minutes at 100-125°C.
7. The component is mechanically pried off the board using pliers to twist the board, or a thin screwdriver can be carefully placed between the component and the board to lift the component away from the board surface without damaging the solder joints.

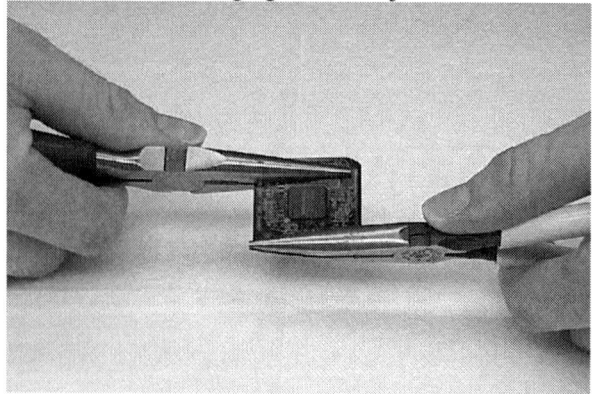

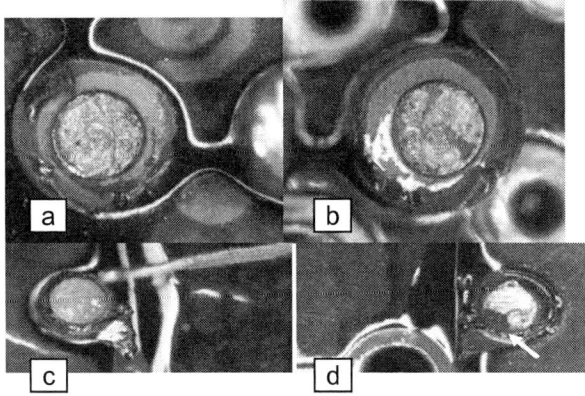

Figure 9, Dye testing results; a, b) board side IMC failure and associated ball removed with component; c, d) PCB pad crater component and ball side.

SEM INSPECTION TECHNIQUES

Most analytical techniques requiring outsourcing will range in cost from several thousand dollars to complete a root cause inspection, to several hundred dollars per hour for use of sophisticated analytical equipment. As an example current, Dual Beam FIB fees can exceed thousands of dollars to analyze a sample. Slightly lower cost techniques such as Scanning Electron Microscopy (SEM) analysis are a useful tool for identifying failure mode conditions, however the inspection can be poorly executed resulting in misinterpretation and confusion.

The most common error in SEM analysis is the use of secondary electron (SE) detectors for metallurgical cross-section inspection of intermetallic. Without getting into technical details, SE is used for imaging topography. Cross-sections are by design flat, so atomic contrast between Ni, Cu, solder and intermetallic is not optimized. Back scatter electron (BSE) detectors provide excellent atomic number contrast and therefore should almost exclusively be used for imaging metallic cross-sections of electronic devices. Examples of BSE vs. SE images for a cross-section are shown in figures 10 and 11.

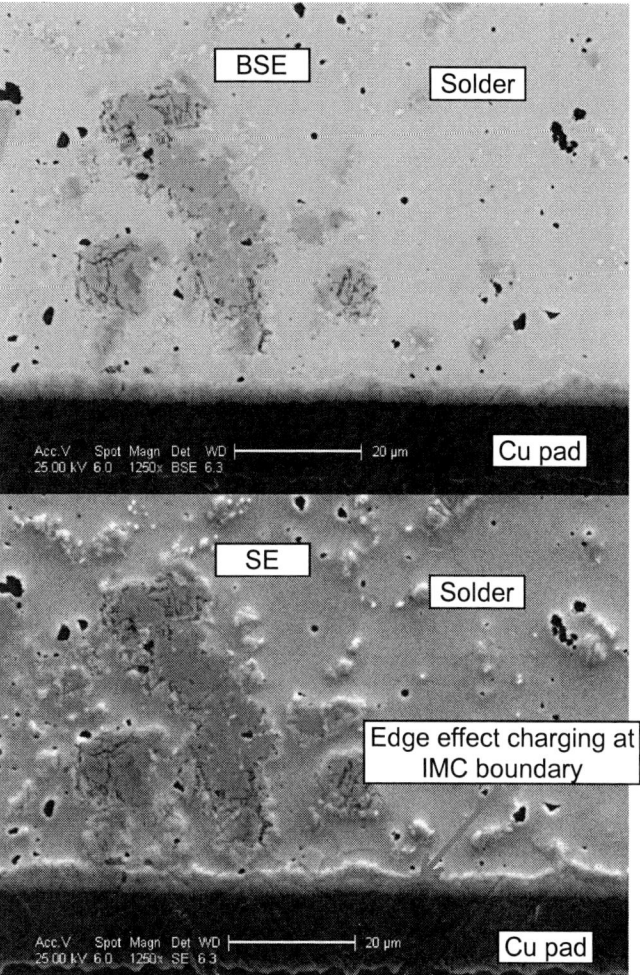

Figure 10, BSE and SE images of identical cross-section locations at 1250x

As can be seen from the images in figure 10 and 11 the ability to distinguish intermetallic regions is compromised using SE detectors, however SE can provide greater detail due to the inherent planarity variation in cross-sections due to the hardness differences in the materials. Harder materials like IMC grind and polish away more slowly than the softer Sn leaving a step between the materials. These steps are highlighted in SE due to edge effect charging.

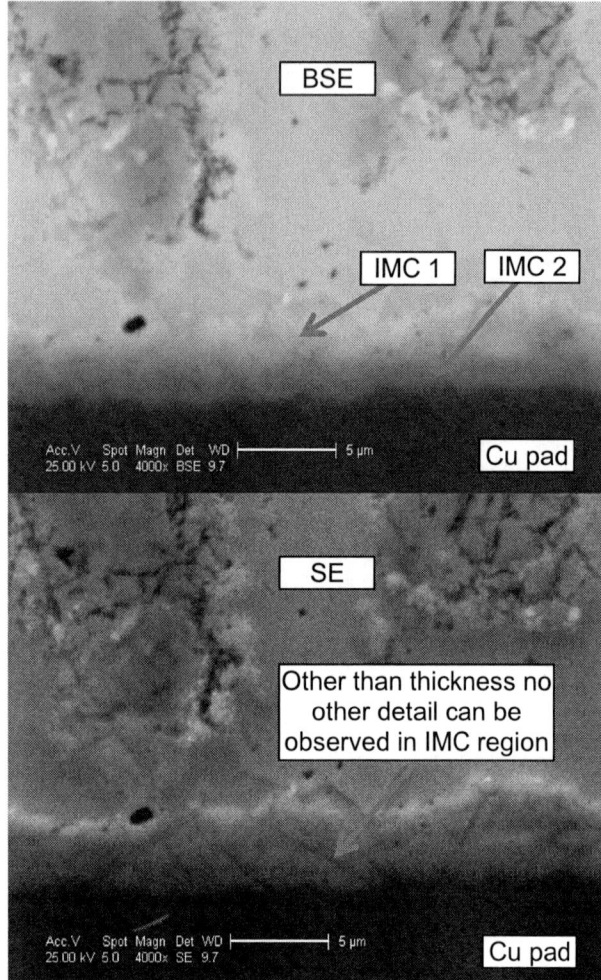

Figure 11, BSE and SE images of identical cross-section locations at 4000x

DETERMINING ROOT CAUSE

Once understandings of the product's use condition, pedigree, and failure mode have been determined the responsible engineer must try to connect the failure mode to the environmental or process condition. This can be accomplished by either comparison to known good product or continued testing. Testing requires materials that may not be on hand and sufficient time to complete test. Both are often not available. Often the most concise conclusions are reached from identification of a clear defect or a dramatic reduction in fallout in the next manufacturing cycle following a corrective action.

CONCLUSION

This paper simply discusses a small fraction of the techniques available to engineers tasked with material assessment. The intent of this discussion was to illustrate the methodology, benefits, and limitations of critical techniques that an engineer may utilize in determining the root cause of a failure. Moreover, any technique used by an engineer has its limitations and requires consideration.

Cost of failure misinterpretation and delay is astronomical and is the cause of significant waste in time and money in an electronics manufacturing factory. With some simple analytical techniques, isolation of the failure and determination of the root cause may be possible. In order to accomplish "root cause" the data collected from analytical techniques discussed in this paper (and others) must be combined with knowledge and experience. Only then can production and field failures be effectively limited and controlled.

ACKNOWLEDGEMENTS

The authors of this paper would like to acknowledge Shantanu Joshi, a graduate student from Binghamton University's System Science and Industrial Engineering department for providing the cross-section samples.

REFERENCES

[1] Singh, A., Meilunas, M., Borgesen, P., Anselm, M., Pitarresi, J., "EFFECT OF STRAIN RATE AND PREDAMAGE ON A PCB USING 4 POINT BEND TEST," 2012 AREA Consortium Report, Universal Instruments Corporation APL, Conklin NY, 2012.

[2] Smetana, J., Coyle, R., Sack, T., Syed, A., Love, D., Tu, D., Kummerl, S., "PB-FREE SOLDER JOINT RELIABILITY IN A MILDLY ACCELERATED TEST CONDITION", APEX 2011

[3] Parrish, M. Et Al, IPC-A-610E-2010, IPC, Bannockburn, Illinois, USA, 2010

[4]www.olympusamerica.com/files/seg_polar_basic_the ory.pdf.

COMPUTED TOMOGRAPHY ON ELECTRONIC COMPONENTS
BETTER WAYS TO DO FAILURE ANALYSIS
PLUS 4D CT THE NEW FRONTIER

Wesley F. Wren
North Star Imaging
Rogers, MN, USA
wwren@4nsi.com or wrenwesley@hotmail.com

ABSTRACT

In the decade past, Computed Tomography has been anunderutilized modality. With the exceptions of failure or quality issues related to safety critical or very expensive components, CT was seldom used. Use of computed tomography systems in the industrial market continues to rise. This recent Shift in paradigm has been attributed to the technologic advances made in digital detectors, new scintillators, better resolution, faster frame rates, and improved bit depth. Most crucial to the growth of computed tomography systems have been the evolutions of better computers, utilization of graphic processing units, along with user friendly software. The proliferation of technology and popularity has allowed the cost of equipment to go down while the acquisition and scan times have dramatically improved.

The new detector technology and computers have allowed for scanning of low density martials inside denser packages that have been almost impossible in the past. This has really opened some doors on what can be done to monitor and understand the integrity of bond wires and via's in electrical components that are being manufactured with aluminum.

The combination of improved computers, detectors, and new software has even allowed the addition of a fourth dimension to a computed tomography data set. Now days a Computed tomography scan on low density material can be done so fast that it allows the element of time to be added in. Making it possible to do a scan of certain parts and watch components move or fluid flow once the 4D computed Tomography scan is reconstructed.

INTRODUCTION

Over the past few decades, Computed Tomography has undergone a significant period of evolution. As recent as five years ago, CT was viewed by many as a slow and cumbersome imaging modality [3]. However, as computational power and speed continue to improve, that perception has begun to change. The ever increasing power of computers has led to faster scan times and reconstruction speeds, as well as more powerful analytical tools. Today, CT has become a proven choice for many applications including failure analysis, metrology, and reverse engineering applications.

The growth in computed tomography has been complemented well by the development of image evaluation software tools. Many of these software tools aid in the process of getting useful data such as volumetric analysis or void calculation. Other software packages paired with quality CT scan data will help cut down on the time to reverse engineer parts. As advancements in computational power continue, 3D software applications and detector technology continue to evolve. These advancements have allowed for inspection of lower density and smaller high density materials that were not previously possible [see figure 6 and 7]. It has also allowed more studies to be performed through reduced scan times and improved accuracy without damaging the integrity of the test specimen. The evolution of 4D CT makes it possible for us to not only study static components, but motion versus time as well. This dynamic volume imaging opens up endless opportunities for expanding our knowledge of product and processes to levels never previously understood or previously proven. This technology has immediate applications in failure analysis and design, and will rapidly be applied in development testing, and inspection areas.

Key words: 4D CT, Computed Tomography, Contrast Sensitivity, and CT

THE SPEED OF COMPUTED TOMOGRAPHY

The first computed tomography system was introduced in 1972 by Godfrey Hounsfield, five year after he had first conceived the idea [1]. Early medical CAT scanners used supper computers where almost a whole room was filled with clusters of central processing units (CPU's) in order to compute the data that was collected, but it was not fast enough for most industrial applications. The lack of speed and great expense were cost prohibitive for most manufactures. However over the last 10 years the return on investment has gone up with the capability of single chassis computer with multiple graphic processing units (GPU's) computing things at speeds that continuing to accelerate. System now days are being used in line with automatic defect recognition on certain parts where data is being collected, reconstructed and software is making the call out in less than 6 seconds [2].

ELECTRONIC COMPONENTS GETTING SMALLER

Electronics components are evolving into smaller packages. This is one of the reasons why high resolution computed tomography images are even more important. With smaller electronic components it is critical that we are able to see inside of the chips and look at the boards before they are destroyed. This allows us to preserve the data and better understand where to look if cutting the part for inspection is even necessary. Figures 1 through 4 show computed tomography results of large and small parts, demonstrating how a CT scan can reveal useful data on a variety of different sized components. As components get smaller they are still being integrated into lager end products. This makes it essential to have the flexibility in the equipment to scan larger and smaller parts with optimum precision, because this allows industrial companies to see faster returns on their investment into this technology by being able to utilize it in a broad array of applications. Figures 1 through 4 are examples of scans that are being performed in less than an hour on various parts when optimized for resolution and contrast. This increase in through put along with the continued improvement of resolution and contrast is making CT technology something that most industrial companies will be able to benefit from now and into the future. These benefits will be realized in reverse engineering, failure analysis, finite element analysis, metrology, CAD to actual comparisons, as well as other areas

faster computers results can be produced in a few seconds on smaller less dense products. When the speed begins to increase down to this level the result will be a loss of some resolution and contrast sensitivity. How ever the data is often times more than adequate as shown in figure 5 on and IC chip that was scanned in under a minute. This makes optimization of a production system product, and criteria dependent, which means most fully automated inline CT application will entail a certain level of software customization.

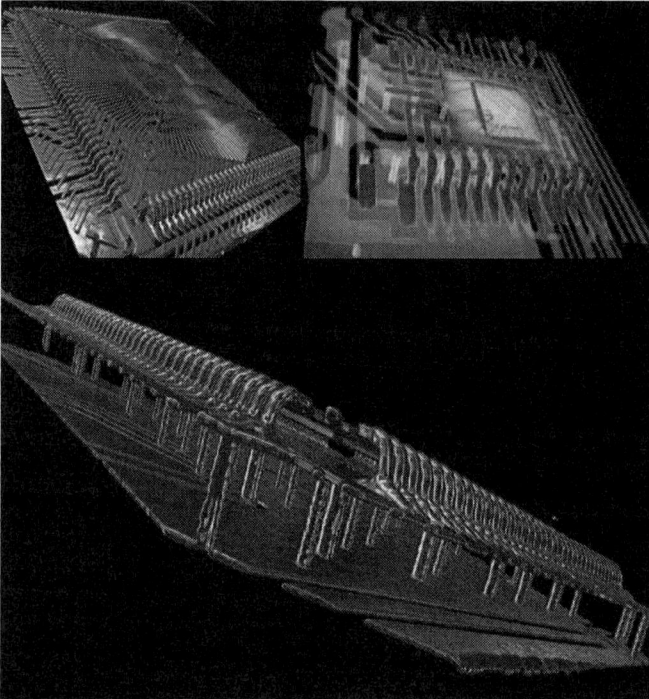

Figure 2. Component Level CT Images

Figure 3. Local CT images of a BGA

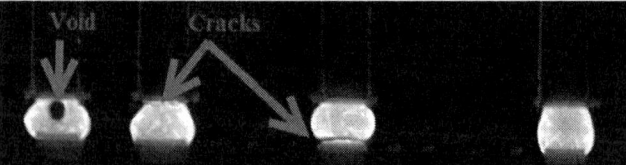

Figure 4. CT Slice Showing a Void and Cracks on BGAs with Enough Senility to also show the vias and pads

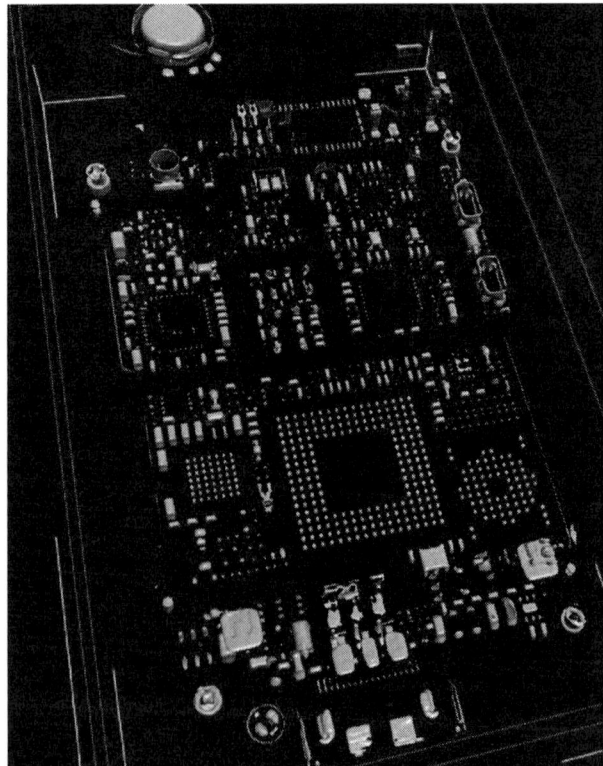

Figure 1. Cell Phone CT Images

Production CT inspection of some components has begun and will continue to utilize CT technology as scan times continue to accelerate [2]. With continuous scanning and

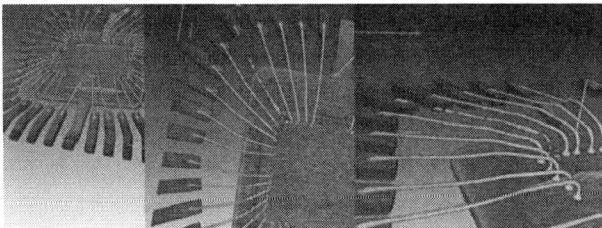

Figure 5. CT Scan Done in Under One Minute Showing a Failure on one of the Bond Wires

Images 2 through 4 illustrate the resolution that can be achieved when imaging smaller objects or local areas with optimum settings. Accomplishing this is possible through increasing the amount of geometric magnification, along with improved contrast sensitivity due to less material being penetrated. Contrast sensitivity and geometric magnification are critical variables in discerning many features of interest.

CONTRAST SENSITIVITY
There are a number of variables that contribute to improved contrast sensitivity in a sample:

1. Material thickness is a vital part in determining what the contrast sensitivity will be. Taking two data sets of the same material, and scanning them at the same resolutions it is clear that the thinner sample would show better contrast sensitivity on a similar void then the thicker sample.

2. Density of material is a key factor in determining whether the contrast sensitivity will be great, average, or poor. Less dense materials show better contrast sensitivity than higher density materials. Highly dense materials require higher energies resulting in a larger percentage of density change in the material in order to see a difference. Whereas low density material like composite or plastics can see subtle changes in the density of the materials.

3. Scanning energy is also a key factor. The lower energy utilized for scanning the sample while still pulling the proper grey scale values the better the contrast will be [5]. One key factor in this is the ability to move the detector closer to the tube, because the numbers of photons that make it to the detector increase by a factor of four. The inverse square law applies to the tube to detector distance [4]. For example if you have 20 inches versus a 40 inch tube to detector distance it will get 4 times the photon count to the detector. Giving the CT technician the chance to either decrease the scan time, or lowering the energy. Lowering the energy will produce better contrast sensitivity. Figure 4 above is a great example of what can be detected when utilizing tube to detector distance reduction to increase contrast. Samples similar to this are often done with a 40 inch or longer tube to detector distance. However, the example shown in figures 6 and 7 were done with less than a twenty inch tube to detector distance.

4. Pixel Pitch is another variable that can affect the contrast sensitivity. The larger the pixel size, the more photons it is able to collect. This results in better contrast resolution while giving up some resolution.

5. Many other variables can play a role in the contrast sensitivity such as BIT depth and scintillators as well.

A technician must consider all of these variables when optimizing a scan for contrast. This is a skill that requires fair amount of training, and practice to optimize your results full potential.

OPTIMIZING CONTRAST FOR ALUMINUM BOND AND TRACE WIRE INSPECTION
When optimizing for the detection of Aluminum Bond wires all of the things discussed our vital in being able to inspect them for failures. Due to the low density of aluminum versus gold wires it is more critical to understand, and utilize all the variables discuss above for superior results. Below in figures 6 and 7 you will find some images of an IC that has aluminum bond wires in it. The bond wires were detectable with longer scan times, shorter tube to detector distance, cesium scintillator was used, and the energies where kept as low as possible while achieving good grey scale values.

The data below clearly demonstrates that utilizing Computed Tomography data in imaging Aluminum Bond wires is possible, but the likelihood of success will depend heavily on the size of the package and density of the outer material. In Figures 6 and 7 you will see that one is illustrated utilizing a black and white histogram and the other was done with a color histogram. Taking a close look at the images one will see how utilizing color can help improve the visualization process.

Figure 6. CT Image Showing Aluminum Bond Wires Inside a Package Utilizing the Black and White Histogram

Figure 7. CT Image Showing Aluminum Bond Wires Inside a Package Utilizing a Color Scale Histogram

When you compare the data from the aluminum bond wire scans in Figures 6 and 7 to the scan of the IC with the gold bond wires in figure 5. It is definitive as to how simple it is to create a good data set on gold bond wires versus aluminum bond wires. The gold bond wires in figure 5 as stated were scanned in under a minute versus the Aluminum Bond wires being scanned in just over an hour.

The IC with the gold bond wires due to the easy of detection could be scanned utilizing the continuous process. The continuous process is when images are collected and the part continually moving versus a step scan were the part will turn, stop, acquire. This allows for some frame averaging which makes for cleaner x-ray images. Better x-ray images with more contrast will help improve the CT scan data. This makes it imperative that components with aluminum bond wires be scanned utilizing the step scan process to obtain optimum results.

This proposes a big question to the industry when it comes to safety critical components. Does it make more sense to build these components with a material that allows their integrity to verified, or is there an advantage of designing components to be manufactures out of aluminum? Building Chips out of aluminum will help ensure that the chips are more difficult to reverse engineer. This could help reduce the number of counterfeit components finding their way to market, which is problem that continues to grow.

4D COMPUTED TOMOGRAPHY

4DCT is a process that takes a number of scans consecutively on a product or sample as some variable of movement is applied. Examples range from fluid being dispersed through a product, a component moving in a part at a relatively slow rate, or a male and female connector being merged together.

As an example application two connectors could be merged together while acquiring the multiple data sets. Most 4D applications require a fair amount of preparation as in this example we may have to manufacture a device to help marry the male and female connectors together at a rate that would be slow enough to gather adequate data as the part is continually. This will ensure the results will be useful and one can see the change occurring upon reconstruction, and the video being made. Once you have the device set up and you begin scanning the connectors as they are slowly being merged you will do multiple scans in arrow. Once the scans are done the reconstruction software allows you to play this in a video format, so you can visualize the interaction or changes occurring in the part or specimen.

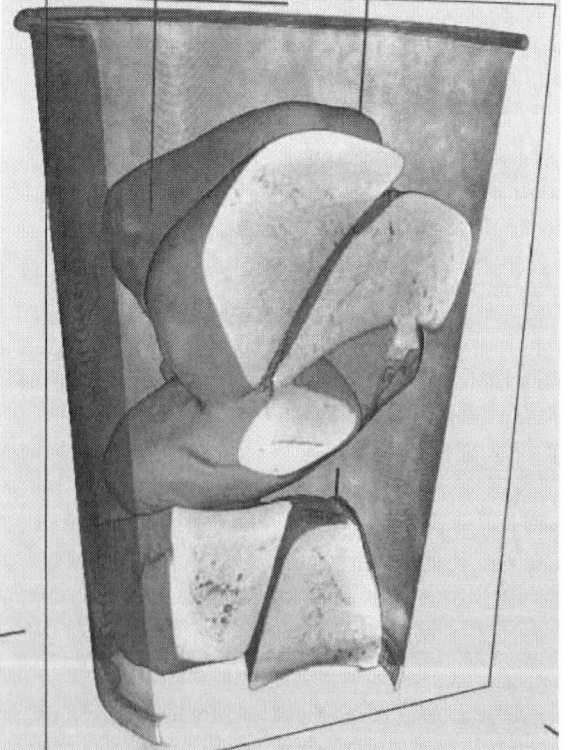

Figure 8. Ice at the Beginning of the 4D CT Scan.

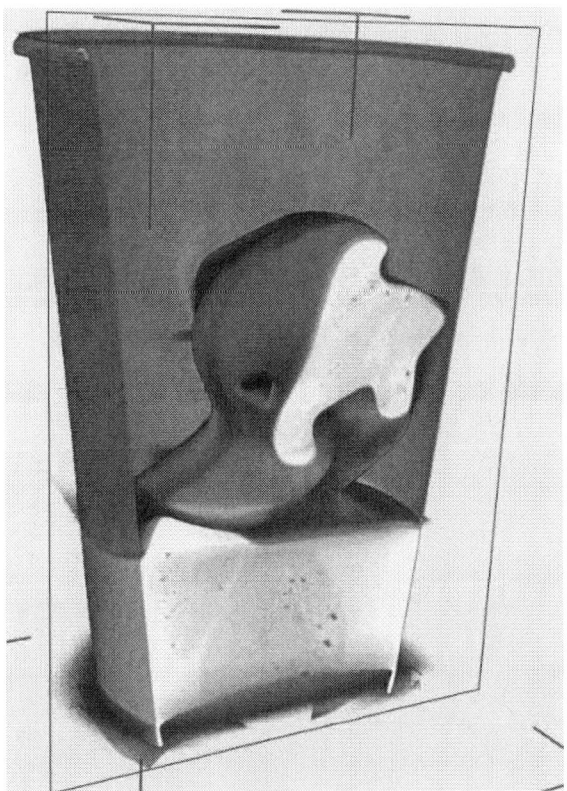

Figure 9. Ice at the Halfway Point of the 4D CT Scan.

4D CT is in its infancy, but could be very useful tool in helping us better understand how certain products interact. Products that lend its self well to 4D CT our typically lower in density and the components of interest usually move at a slow speed. Some application that fit the build would be a slowly occurring chemical reaction, fluid distribution at a controlled rate, switches, or any small component with moving parts. As this technology continues to evolve, so will the possibilities and ideas for new applications.

Figure 8 and 9 above show a plastic cup filled with ice that is being scanned while it was melting. The images only give you part of the story. To really understand how valuable this data can be one must see it in a video where you can witness the ice melting.

In figure 10, 11, and 12, you will see one more sample of 4D CT that demonstrates the versatility of the various applications. In this component it is easy to see how monitoring the interaction of parts could be useful in understanding ways to improve products and resolve issues that are related to failure.

FORESEEN ADVANCEMENTS IN TECHNOLOGY
Two advancements that appear to be coming down the road that will further the capabilities of computed tomography are faster computers and detector improvements. With the advancements in detectors having better resolution, bit depth, and speed there is a growing need for more storage space and faster computation. Luckily computational speeds and storage devices continue to grow. Presently, it is

not uncommon for a single high resolutions data set to be more than 50 gigabyte in size. This will continue to be a big concern due to some of this data needing to be saved for the life cycle of the part. This current environment we work in storage capabilities do not be seems to be growing fast enough to keep up.

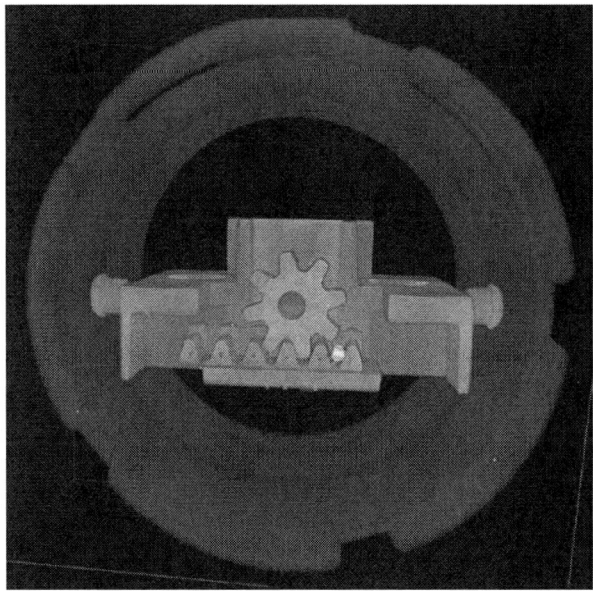

Figure 10. View 1 of 3.

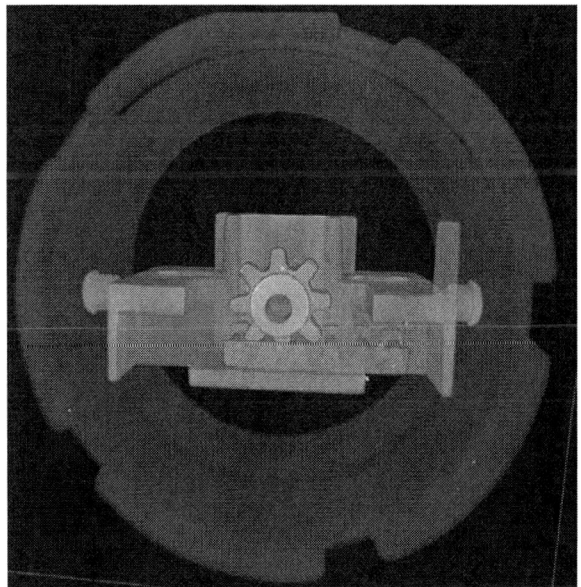

Figure 11. View 2 of 3.

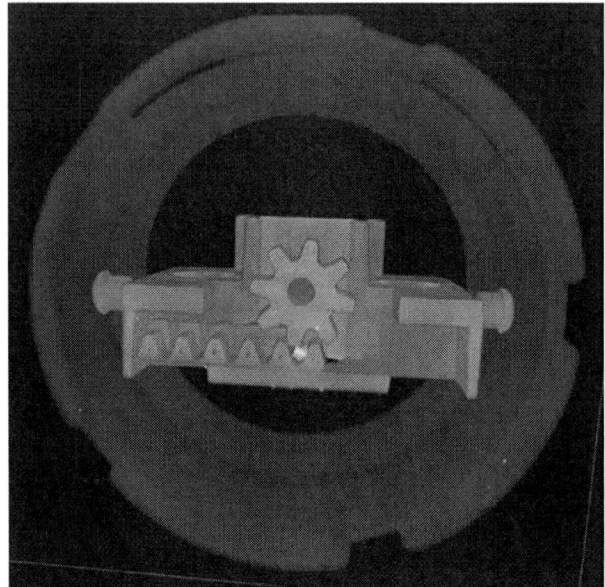

Figure 12. View 3 of 3.

CONCLUSIONS

Computed tomography is a valuable tool that is growing in use and functionality for many applications. With the ever increasing capabilities of computers and the advancements of the detector technology it is clear that CT will continue to see increases in speed, contrast sensitivity, and special resolution. These improvements will allow for better imaging of smaller products with a faster through put time. Do to this many industries and companies will be looking into CT technology as the return on investment improves with more and better data.

4D CT will begin to play a much larger role in the learning more about the functionality of parts, and different processes. Applications in areas never thought of before will begin to be explored with this technology.

ACKNOWLEDGEMENTS

The author would like to thank North Star Imaging for providing the equipment time and scan data for this paper especially Brett Muehlhauser, Shaun Coughlin, Nick Brinkhoff, Julien Noel. The author would also like to thank John Clark for his time in proof reading this paper.

REFERENCES

[1] Kaiser, C. P. (2004, August 19). CT Inventor Godfrey Hounsfield Dies. In Diagnostic Imaging Online. Retrieved November 24, 2012, from http://www.diagnostic imaging.com/dimag/legacy/dinews/2004081901.shtml

[2] EISENBERG, A. N. N. E. (2011, January 2). A C.T. Scan for Hairline Fractures (in Industrial Parts, That Is). The New York Times, p. BU3.

[3] Kim, H., PHD. (2007). Software and Hardware Cooperative Computing. In Georgia Institute of Technology. Retrieved 2012, from http://bcs.bedford

stmartins.com/bbibliographer/bbib_frameset.htm?uid=7951 593&rau=7951593

[4] Radiographic Inspection - Formula Based on Newton's Inverse Square Law. (2012). In NDT Resource Center. Retrieved 2012, from http://www.ndt-ed.org/GeneralResources/Formula/RTFormula/InverseSquar e/InverseSquareLaw.htm

[5] X-ray. (n.d.). In Wikipedia. Retrieved 2012, from http://en.wikipedia.org/wiki/X-ray

ACOUSTIC MICRO IMAGING ANALYSIS METHODS FOR 3D PACKAGES

Janet E. Semmens
Sonoscan, Inc.
Elk Grove Village, IL, USA
Jsemmens@sonoscan.com

ABSTRACT

Earlier studies concerning evaluation of stacked die packages using Acoustic Micro Imaging (AMI) demonstrated the feasibility of using AMI to analyze 3D devices. The construction of the devices evaluated in the studies were typically stacks of silicon chips bonded with an adhesive and using wire bonding for the interconnections. More recently 3D processes include stacked flip chip, silicon interposer, and TSV (Through Silicon Via). The various methods to achieve 3D integration provide challenges to the inspection of the devices using AMI. These challenges include multiple layers, thin silicon layers, different layer thicknesses, varying material properties, small feature sizes and in some cases the devices require analysis post encapsulation.

AMI (Acoustic Micro Imaging) is a non-destructive test method that utilizes high frequency ultrasound in the range of 5 MHz to 500 MHz. Ultrasound is sensitive to variations in the elastic properties of materials and is particularly sensitive to locating air gaps (delaminations, cracks and voids). There is a direct relationship between frequency and resolution in AMI. Higher frequencies have shorter wavelengths and therefore provide higher resolution. Lower frequencies, which have longer wavelengths, provide better penetration of the ultrasound energy through attenuating materials, thicker materials or multiple layer assemblies. Generally a compromise is found between sufficient resolution and maintaining satisfactory penetration and working distance for a given application. In order to accommodate the changing designs in 3D devices AMI technology is evolving. This includes different frequency transducers, and enhanced software tools and imaging techniques to aid in the detectability of features and assist in locating the sequential levels in the devices.

This paper will demonstrate how recent developments in AMI analysis methods can facilitate evaluation of various types of 3D devices.

Key words: Acoustic Micro Imaging (AMI), Flip Chip, Stacked Die, 3D Packaging

INTRODUCTION

There are a number of methods being proposed and investigated to achieve 3D integration in devices. With varying assemblies and materials and with many device types still in the development phase it can be a challenge to find the optimum evaluation method using acoustic micro imaging. However many 3D package types incorporate technologies used in single layer IC packages and there is much past experience working with these related devices such as flip chips. In addition evaluations of 3D devices that are currently in production such as stacked die parts lend information that can be applied and /or modified to analyze the devices that use different approaches to implement 3D integration or devices that are in development.

This paper will present a review of AMI methods that are applicable to analyses of 3D devices and show example applications.

REVIEW OF AMI PRINCIPLES

AMI (Acoustic Micro Imaging) is a non-destructive test method that utilizes high frequency ultrasound in the range of 5 MHz to 500 MHz. Ultrasound is sensitive to variations in the elastic properties of materials and is particularly sensitive to locating air gaps (delaminations, cracks and voids). There is a direct relationship between frequency and resolution in AMI. Higher frequencies have shorter wavelengths and, therefore, provide higher resolution. Lower frequencies, which have longer wavelengths, provide better penetration of the ultrasound energy through attenuating materials, thicker materials or multiple layer assemblies.

A-Scan and - C-Scan (Interface Scan)

In reflection mode Acoustic Micro Imaging the fundamental information is contained in what is called the A-Scan. The A-Scan displays the echo depth information in the sample at each x, y coordinate. Echoes displayed in the A-Scan correspond to different interfaces in the device being examined. There is a time - distance relationship between the echoes related to their depth in the device and the ultrasonic velocity in the materials. The amplitude and phase (polarity) information of the echoes is used to characterize the condition at the interface and is dependent on the acoustic impedance value of the materials involved.

The "interface scan" is the most common imaging method used to evaluate devices for voids and delaminations between layers. This method involves gating the A-Scan signal for the appropriate echo from the interface to be investigated. The gate corresponds to a time window that is selected and applied to each x-y position for the scan. The geometric focus of the acoustic beam is optimized for the interface as well. At each x-y position only the peak intensity value and the polarity of the echo within the gate

145

are displayed. The equation that describes the pulse reflection at an interface between materials is as follows:

$$R = I \; \frac{Z2 - Z1}{Z2 + Z1}$$

Where R is the amplitude of the reflected pulse, I is the amplitude of the incident pulse, Z1 is the intrinsic acoustic impedance of the material through which the pulse is traveling and Z2 is that of the next material which is encountered by the pulse.

As the equation indicates the greater the impedance difference between materials the stronger the reflection at the interface. Whereas bonded areas between similar materials or materials with similar impedances (such as solder die attach) show very little signal reflection at a bonded interface die attach using epoxy bonding shows a significant reflected echo even in the bonded areas. Also multiple reflections for the same interface occur periodically at regular intervals based on the thickness to the interface. The magnitude of these echoes maximizes at deeper focus levels in the sample and often is coincident with the time position on the A-scan and focus for actual subsequent interfaces in the sample (Figure 1).

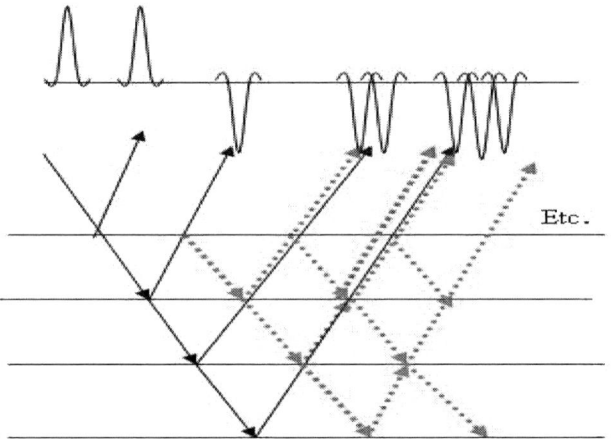

Figure 1: At each interface some of the signal transmits across the boundary (if there is no air gap) and some is reflected at the interface for both the incident and returned signal. This causes multiple reflections from the different interfaces that can interfere with the reflections from the actual levels of interest.

There is a time/distance relationship between the echoes based on the acoustic velocities in the materials that can be used to predict the positions of the interface echoes for the various levels.

Velocity = 2 x depth/time

However, in multilayer silicon devices, typically the layers are very thin relative to the wavelength of the frequency needed for inspection. In some instances the echoes from the various levels may not be completely separated from one

another on the A-scan and this causes interference effects that can be difficult to interpret.

In the reflection mode simultaneous sequential gates can be used to image the separate layers. The location of the gates for the different interfaces is based on the acoustic velocity and the thickness of the material(s). But depending on the thickness (or thinness) of the layers and the influence of multiple reflections that can be overlapping information from previous levels can repeat in images of subsequent interfaces.

Information from the multiple reflections (resonances) has proved useful to examine the individual layers. The location of the different interfaces is determined empirically at this time but will be consistent within a part type.

Resolution: Higher frequency, Lower F#, Shorter Focal Lengths, and Heated Fluid Couplant
Experience with applications such as flip chip evaluation has shown that there are a number of factors that can be manipulated to increase the resolution capabilities. The frequency of the transducer is the most obvious factor in improving resolution. In general, the higher the ultrasonic frequency the higher the resolution possibilities. At present flip chip devices are routinely evaluated using frequencies of 230 MHz to 300 MHz.

However there are other design factors that affect the resolution at a given frequency. The water path from the transducer to the sample at the point of focus and the interface of interest is one important factor. A shorter fluid path will cause less attenuation of the high frequency portion of the transducer bandwidth and therefore allow for the best resolution in the sample. Shorter focal length transducers can accomplish this but the initial focal length of the transducer has to be sufficient to allow for refraction in the sample and to be able to reach the interface of interest with optimum focus.

The F# of the transducer also affects the resolution. This is the relationship of the transducer element size to the transducer focal length (F# = Focal length/diameter of beam). The F # of a lens is used as a measure of the degree of focusing achieved by the lens. For example, two transducers having the same frequency characteristics but different focal lengths will exhibit the same resolution if their F # s are identical (disregarding the attenuation in the fluid couplant).
A smaller F # results in a more highly focused ultrasonic beam and a better resolution when transducers are focused in a couplant such as water [1].

Another factor that controls resolution in broadband acoustic microscopy systems is frequency downshifting due to attenuation in the water path and material. Most acoustic microscopes employ ultrasound in the frequency range of 15 to 300 MHz. The ultrasound is emitted by a piezoelectric element as a short duration pulse. The finite duration of the pulse results in the ultrasound having a broad range of frequencies whose distribution is similar to a Gaussian function (bell curve). The central peak is usually close to the transducer's rated frequency. As the ultrasonic pulse propagates from the transducer through the water couplant into the device and back, the higher frequencies in the incident pulse suffer more attenuation (reduction) than the lower frequencies. The net effect is that the peak in the spectrum shifts to lower frequencies. In other words, an incident pulse with a center frequency of 50 MHz might resemble, after reflection from the target, a pulse from a 30 MHz transducer. This downshifting can cause a significant reduction in the resolution afforded by a high frequency transducer. It has been shown that a shorter focal length transducer will yield better resolution than a longer focal length transducer because the water path between the transducer and sample surface is smaller.

The images shown in Figures 2a, b, and c illustrate the effect of F# and focal length in the acoustic image. All three images were made using the same flip chip sample. Figure 2a displays a 230 MHz image using a transducer with F# 2 and a 9.5 mm focal length. White features are present in the image which corresponds to voids at the chip/bump level. Voids in the underfill are also present. Figure 2b is also a 230 MHz image using an F2 transducer but the focal length in this case is 3.8 mm. Notice that the appearance of the voids is more defined in the image. Figure 2c shows a 230 MHz, 3.8 mm focal length image but in this instance an F# 0.8 transducer was used. This image shows the best resolution of the features and additional small voids can be seen when compared to the other images.

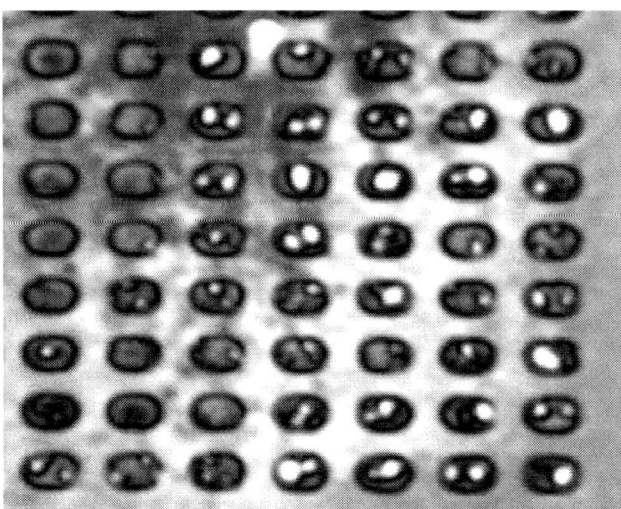

Figure 2a: 230 MHz, F# 2, 9.5 mm fl

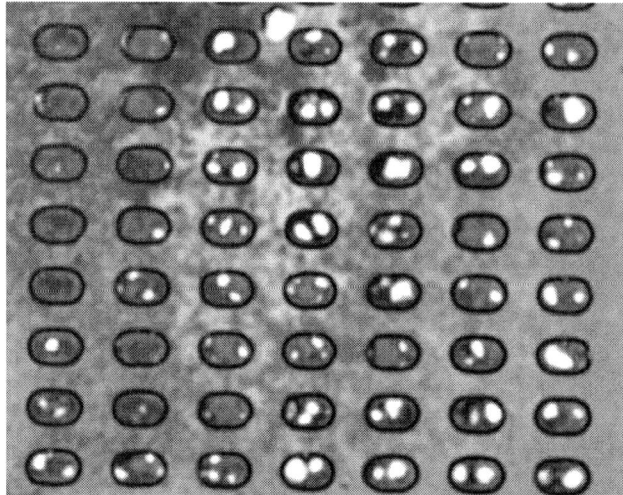

Figure 2b: 230 MHz, F# 2, 3.8 mm fl

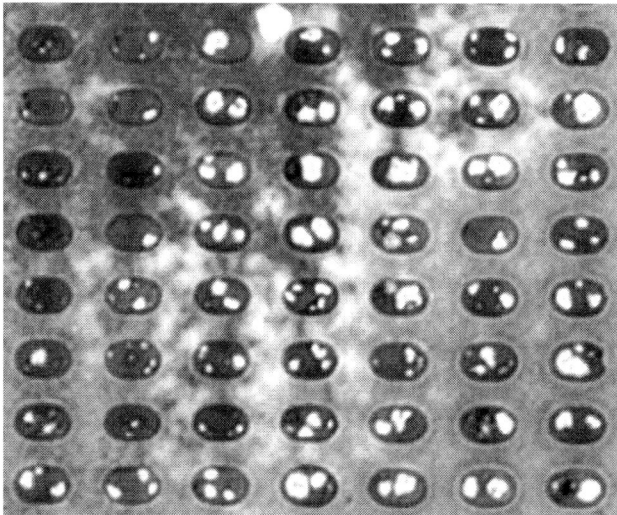

Figure 2c: 230 MHz, F# 0.8, 3.8 mm fl

Heating the water couplant to 40-50 degrees Centigrade has also shown improvement in the resolution in acoustic images. There is less attenuation of the high frequency portion of the signal in water at higher temperatures.

Plastic materials are more attenuating to high frequency ultrasound than water. However even though heating water improves the acoustic transmission through the couplant it has the opposite effect on polymer encapsulant materials. Therefore it is not always advisable to use a heated fluid couplant on encapsulated devices.

Shorter focal length transducers still provide an advantage as there is less attenuation from the water path before reaching the part but the smaller working distance from the parts can be a clearance issue for production screening of an entire strip or batch of parts at a time.

Frequency Domain (FFT) Imaging
Currently a method is used that stores the A-scan information for each x-y point in a scan. From the stored information images of depths within the device not included

in the original gate for the image can be recreated and/or waveforms (echoes) can be viewed for analysis without rescanning the sample. In addition to this the echoes can be digitally processed, frequency filtered, etc., to extract further information about the condition of the sample, or extract information at or slightly beyond the limits of conventional AMI.

Frequency Domain imaging is one method that can extract further information by using the frequency content of the signal. In this technique each A-scan of the image relates to the localized frequency response of the corresponding pixel in the sample. For reference, the conventional image is a time domain image in which each pixel relates to the magnitude of a return echo [2].

Transducers typically used in AMI have highly damped waveforms in order to achieve better resolution, both spatial and axial, using time domain imaging. Figure 3 displays an A-scan with typical echoes (pulses) as seen in the time domain. However these highly damped waveforms contain broad-spectrum frequency information that can be displayed in the Fourier (frequency) domain.

Because the A-scans for each point in the image are collected with the image changes in frequency that may occur during reflection can be analyzed. The gated echo(es) from the stored A-scans can be filtered by means of a Fast Fourier Transform (FFT), also called a Frequency Domain algorithm, to isolate a given frequency. The FFT identifies the different frequencies present in the bandwidth and their respective amplitudes. Figure 4 shows the frequency content distribution of the gated echo shown in Figure 3 in the time domain. Images can then be reconstructed from components of the frequency information.

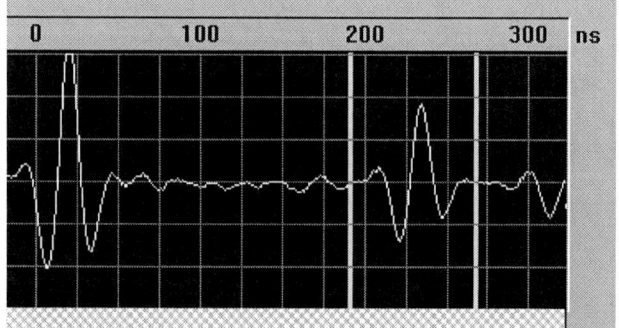

Figure 3: A-scan displaying typical waveforms (pulses) in the time domain.

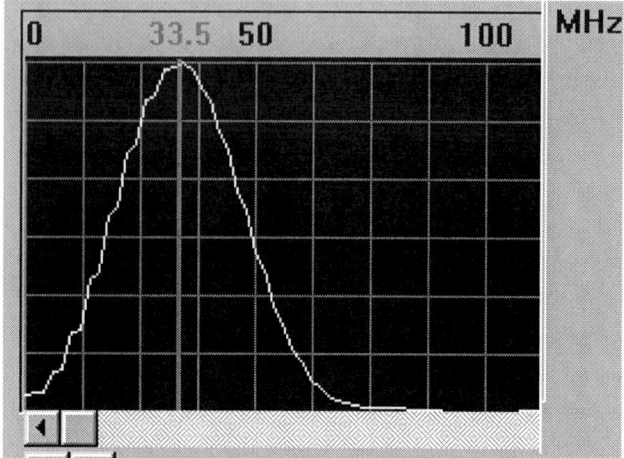

Figure 4: Broadband pulse content in the frequency domain for the echo within the gate on the A-scan in Figure 3.

Frequency domain imaging can be used to improve detection/resolution in the lateral dimensions by removing the low frequency component from the image. Conversely by selecting a lower frequency component of the bandwidth features that were masked by the high frequency portion of the signal have been detected.

APPLICATIONS
Stacked Die Packages
In this example an un-encapsulated six die stack is shown using reflection mode C-Scans [3]. As this sample was un-encapsulated it could be evaluated at a frequency of 230 MHz. In Figure 5 the die attach interface of the first to second die is shown. A large delamination at the lower left corner of the die attach is present in the image as well as a band of smaller voids. The corner delamination corresponds to the position of one of the dark areas in the through scan image but a more central dark area in the through scan is not accounted for at this level. At a subsequent level in the stack a large void area is detected that corresponds to the more circular area toward the center of the part in the through scan (Figure 6). Please note that the orientation of alternating dies was rotated 90° creating an area of intentional disbond at the edges of the stack which also shows up in the images.

Figure 5: Reflection mode C-Scan image of the first die attach level in a stacked die part. A large corner void and additional smaller voids (white areas) are present in the bond.

Figure 6: Reflection mode C-Scan of a die attach level deeper in the stack revealing another large void (white area). The shadows from the voids at die attach level one are also present in the image.

In Figures 7, 8, and 9 voids were detected in an encapsulated four stack die package with 80μ nominal die thickness using through transmission imaging. However, initial reflection mode scans using 75 MHz did not reveal the presence of several of the defects. Subsequent scans at 100 MHz did show the presence of the defects at the deepest die attach level (die 4 to substrate). Using Frequency Domain imaging the appearance of the voids in the image was much improved when the image was reconstructed using a single frequency of 51 MHz.

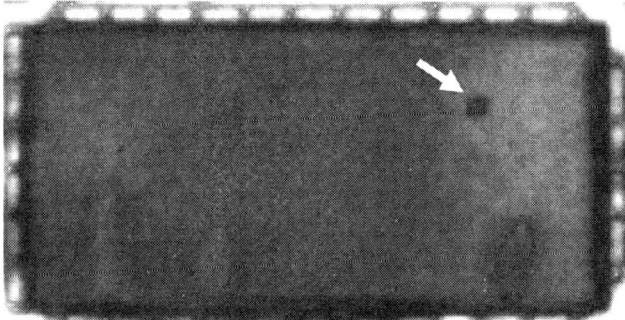

Figure 7: 75 MHz acoustic image of deepest die (die 4) attach to substrate. The shadow of one void from a previous level is seen (white arrow).

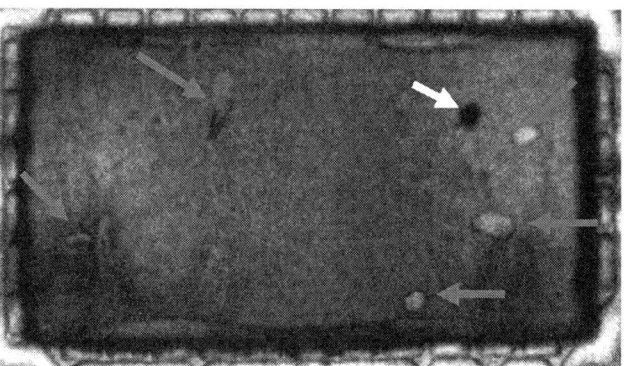

Figure 8: 100 MHz acoustic image of die 4 to substrate. In addition to the void at a level above the interface (white arrow) additional voids are revealed at the interface (red arrows).

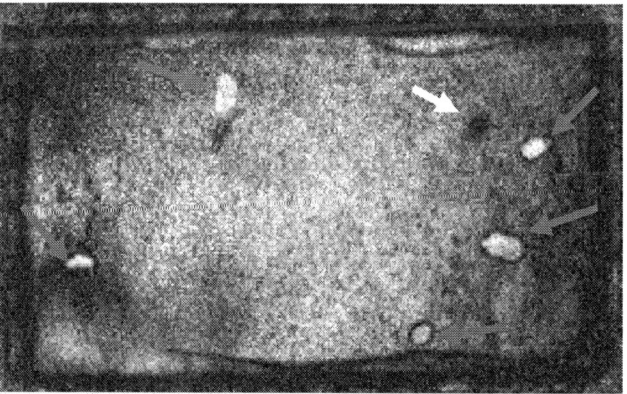

Figure 9: 51 MHz Frequency Domain image of the same die attach level as shown in Figures 8 and 9. The voids in the die 4 attach (red arrows) are much more evident in this image. The shadow of the void from an earlier interface is still present (white arrow).

In this case the interface and defects of interest required sufficient frequency content in the bandwidth of the transducer around 50 MHz to be detected. However it is not as easy as simply using a 50 MHz transducer to begin with. Due to downshifting of the transducer center frequency in the fluid path and in the molding compound 50 MHz and 75 MHz transducers no longer contained sufficient frequency response at 50 MHz to detect the defects.

Micro Bump Flip Chip and Silicon Interposer
This example of a stacked flip chip application involves the top silicon die bonded using "micro bumps" to a silicon interposer and the interposer is bonded to the substrate using larger size bumps. Analysis of micro bump attach is similar to typical flip chip evaluation. It necessitates high frequency transducers and minimizing the fluid path to achieve the best resolution in the images of the bumps. However the optimum transducer for evaluating the top die usually is not suitable for accessing the bump bond interface of the lower interposer and a slightly lower frequency and greater focal length/working distance is required. Figure 10 shows a 300 MHz image of 20μ micro bump bonds of the top die in a silicon interposer sample. Figure 11 shows a 230 MHz image of the larger interposer bump bonds.

Similar to analysis of standard flip chip devices, other factors that affect the analysis of stacked flip chips are the thickness of both silicon layers and the type of die coating or dielectric materials being used.

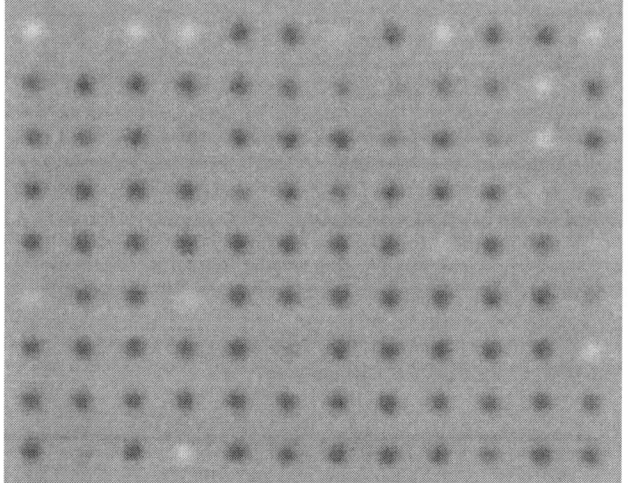

Figure 10: 300 MHz image of 20μ flip chip micro bumps. Disbonded bumps appear white or absent in the image.

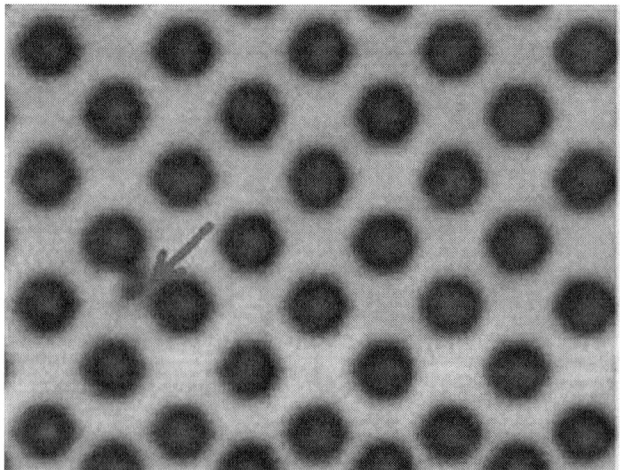

Figure 11: Bump bonds of silicon interposer. A small defect in the underfill is seen adjacent to one of the bumps.

Cu-Cu Wafer Bonding
Another method of stacking silicon die is to bond silicon wafers using Cu-Cu thermo compression bonding. The Cu metallization features provide the interconnections between the two wafers. Usually one of the wafers is back thinned and can be as thin as 25μ. Again here, similar to flip chips, the challenge is to find a transducer that can provide adequate resolution and still penetrate through the thickness of the silicon. Figure 12 displays an area of a Cu-Cu bonded wafer with resolution test features at the interface. The smaller copper bumps are approaching the resolution limits for this transducer however a missing bump is still detected.

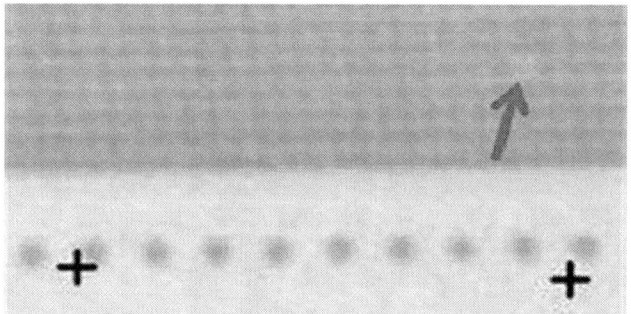

Figure 12: 230 MHz image of Cu-Cu thermo compression bonded test features in a bonded wafer. The distance between the cursors is 200μ.

CONCLUSION
During the development of a 3D device construction and materials are changed or modified. Design and construction varies significantly between types of 3D devices and can vary between manufacturers for the same type of device. Fortunately many of the methods that are currently in use to evaluate standard device types can be used or modified to analyze 3D packages. The diversity of 3D device types also necessitates continuing experimentation and development of acoustic methods and transducers to accommodate evolving device technology.

REFERENCES
[1] Shriram Ramanathan, Gerald Leatherman, Janet E. Semmens, and Lawrence W. Kessler, "High-Frequency Acoustic Microscopy Studies of Buried Interfaces in Silicon", ECTC 2006, San Diego, CA

[2] Janet Semmens and Lawrence Kessler, "Acoustic Micro Imaging in the Fourier Domain for Evaluation of Advanced Packaging", Proceedings of Pan Pacific Conference, Maui, Hawaii, 2002.

[3] Janet Semmens, "Update on the Evaluation of Stacked Die Packages using Acoustic Micro Imaging", Proceedings of Pan Pacific Conference, Kauai, Hawaii, 2008.

SILICON V-GROOVE ALIGNMENT BENCH FOR OPTICAL COMPONENT ASSEMBLY

Terry Bowen
TE Connectivity
Harrisburg, PA, USA
Email(s) tpbowen@te.com

ABSTRACT

One of the primary technical challenges associated with the manufacture of optical assemblies, especially systems with higher levels of integration, is component to component optical alignment. Components must be brought into precise spatial relationship and the precision of this alignment must be captured and maintained throughout the useful lifetime of the assembly. The alignment step can be done actively or passively. The active approach is more complicated involving powering up devices and measuring optical coupling results while moving the components into the aligned position. The passive approach relies on building precision into the parts so that they can be directly assembled into the aligned position without activating the components. This paper describes the use of a large V-groove wet etched into a silicon wafer to form an alignment bench component, which can be used in combinations with silicon based interposer / photonic integrated circuit (PIC) die and either additional similar die, or optical fibers assembled with cylindrical ferrules such as the LC connector ferrule. The silicon interposer die / PIC die are constructed with wet etched v-grooves along the locations where the die will be diced. This allows the sidewalls of these edge locator grooves to be used in combination with the alignment bench to precisely position features on the die relative to other similarly constructed components or optical fibers. The accuracy achieved by the photolithographic processes employed allow passive alignment to be used for constructing these asscmblies.

Key words: Active Alignment, Passive Alignment, Silicon Interposer (SI), Photonic Integrated Circuit (PIC), Wafer-Scale Processing, Wide Alignment Groove (wag), LC Connector Ferrule, lC Connector Assembly, Laser Fiber Endface Cutting, Through Silicon Via (TSV), Chip-to-world Interconnect, Bi-directional Optical Transceiver

INTRODUCTION

Photolithographic methods provide extremely precise features on parts produced in foundries established for the processing of silicon wafers. Features such as solder pad arrays that are designed for flip-chip mounting can provide accurate positioning of electrical and optical die in addition to electrical and thermal interconnections to these die. Silicon Interposer (SI) die or Photonic Integrated Circuit (PIC) die incorporating such features can provide a number of sophisticated functions per die. In this paper, an approach will be prsented that uses wet etched v-grooves along the sides of Silicon Interposer / PIC die. These side locator alignment surfaces are used when assembling these die to a standard optical fiber connector front end. A precision alignment bench formed as a Wide Alignment Groove (WAG) that has been wet etched from a crystalline silcon wafer forms the base for the assembly. In this paper, a Silicon Interposer is described that provides electrical interconnects to an edge emitting laser die and a laser monitor detector die. The electrical interconnects for the SI described in this paper are formed using a process developed at MA/COM for their glass microwave integrated circuits (GMIC process). It involves etching pockets into the silicon wafer and then filling the pockets with glass frit material. Further processing of the wafer produces a surface with areas of glass and areas of silicon. The metallic electrical traces for this SI are run on the glass material areas in order to provide high frequency impedance control and dielectric isolation from the silicon substrate. Through Silicon Vias (TSVs) are used to interconnect the top surface pads of the SI to the bottom surface ground layer to provide top surface grounds where needed.

The electrical connections on the back end of the SI can be bonded to an electrical flex circuit for interconnection to a printed circuit board. The laser die on the front end of the SI is passively aligned to the solder pad array features when the solder is reflowed. The solder pad array features are accurately positioned relative to the SI wet etched sidewall alignment features. These sidewalls are aligned to a standard optical fiber connector sub-assembly by the WAG. In this case, the connector is an LC connector. The resulting assembly forms a chip-to-world interconnect that is small in size, offering a high speed packaging approach. The Silicon Interposer / PIC to standard LC connector receptacle sub-assembly alignment approach is shown in Figs. 1 & 2.

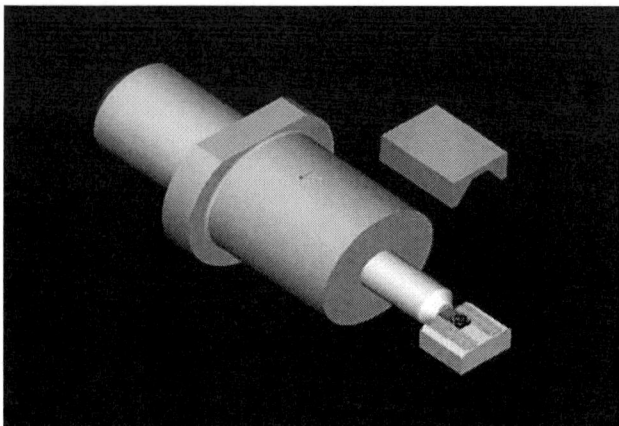

Figure 1. Silicon Interposer / PIC to standard LC connector receptacle sub-assembly alignment using Wide Alignment Groove (WAG).

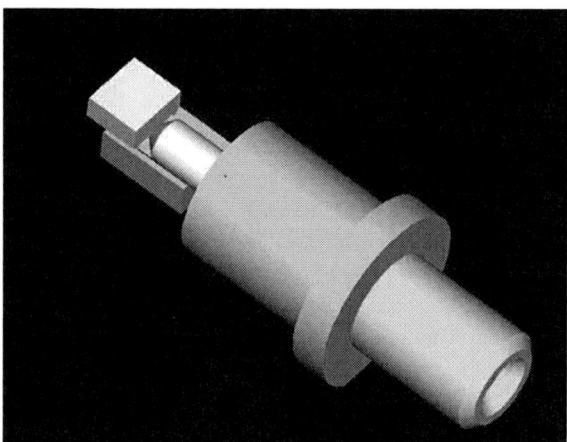

Figure 2. Wide Alignment Groove (WAG) is used to accurately position and provide stability to assembly.

SILICON INTERPOSER (SI)

TE Connectivty, (formerly Tyco Electronics, a part of Tyco International) acquired AMP Inc. in Harrisburg, PA & Somerville, NJ and M/A-COM in Lowell & Burlington, MA) who developed a manufacturable process for mounting opto-electronic flip chip die onto GMIC silicon wafers. This work was presented at the Electronic Components and Technology Conference in 2001 [1]-[2]. US Patent 6,625,357 assigned to Tyco Electronics Corporation claims methods to fabricate devices having either mechanical seating or visually aligned fiducials. It also claimed the method of using matched solder metal pad arrays on the flip chip laser or detector die and on the silicon interposer for passive solder reflow alignment [3]. This invention was extended to include additional methods as well as other product configurations that use matching solder metal pad arrays on the flip chips and the silicon wafer as described in US Patent 6,933,536 assigned to Tyco Electronics Corporation [4]

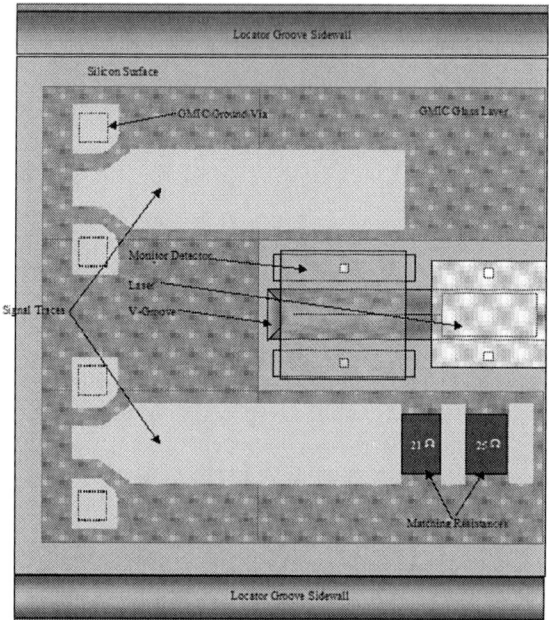

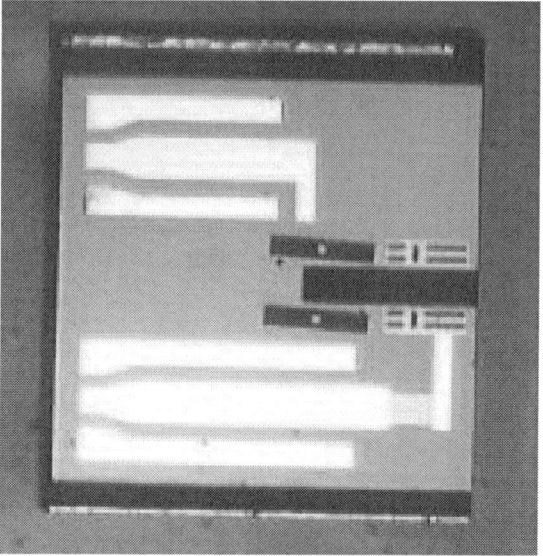

Figure 3a and 3b. Silicon Interposer for laser and monitor detector with locator groove sidewalls.

Passive die alignment using Gold / Tin solder reflow was presented at the 2003 M/A-COM Engineering Conference as an internal Tyco Electronics publication [5]. Figure 3 shows the metal pad array configuration for an edge emitting laser die and a laser monitor detector die. The signal traces for the high speed laser die include impedance matching resistors and are placed onto the glass sections of the GMIC wafer. The laser is flip chip mounted to the metal pad array in a silicon section of the GMIC wafer. Through silicon vias connect the top side ground pads to the ground plane back side of the SI. The LC connector sub-assembly as shown in Fig. 4 can be prepared with a fiber endface that has been polished or prepared by laser cutting as described in ref.[6].

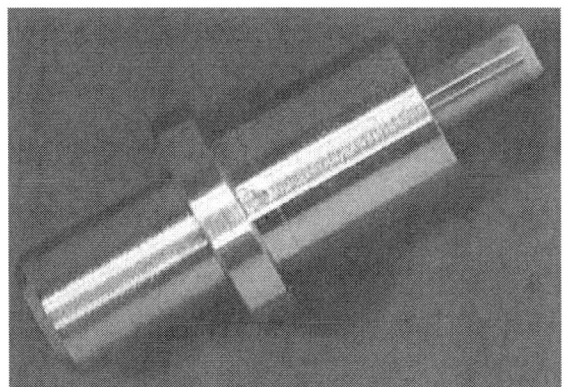

Figure 4. LC Connector Sub-Assembly

The Wide Alignment Groove (WAG) is produced by wet etching arrays of grooves into a crystalline silicon wafer as shown in Fig 5 and dicing into individual parts as shown in Fig 6. It is very important to accurately align the etching mask to the crystal axis in the wafer in order to avoid the production of crystal plane steps on the large sidewalls of the WAG [7]. It is also important to control the etch process parameters in order to avoid the production of hillocks on the surface of the etched silicon for such large sidewalls.

Figure 5. Wide Alignment Groove (WAG) array

Figure 6. Diced Wide Alignment Groove (WAG)

ALIGNMENT OF LC FERRULE TO SI USING WAG
As shown in Fig 7, an LC connector sub-asembly is placed into the large WAG and the exposed LC ferrule contacts the WAG sidewall surfaces in 2 lines that accurately position the center axis of the LC ferrule in the WAG. The SI shown in Fig 8 is constructed with it's alignment sidewalls accurately placed about the metal pad array that positions

the edge emitting laser die. When the spacing of the alignment sidewalls is chosen correctly, the centerline axis of the edge emitting laser die is positioned co-incident to the center axis of the LC ferrule. This alignment can be achieved to high precision using the crystal plane wet etching of the two silicon parts. This alignment mechanism is described in detail in US 7,511,258 assigned to Tyco Electronics Corporation [8]. This true position optical bench interconnection approach enables optical fibers to be interconnected to electro-optical chips that are flip chip mounted onto SI substrates with a high degree of precision. This type of interconnection has been termed Chip-to-World Interconnect since the distances of transmission enabled by the optical fiber can in fact span the world.

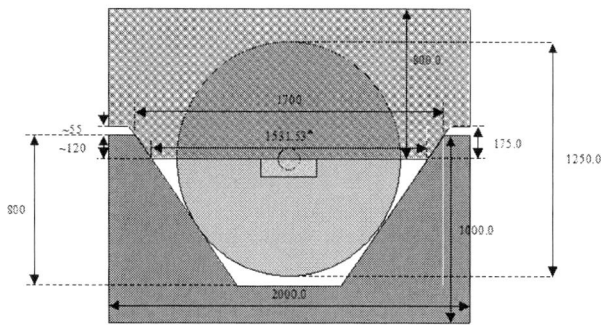

Figure 7. LC Ferrule to SI Locator Groove Sidewalls aligment using WAG

Figure 8. LC Ferrule Sub-Assembly aligned to SI using WAG

THE LC CONNECTOR-TO-SI ASSEMBLY CAN REDUCE THE SIZE OF TRANSCEIVERS CURRENTLY BUILT WITH TOSA AND ROSA ASSEMBLIES
The LC connector-to-SI assembly approach can bring a significant reduction in size to an optical transceiver when compared to optical transceivers built using conventional TOSA and ROSA assemblies that are based on hermetic TO-Can packaging. The electrical connections on the back end of the SI can be bonded to an electrical flex circuit for connection to a printed circuit board. The laser die on the front end of the SI is aligned to a standard optical fiber connector sub-assembly. In this case, the standard

connector is an LC connector interface. The result is a chip-to-world interconnect that is small in size and low in cost while offering a high speed package with improved optical coupling efficiency.

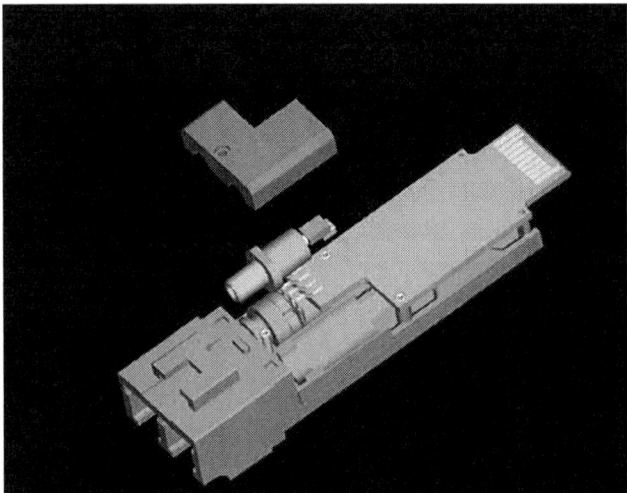

Figure 9. LC Connector-to-SI Assembly comparison to TOSA and ROSA based transceiver packaging

The exploded view of the LC receptacle assembly as shown in Fig. 10 uses a conventional split sleeve to align the LC ferrule component included in the receptacle assembly to the ferrule from the LC cable connector. The standard LC cable connector is shown in Fig 11. A typical optical transceiver will use 2 separate fibers for the send and receive directions, so there are 2 LC cable connector receptacles on the front end of the transceiver as shown in Fig 9. If the SI / PIC includes the capability to do wavelength division splitting and routing, a single Bi-Directional (Bi-Di) optical transceiver device can be formed. In this case, only a single LC connector front end receptacle is required per Bi-Di transceiver.

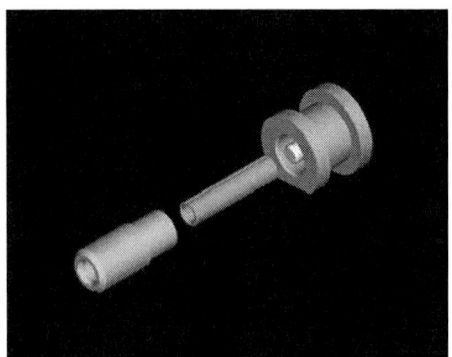

Figure 10. Exploded view of LC receptacle assembly

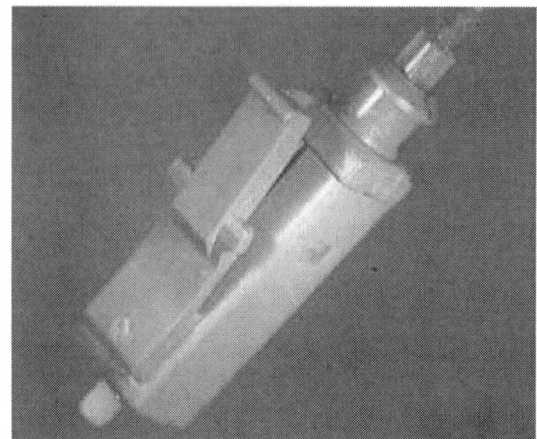

Figure 11. Standard LC cable connector

SILICON INTERPOSER-TO-SILICON INTERPOSER SILICON INTERPOSER-TO-PIC INTERCONNECTS

The WAG can also be used to align 2 or more SI components to each other to construct a complex device, or it can be used to align a SI component to an already complex PIC component. The only requirement is that the SI and PIC components must be produced with corresponding locator groove sidewall width and orientation to the coupled optical object positions. As an example of one of these additional SI configurations, arrays of optical waveguides on a first SI component can be aligned to a second array of optical waveguides or an array of optical fibers as shown in Fig 12. Such array Sis can greatly increase the interconnect density between components. When combined with dense wavelength division multiplexing on each optical waveguide, it is possible to envision 10's of Terabits of bandwidth interconnect between such components using this approach.

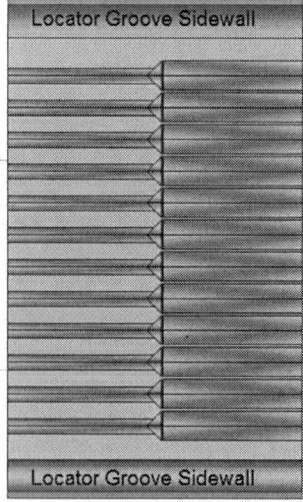

Figure 12. SI to position an array of optical fibers

154

CONCLUSION

The use of a Wide Alignment Groove (WAG) that is wet etched from a silicon wafer to form an alignment bench component has been presented. The WAG, is used to align a combination of components such as a Silicon based Interposer (SI) / Photonic Integrated Circuit (PIC) die to either additional similar die, or to cylindrical ferrule connector receptacle sub-assemblies such as the LC connector ferrule receptacle sub-assembly described here. The silicon interposer die / PIC die are constructed with wet etched locator v-groove sidewalls along the locations where the die are diced. The sidewalls are used in combination with the WAG to precisely position features on the die relative to features on other similarly constructed die or to optical fibers. The accuracy achieved by the photolithographic processes employed enable passive alignment to be used for the construction of these assemblies.

The Silicon Interposer described provided electrical interconnects to an edge emitting laser die and a laser monitor detector die. The electrical interconnects were formed using an advanced electrical interconnect process developed at MA/COM for their glass microwave integrated circuits (GMIC process). It involved etching pockets into the silicon wafer and then filling the pockets with glass frit material and processing the wafer to produce a surface with areas of glass and areas of silicon. The metallic electrical traces for this SI were run on the glass material areas in order to provide impedance control and dielectric isolation from the silicon substrate. Through Silicon Vias (TSVs) were used to interconnect the top surface pads of the SI to the back surface ground layer in order to provide top surface grounding.

One of the primary technical challenges associated with the manufacture of optical assemblies, especially systems with higher levels of integration, is the component to component optical alignment. The alignment step can be done actively or passively. The active approach is more complicated involving powering up devices and measuring optical coupling results while robotically moving the components into the optimum aligned position. The passive approach as described in this paper, relies on building precision into the parts using wafer scale photolithography. This approach provides component parts that can be directly assembled into the aligned position without activating the components. Once the components are brought into precise spatial relationship the precision of this alignment is captured and maintained throughout the useful lifetime of the assembly.

The resulting assembly provides a chip-to-world interconnect that can be automated to produce high volumes.

ACKNOWLEDGEMENTS

AMP, M/A-COM, Tyco Electronics, and TE Connectivity are trademarks of Tyco Electronics Corporation

REFERENCES

[1]   J. Goodrich $, "A silicon optical bench approach to low cost high speed transceivers" *Electronic Components and Technology Conference 2001 Proceedings*, 51st Volume, issue 2001, Page(s) 238-241

[2]   J. Breedis †, "Monte Carlo tolerance analysis of a passively aligned silicon waferboard package" *Electronic Components and Technology Conference 2001 Proceedings*, 51st Volume, issue 2001, Page(s) 247-254

[3]   T. Bowen †, W. Ring #, C. Jiang #, R. Wilson #, M. Soler #, J. Breedis †, R. Anderson *, "Method for fabricating fiducials for passive alignment of opto-electronic devices" *US Patent 6,625,357*

[4]   T. Bowen †, W. Ring #, C. Jiang #, R. Wilson #, M. Soler #, J. Breedis †, R. Anderson *, "Method for fabricating fiducials for passive alignment of opto-electronic devices" *US Patent 6,933,536*

[5]   J.Breedis,† T.Bowen,† R.Perko,† R.Anderson,* J. Goodrich$ and R.Thompson#, "A reliable process for obtaining micrometer-scale passive die alignment in Au-Sn solder" *M/A-COM Engineering Conference 2003, Tyco Electronics Internal Publication*

[6]   H. Vergeest %, "Process for cutting an optical fiber" *US Patent 6,246,026*

[7]   S. Tan †, J. Schramm †, "Method for precise crystallographic semiconductor wafer alignment orientation" *US Patent 6,007,951*

[8]   T. Bowen † R.Perko† J. Breedis† "Optical Bench having v-groove for aligning optical components" *US Patent 7,511,258*

† TE Connectivity, * M/A-Com Corporate R&D, $ M/A-Com Microwave Solutions Business Unit,
# Tyco Electronics Fiber Optics Business Unit, % TE Connectivity Fiber Optics Advanced Development Europe

THREE DIMENSIONAL INTEGRATION RESEARCH FOCUSING ON DEVICE EMBEDDED SUBSTRATE

Hajime Tomokage
Department of Electronics Engineering and Computer Science
Fukuoka University
Fukuoka, Japan
tomokage@fukuoka-u.ac.jp

ABSTRACT

The national research project on 3D integration technology had been carried on in Fukuoka, Japan from 2002 to 2012. The system-in-a-package (SiP) design tools STEERSIP and STEERMEMS, test element group (TEG) chips for evaluating the assembling process, and the evaluation equipment such as scanning electron and laser beams induced current (SELBIC) measurement system have been developed. In 2011, a new research center for 3D semiconductors was constructed, where the main research is on device embedded substrate and silicon interposer with through silicon via (TSV). According to the Japan Electronics Packaging and Circuits Association (JPCA) standard on device embedded substrate EB01 and EB02, the evaluation kits for device embedded substrate are developed in order for device companies to perform function test of embedded devices with the common substrate structure.

INTRODUCTION

Device embedded substrates is believed to be a key technology for 3D integration as well as TSV technology[1]. Active and passive devices are embedded inside the substrate during the PCB process and connected three dimensionally. Then surface patterns are formed, and devices are mounted on the surface.

The national research project on system integration platform had been carried on from 2002 to 2012 in Fukuoka, Japan. There had been three main fields; one is to develop 3D system-in-a-package (SiP) EDA tool. The other was to develop test element group (TEG) chips and reference substrate (RS) for evaluating assembling process, and to develop evaluation equipment. EDA tools named STEERSIP and STEERMEMS were developed, and are now commercially available in Japan. TEG chips specially designed for assembling process have been used by many Japanese companies. In March, 2011, a new research center for 3D semiconductors was constructed in Fukuoka.

On the other hand, the Japan Institute of Electronics Packaging (JIEP) organized a technical meeting named Embedded Passive and Active Devices (EPADs) in April,

2007, and the academic discussion on the embedding technology started in Japan. Then the Japan Electronics Packaging and Circuits Association (JPCA) organized the standard committee on device embedded substrate in March, 2008. In June, 2008, the first standard on device embedded substrate EB01 (Edition 1) was published, and since then every year EB01 has been revised, and EB01 (Edition 5) was published in June, 2012[2]. In the standard, TEG chips developed in Fukuoka are used for reliability test. In terms of design format, JPCA published EB02 (Edition 1) in November, 2011[3],[4]. Currently EB02 is available by three CAD vender tools in Japan[5].

In this paper, EDA tools, TEG chips, and several evaluation equipment developed in Fukuoka are explained. The evaluation kits for device embedded substrate developed at the research center and the collaboration with JPCA on the evaluation method are also explained.

EDA TOOLS

In order to design 3D SiP structure, not only the electrical connection but also thermal analysis and electromagnetic analysis are needed and the co-design is important to shorten the development time. The fully 3D EDA tool named STEERSIP was developed. It has the interface to IcePAK for thermal analysis, and the interface to HFSS for electromagnetic analysis. The feedback of simulated data to STEERSIP is possible, making co-design system.

For micro-electromechanical systems (MEMS) device SiP, the stress analysis software ADVENTURE Cluster are added to STEERSIP, and finally MEMS/SiP tool named STEERMEMS are developed. Figure 1 shows the stress analysis of 3D acceleration device with STEERMEMS.

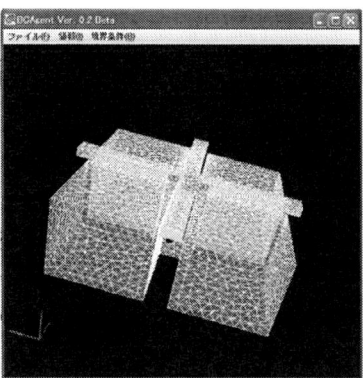

Figure 1. Stress analysis of 3D acceleration device

TEG CHIPS

In order to evaluate the assembling process such as dicing, bonding, and molding, TEG chips are used commonly. Figure 2 shows the TEG chips with low-k insulating layer. Since the low-k materials are fragile against the mechanical stress, defects are introduced sometimes during assembling processes such as dicing and bonding, and molding. The assembling process is evaluated from the change in capacitance and resistance. There are several capacitors near the dicing street, for example. If the low-k layer is damaged by the dicing, the change in capacitance must be measured. TEG wafers are 300 mm in diameter, and fabricated by 90 nm and 130 nm process.

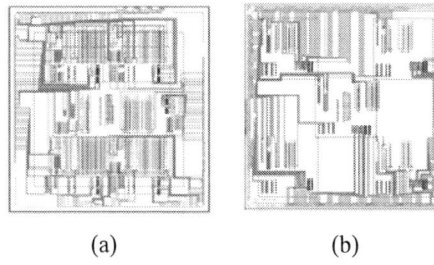

(a)   (b)

Figure 2. Low-k TEG chips with 90nm node (a) and 130 nm node (b) for evaluating the assembling process

EVALUATION EQUIPMENT

Figure 3 shows the scanning electron and laser beams induced current (SELBIC) measurement system[6]. Both of electron and infrared laser beams irradiate the sample at the same time from both sides of the sample, and can be scanned. The cold cathode field emission gun and laser sources with two wavelengths of 1064 nm (YAG laser) and 1400 nm (laser diode) are installed coaxially, and the vacuum chamber is sandwiched between an electron microscope positioned above and an inverted-type laser microscope positioned below. It is possible to fix an electron beam and to scan the optical beam. It is effective to supply electrons to an electrode in

order to obtain the image induced by the optical beam.

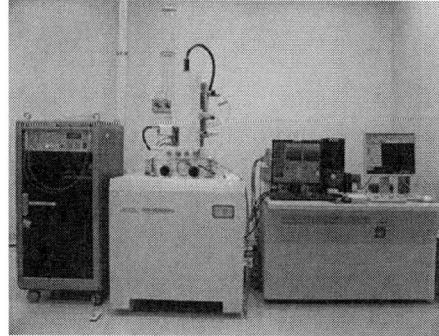

Figure 3. Scanning electron and laser beams induced current (SELBIC) measurement system

Figure 4. Current image of TSV obtained by scanning 1.064 μm laser

Figure 4 shows the current image of TSV obtained by scanning 1.064 μm laser. It is seen that the right hand side of one TSV showed the high current. It suggests that the leakage point in the insulating layer exists inside the via wall. Since nondestructive and electrical measurements are available, this method is effective for the failure analysis of TSV.

Usually the usage of electron beam requires the vacuum chamber, and the size of chamber must be big if 300mm wafer is tested. The scanning laser beam induced current (SLBIC) system was also developed, as shown in Fig.5. Two laser beams without electron beam can irradiate 300 mm wafer from the bottom, and the current image can be obtained by moving the wafer stage after scanning the laser.

Figure 6(a) shows the time domain reflectmetry (TDR) measurement system developed for detecting failure points inside a device embedded substrate. It consists of TDR oscilloscope and semi-automatic prober. The step voltage with a rise time of 15ps is applied to the substrate, and reflection signal is measured. Figure 6(b) is the stacked TEG chips with TSV on the silicon interposer. Intentionally solder bump between TEG chips was removed in order to make disconnections, and TDR signal was measured in order to detect the failure points. The reflection signal was differentiated with respect to time, and the peak position was obtained in order to calculate the electrical length from the input terminal. Figure 7 is the plot of electrical length vs. distance. The distance between the input terminal and disconnection point was measured by observing the cross-sectional view. It is seen that electrical length is proportional to distance. It means that from the TDR signal it is possible to evaluate the failure point for stacked chips.

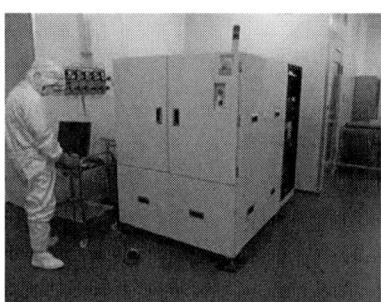

Figure 5. Scanning laser beam induced current (SLBIC) measurement system

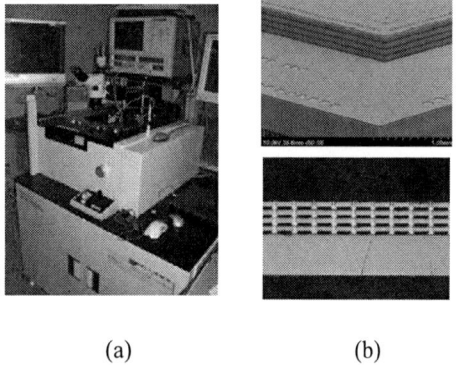

(a)   (b)

Figure 6. Time domain reflectmetry (TDR) system (a) with 15ps rise time was applied to stacked TEG chips (b) with TSV in order to detect the failure point of TSV interconnection

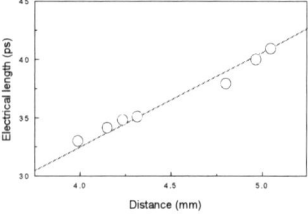

Figure 7. Electrical length vs. distance obtained from TDR measurements

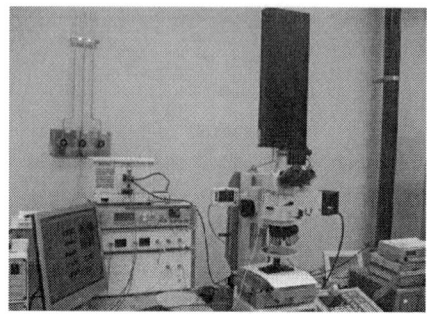

Figure 8. MEMS wafer test system with multimode infrared laser of 1.55 μm in wavelength

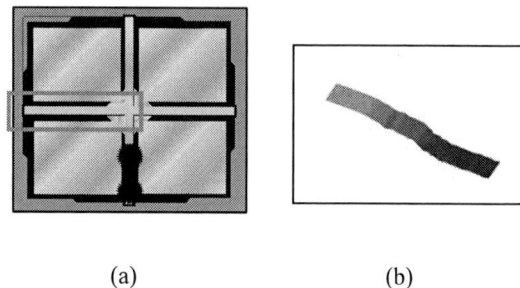

(a)   (b)

Figure 9. 3D acceleration device (a) and the vibration with time (b) measured at the red square region of (a)

Figure 8 shows the MEMS wafer test system. The multimode infrared laser with central wavelength of 1.55 μm has about 60 different wavelengths. After passing the spectrometer, 60 laser points irradiate the sample at the same time. Figure 9(a) shows the 3D acceleration device, and laser beams irradiate the red square region at the same time, for example. Doppler effect enables to measure the motion, as shown in Fig.9(b). Fast measurements in μs range are possible. Since it can measure the motion very fast, the failure due to deep reactive etching can be eliminated by the scanning the laser on the wafer.

EVALUATION KITS OF DEVICE EMBEDDED SUBSTRATE

In March, 2011, a new research center was constructed in Fukuoka, Japan. Starting from design, analysis, prototyping and test of device embedded substrates are possible in the volume production level. Also 8" silicon wafer process to make TEG chips with TSV structure is available in the center. According to JPCA standard on device embedded substrate EB01 (Edition 5), electrical test boards have been developed. Figure 10 shows structure of device embedded substrate for evaluation. TEG chips and discrete devices 1005 and 0603 are embedded, and connected through vias.

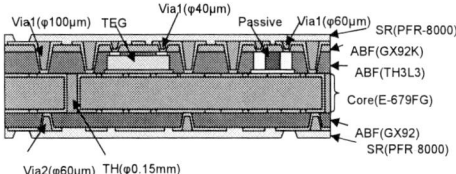

Figure 10. Cross section of device embedded substrate

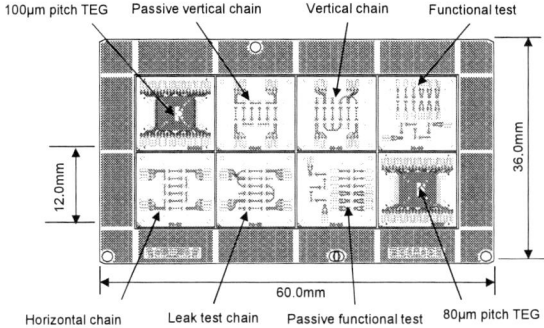

Figure 11. Evaluation kit of device embedded substrate SIPOS_EB01

Figure 11 shows the evaluation kit SIPOS_EB01 for device embedded substrates. This kit has been used in order for device companies to perform the functional test of discrete devices embedded inside the substrate with common substrate structure.

ACKNOWLEDGEMENTS

This work was supported in part by a grant from the Fukuoka Project of the Cooperative Link of Unique Science and Technology for Economy Revitalization (CLUSTER) by the Ministry of Education, Culture, Sports, Science and Technology (MEXT) of Japan. The author would like to thank researchers who joined MEMS/SiP project for ten years. The author also would like to thank Mr. Uranishi and other members of the Japan Electronics Packaging and Circuits Association (JPCA) standard committee on device embedded substrate.

REFERENCES

[1] H. Tomokage, "Design, Evaluation and Analysis Technologies of High-Frequency System-in-a-Package (in Japanese)", IEICE Trans. Electron., **J89-C**, 2006, pp.751.

[2]Japan Electronics Packaging and Circuits Association (JPCA), Standard on device embedded substrate EB01, Edition 5, June, 2012.

[3] Japan Electronics Packaging and Circuits Association (JPCA), Standard on device embedded substrate EB02, Edition 1, November, 2011.

[4] H. Tomokage and M. Kawase, "A New Data Format FUJIKO for Designing Device Embedded Substrates (in Japanese)", IEICE Trans. Electron., **J95-C**, 2012, pp.160.

[5] H.Tomokage, "Standardization of Data Format for Designing Device Embedded Substrates (in Japanese)", Japan Institute of Electronics Packaging, **15**, 2012, pp.515.

[6] H. Sueyoshi, S. Takasu, W.Choi, and H. Tomokage, "Scanning Electron and Laser Beams Induced Current (SELBIC) Method for Observing Failures in GaAs High Electron Mobility Transistors", Superlattices and Microstructures, **45**, 2009, pp.249.

THE CHALLENGES OF LGA SERVER SOCKET TRENDS

Jackson Chang, Michael Hung, Bono Liao, and Nick Lin
Foxconn Electronics, Inc.
Tu-Cheng, Taipei Hsien, Taipei R.O.C.
jackson.cy.chang@foxconn.com, michael.tu.hung@foxconn.com, bono.fj.liao@foxconn.com,
nick.lin@foxconn.com

Andrew Gattuso
Foxconn Electronics, Inc.
Chandler, AZ, USA
andrew.gattuso@foxconn.com

Bob McHugh
Foxconn Electronics, Inc.
Evergreen, CO, USA
bmchugh@foxconn.com

ABSTRACT

The high end server market trends present several manufacturing and technical challenges to socket manufacturers. There are five notable trends that present challenges:

1. There is a trend towards lower contact resistance. Contact resistance has a negative impact on current carrying capability and leads to higher socket operating temperatures. Given space constraints, the best approach to lowering contact resistance is by selecting materials with higher electrical conductivity. However, the tradeoff to higher conductivity materials is reduced mechanical performance.
2. There is a trend towards increasing pin-counts coupled with a need to maintain or lower the socket's overall loading force. Larger loading forces can impact the socket's mechanical integrity and create ergonomic challenges. One way to achieve this is by reducing contact normal force.
3. There is a trend towards contacts having shorter electrical paths. As server frequencies increase, signal becomes more sensitive to noise. Reducing the seating plane height forces a shorter contact signal path and reduces the potential for noise in the circuit.
4. There is a trend towards higher contact density. The need for maintaining legacy real estate requirements along with the need for higher pin-counts results in the need for contact pitch reduction.
5. Due to the higher pin counts and smaller contacts, the trend in manufacturing is towards robotic assembly lines to make the production process more stable and increase yield rates.

This presentation aims to discuss the challenges and some potential solutions for each of these trends.

INTRODUCTION

With each succeeding year, as the technology of high end servers advances, the requirements for the LGA sockets used in these servers becomes more varied and restrictive. As mentioned in the abstract, there are five notable trends in connector requirements. As can be seen in Chart 1 below, the requirement for LGA server sockets is constantly evolving, with changes in pitch, pin counts, style of contacts, and seating plane height.

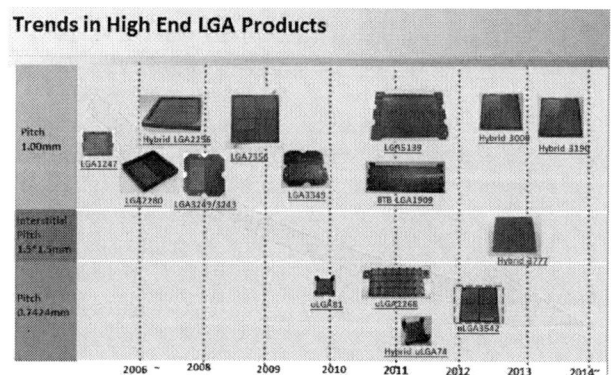

Chart 1. Recent trends in LGA Server Sockets

Thus, it is necessary to be constantly working on developing new stamping, molding, and assembly technologies. Additionally, it is necessary to work with the material suppliers to provide contact materials that have higher conductivity without compromising strength properties and thermoplastics that have better flow properties and maintain high strength with thinner walls.

LGA SOCKET TECHNOLOGY DEVELOPMENT
Reduced Contact Resistance
There is a need to be able to carry higher current levels thru the LGA sockets as the processing power increases. In order to accomplish this, the bulk contact resistance needs to be reduced. There are several ways to accomplish this task. The contact length could be made shorter, the contact material width could be increased, the contact material thickness could be increased, the conductivity of the material could be increased, or a combination of the above. Each of these changes has its limitations, but generally due to space constraints finding a higher conductivity material is the preferred solution. The tradeoff, however, is that the higher conductivity materials typically have lower material strength properties, thus adversely affecting the mechanical performance of the socket. Graph 1 below shows how the material selection between conductivity can change with respect to tensile strength, illustrating the tradeoff that occurs.

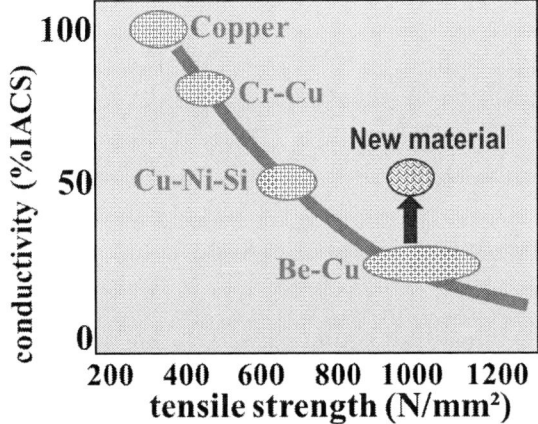

Graph 1. Contact material conductivity versus tensile strength

The challenge for the contact material industry is to develop new materials that have higher conductivity without a significant loss in material strength properties. We are actively working with our contact material suppliers to find new materials that will meet these requirements.

Reduced Contact Normal Force
Another way to try combating the loss of material strength with the higher conductivity material is to lower the normal force requirement for the contact without increasing the contact resistance. This is also being driven by the higher pin count requirements for the sockets. With higher pin counts the loading force required to activate the sockets increases, which increases spacing requirements for loading mechanisms, as well as, the integrity of the socket body. For example, if you have a 3000 pin LGA socket and the nominal contact normal force is 36gf, then the nominal loading force required to activate the socket is 108 kgf. Now if the pin count increases to 3777, the loading force will increase to 136 kgf, a significant jump in the amount of force that must be applied and maintained.

The contact normal force can be reduced if a high hertz stress is maintained. We have developed a contact to achieve this by employing a proprietary thin gap blanking technology to reduce the width of the contact by 28%, which allowed us to reduce the contact normal force by 47%, while maintaining a high hertz stress, reference Pictures 1 and 2.

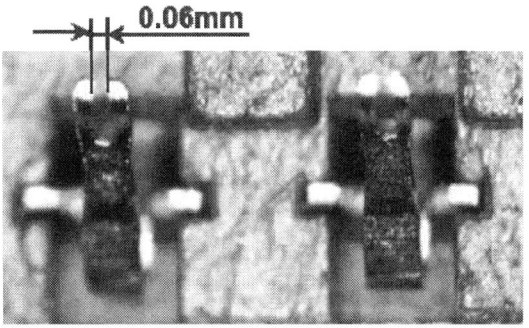

Picture 1. Top view of contact with smaller gap

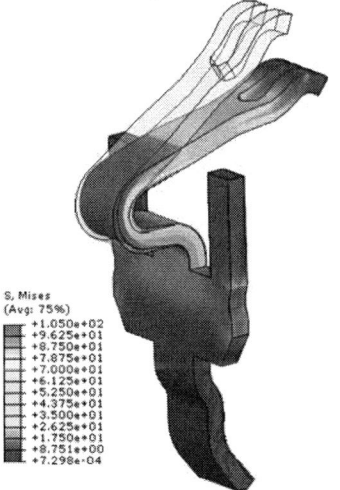

Picture 2. FEA deflected stress model of redesigned contact

The contact normal force went from 36gf in our current contacts to 19gf in the redesigned contact, as shown in Graph 2.

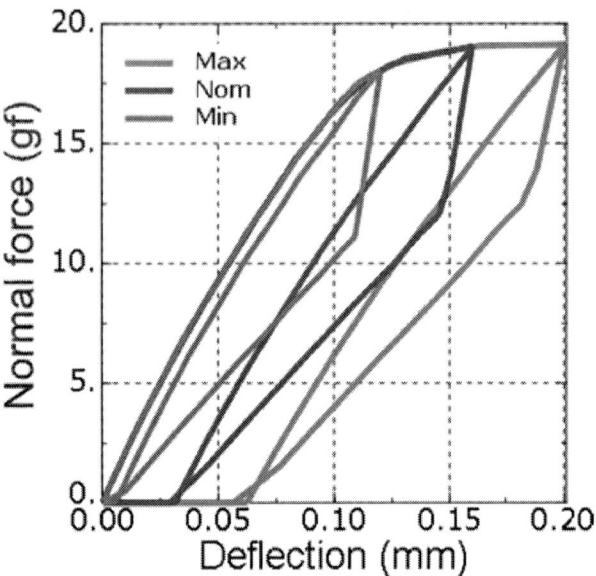

Graph 2. Normal force versus deflection of redesigned contact

As can be seen in the Graph 3 below, the LLCR on the contact begins to stabilize at the 10gf level.

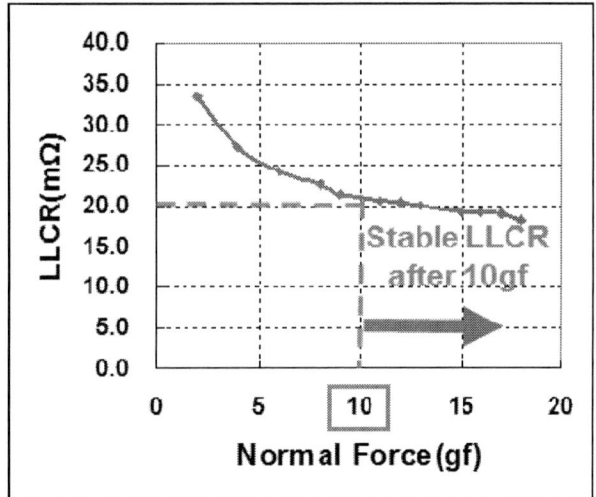

Graph 3. Contact LLCR versus contact normal force

Thin Blanking Technology
As mentioned above, thin blanking technology was utilized to decrease the space requirements for the contact, and keep the hertz stress high while lowering the contact normal force. The thinner gap requires a thinner blanking punch, which would shorten the life of the tool due to higher tool breakage.

We internally developed a tool punch using FEA simulation to optimize the tool geometry. With the redesigned punch, and optimizing our stamping process, we were able to thin the punching area by 40% as shown in the first two sets of pictures in Picture 3. We strengthened the punch by adding two reinforcing side ribs, as shown in the last two set of pictures of Picture 3. [1 and 2]

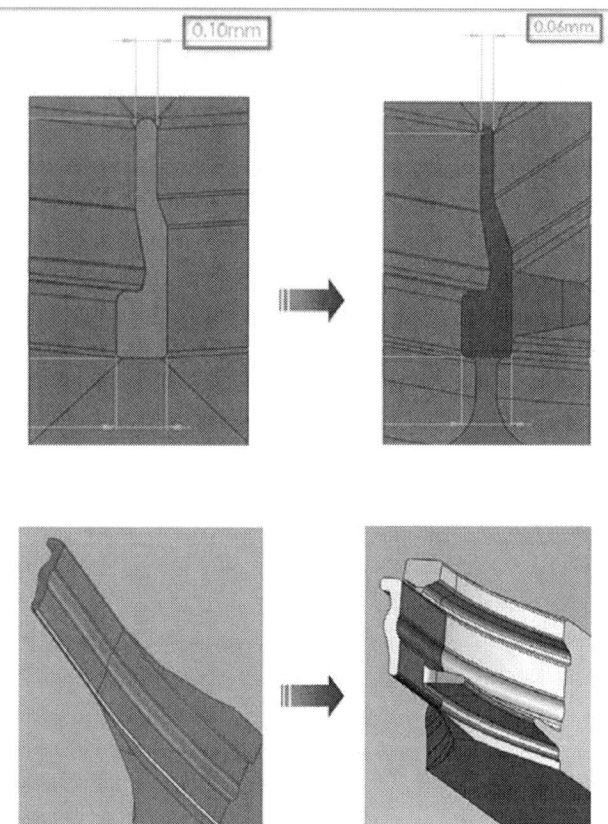

Picture 3. Standard Blanking tool on left side versus optimized thin blanking tool on right side

With the new punch design, we were able to reduce the blanking gap from 0.10 mm to 0.06mm, without reducing the tool life.

Reduced Seating Plane Height
As the processor's speed increases, there is a need to reduce the electrical length of the contact. If the electrical length is too long, there will be more noise in the system with the higher speeds. This requires reducing the seating plane of the socket (the height from the top of the motherboard to the bottom of the processor package). One method that can be employed to reduce the seating plane height is to use thinner material so the deflection range of the contact can be maintained with a shorter spring beam length. Employing a proprietary thin metal stamping technology, we have been able to design a contact that has a lower profile and reduces the seating plane from the current 1.6 mm to 1.0 mm as shown in Pictures 4 and 5.

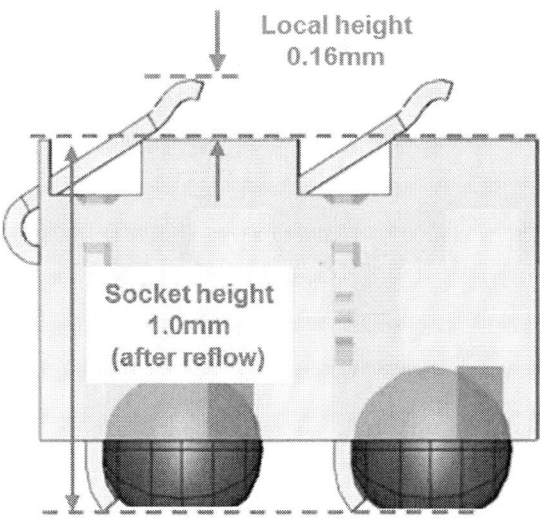

Picture 4. Drawing side view showing seating plane

Picture 5. 1.0 mm seating plane height socket

Thin Metal Stamping Technology

Progressive die stamping is the preferred method for producing LGA contacts. In a progressive die, a strip of metal is horizontally fed into a long die that has a series of stations for punching, swaging, blanking, shearing, and bending to produce the contacts. However, one of the problems with using thinner metal for progressive die stamping is that the carrier strip strength becomes significantly weaker. This makes it difficult to properly feed the material in the progressive die. [3 & 4] We were able to optimize our contact carrier design, and enhance the strength of the pilot hole, allowing us to reduce the material thickness of the contact strip by 33% from 0.06 mm to 0.04 mm, while maintaining good manufacturability, as shown in Pictures 6 and 7.

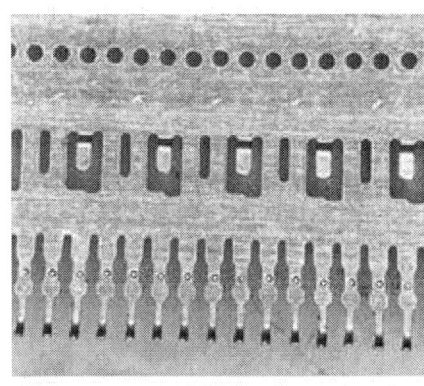

Picture 6. Standard contact carrier

Picture 7. Optimized contact carrier

Reduced Contact Pitch

As the processors become more powerful, there is a need for more pins to utilize this capability. However, it is also desired to not take up more of the board real estate to achieve the higher pin count, as the board also needs more components to maximize the processor capability. Thus, methods need to be found for reducing the contact pitch. We used a combination of thin metal stamping technology, thin wall injection molding technology, changing the method of solder ball attachment, and FEA to design a socket with a pitch of only 0.40 mm from the current pitch of 0.7424 mm. Pictures 8 & 9 show the socket.

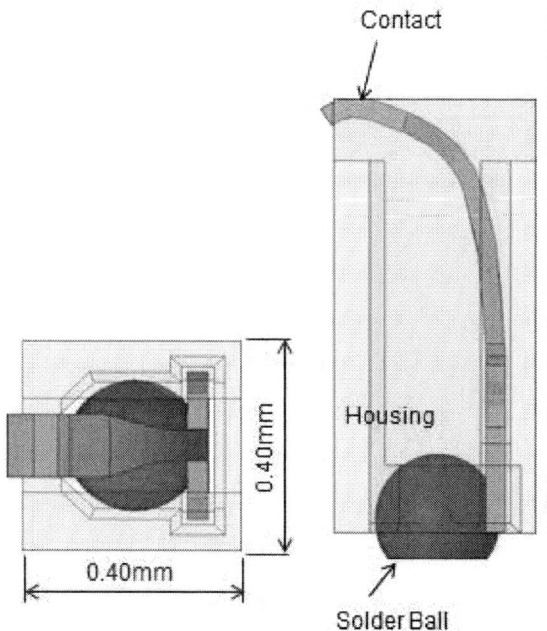

Picture 8. Top and side view drawings of contact

Picture 9. Top view of 0.40 mm pitch LGA socket

Thin Wall Injection Molding

Thin wall injection molding requires higher pressures and speeds than more conventional molding. There are many design considerations to take into account, such as harder steel for the molds, gate and runner sizes, core pin and ejector pin designs, heating and cooling systems, and venting [5 & 6].

Thin wall molding can also create issues with maintaining true position of contact cavities and warpage of the socket body. For a 0.74 mm pitch socket, the TP is held to 0.25mm, so for a 0.40 mm pitch socket it might now be necessary to hold it to 0.125 mm or less depending on contact design, alignment pins, and pad design. Warpage control is also harder. Socket flatness is typically held to 0.20 mm. With the smaller pitch, socket walls are not as stiff and there exists fewer areas to employ coring to control warpage.

We were able to employ thin wall injection molding technology to reduce the wall thickness on a socket from 0.25 mm down to 0.15 mm. In particular, we utilized mold flow analysis to optimize the core pin and gate designs, optimized processing parameters, heating and cooling of the mold, and used a small form factor ejector pin system. As shown in Pictures 10, 11, and 12.

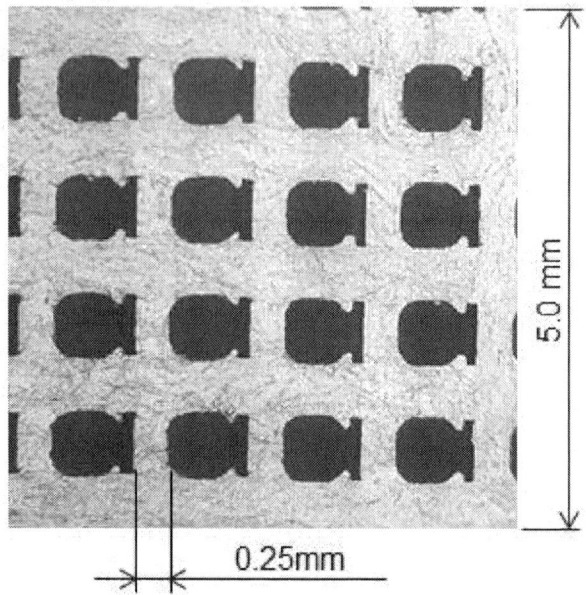

Picture 10. Top view of standard injection molded housing

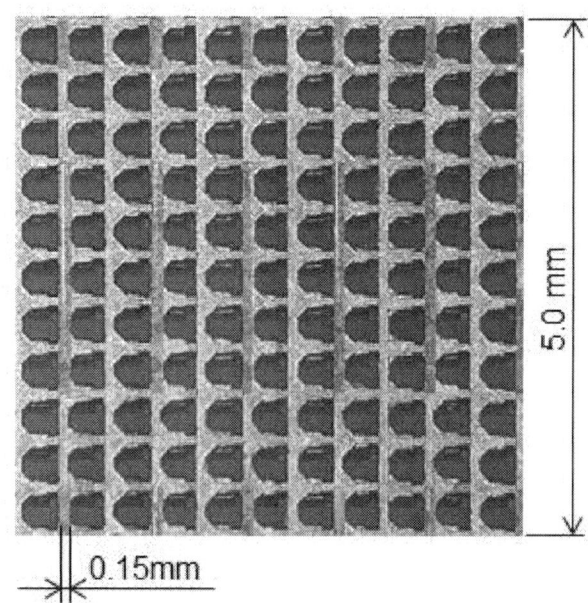

Picture 11. Top view of thin walled injection molded housing

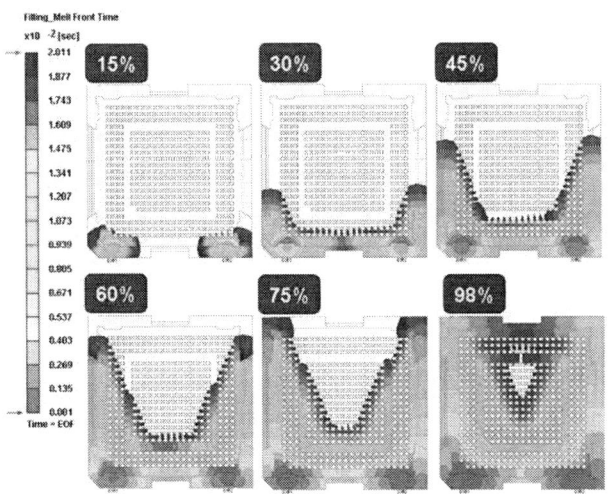

Picture 12. Mold flow analysis showing fill pattern of housing

Robotic Assembly of LGA Sockets

The above trends that have been discussed have put more pressure on efficiently assembling the sockets. As the sockets become smaller, it becomes difficult to assemble the sockets using the traditional methods. It becomes necessary to remove as much human interaction as possible from the socket assembly process. Thus, we have developed a completely self-contained automated robotic process for assembling high pin count LGA sockets. The system utilizes a robot to move and precisely position the socket assembly from station to station. Thus, operator error is removed from the process. The contacts are fed and cut from a stamping reel and inserted into the socket body. The socket bodies are fed into the robotic line from a tray. A high resolution camera, with 15 million pixels, is used to inspect the sockets after assembly to verify the contacts have been inserted correctly, before pressing them to their final position. The sockets are then routed thru automated electrical testers to verify the socket is functioning properly. Defective sockets are automatically sorted out and the good sockets are tray packaged for shipment. A descriptive picture of the robotic assembly cell is shown below. Yield rate was increased from 85% to 95%. The output volume was not increased, but the number of operators required for assembling was reduced by 75% versus the old assembly method. Picture 13 shows the robot assembly line and the flow of the product.

CONCLUSIONS

As can be understood from the preceding discussion, there is much work in many areas that needs to be done to continue to meet the difficult requirements of future high end LGA sockets. Not only does the manufacturing technology need to continue to be optimized and improved, but also the raw materials used to make the sockets needs to also be developed. In particular, contact materials with high strength and better conductivity are needed to meet the increased current requirements on tighter pitches, and thermoplastics that have good flow characteristics, while maintaining high strength with thinner walls.

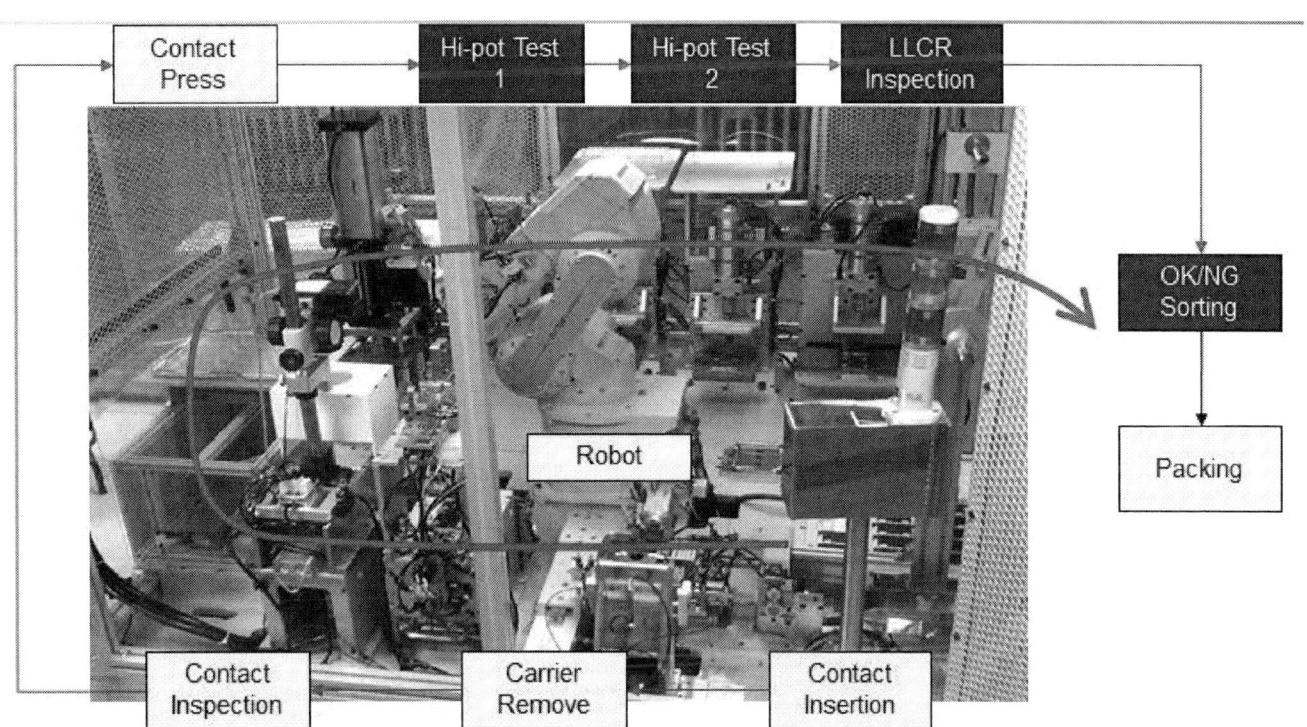

Picture 13. Robotic cell used to assembled LGA socket

ACKNOWLEDGEMENTS

The authors would like to thank the material suppliers, who provided support for this paper. We would also like to acknowledge the many Foxconn Electronics, Inc. mold, die, and assembly engineers, and FEA analysts, who provided valuable information and insight.

REFERENCES

[1] Kalpakjian & Schmid, "Manufacturing Processes for Engineering Materials," 5th ed, Chapter 7, Pearson Education, 2008.

[2] Sue Roberts, "Tips for Punching Thin Material", Canadian Industrial Machinery, August, 2012.

[3] Art Hedrick, "Die Basics 101", TheFabricator.com, August, 2005.

[4] Wikipedia, "Progressive Stamping", October, 2012.

[5] Kurt Weiss, "Secrets of Successful Thin-Wall Molding", Plastic Technology, February, 2002.

[6] Donna Bibber, "Secrets of Success in Micro Molding", Plastic Technology, March, 2012.

THE QUEST FOR RELIABILITY STANDARDS

Dieter W. Bergman
IPC Inc.
Bannockburn, IL, USA
dieterbergman@ipc.org

ABSTRACT

As electronic assemblies become more complex each Original Equipment Manufacturer (OEM) struggles with the question as to whether the product will work in the intended environment, for the length of time expected by the user. Many methods have been developed to simulate the end-use characteristics of equipment in the field, but a clear understanding is required to differentiate between the methods used to assess quality and those used to validate reliability. The considerations and relationships between the customer and supplier are also now a major factor, especially since most OEMs outsource not only the fabrication, but also the design of the product.

The industry trade associations have written many standards which define methods for both quality conformance and reliability testing. Sometimes these methods are misapplied; however there should be no misunderstanding about the fact that poor quality can never achieve the intended reliability of the product. Quality conformance is meeting the requirements of the customer. Since each end-use varies classes or levels of complexity and performance were developed in most of the performance standards in order to simplify the procedures for determining whether the finished goods meet the customer requirements. There is no doubt that the requirements for a hand-held commercial product are different than those for an aerospace application. Also the degree of validation of both quality and reliability differs, as does the cost of producing the item.

Some of the existing reliability methods have become suspect in simulating the end-use, so some OEMs are requiring new methods in order to accept delivery of those products indicated in their fabrication documentation package. Reference to classes or levels of quality are not a guarantee that the product will perform according to its reliability assessment. An example is one where an OEM has developed a product and the prototype version has been tested to validate the reliability of the unit in its intended use. The production version is then outsourced to a new fabricator and it is unclear as to whether the new supplier's process is robust enough to achieve the same reliability as the original prototype.

The dilemma facing the industry is that the supply chain is hard pressed to keep up with all the new "home-grown" methods and new stress techniques being developed. The concepts badly need industry consensus on robustness and reliability assessment methods linked to the end-use environment. This paper explores the various alternatives.

BACKGROUND

The idea of testing a product in order to establish a form of reliability is not new. Many companies that manufacture products for use in the industry want to establish some form of verification that the product will work for the time expected by the customer. These tests have been developed in order to stress the product to the point where a determination can be made as to the products' ability to survive under extreme conditions. Highly Accelerated Life Testing (HALT) has been in existence for over 40 years. Based on the simple idea that if the product was tested beyond specifications the developer would better understand the design margins and improve the products' robustness in the field where it will be used.

Hewlett Packard started life testing in the early seventies. It was David Packard who believed that "Reliability cannot be achieved by adhering to detailed specifications. Reliability cannot be achieved by formula or by analysis. Some of these may help to some extent, but there is only one road to reliability. Build it, test it and fix the things that go wrong. Repeat the process until the desired reliability is achieved. It is a feedback process and there is no other way". Some think that the work of HP in the 70's was a predecessor to HALT testing.

A good combined definition for HALT is that it is a Design technique used to discover product weaknesses and improve design margins. The intent is to systematically subject a product to stress stimuli well beyond the expected field environments in order to determine and expand the operating and destruct limits of the product. As such many pieces of equipment have been developed that can provide these stresses which include cold temperature, hot temperature, rapid thermal transitions, vibration and shock stress, and combine thermal and vibrational environments. There is a similarity between Design for Robustness and HALT except that the latter is intended to take the product to failure during testing, understanding the failures, and improving the design of the product.

In order to expose the product to the HALT conditions requires that the entire product is subjected to the stresses intended to simulate the severity of the environments that might be encounter. In many instances many aspects of the product is able to survive the stresses, and the redesign normally concentrates on those aspects that are the subject of the failure. If the vibrational stresses caused the problem it is a simple matter to add mechanical structures to enhance the capability of the product to resist the impact of the vibration or shock. Thermal transitions however are a little more difficult to isolate especially if these are severe and cover a large range between hot and cold.

The Weakest Link

The electronic assembly within any container has always been the subject of reliability concern. Since the functions of the product would be severely impaired, some designers build redundancy into the electronics. If a circuit was going to fail an alternate would take over to accomplish the same task or to provide a warning to the user that something needed to be replaced or recharged or whatever it took to get the product to work correctly. These solutions were usually not sufficient in order to establish a high level of confidence by the customer.

Once the idea of using multilayered circuitry was established in the 1970s, the interconnection between layers was determined to be the weakest link of the electronic assembly. Electronic components were tested and after the assembly, were subjected to Burn-In exposure. If they passed these conditions their failure rate was minimized for a long period. Through-hole attachment and good solder joints provided sufficient mechanical strength to offset some of the vibrational stresses, and the thermal characteristics imposed on the assembly were not sufficiently detrimental to offset the pretesting accomplished by the component supplier to prove that the components would be reliable in many different environments.

The concern was for the plating in the plated-thru holes of the multilayer board. The only electrical connection made between the inner layers of circuitry was the plating that served to coat the inside of the hole. The process was one with multiple steps each of which could cause serious problems in performance within the field. After the assembly saw many thermal cycles of hot and cold, during the normal product operation, the concern was that the plating attached to a particular layer would come apart. In addition, with thousands of holes in any multilayer panel, or for that matter large backplane, it was not possible to examine and certify that each hole would survive the thermal stresses of normal assembly operation. The process needed to produce a reliable product.

Once the various layers had been prepared with their defined circuitry, they were laminated into a homogeneous structure. This effort was normally accomplished in a panel format with the individual boards outlined to be later sectionalized as they become separated boards or assemblies. The plated through-hole drilling was the most critical part of the process, and if not accomplished properly would eventually lead to product failure due to the following steps not accomplishing their goals of interconnection attachment. Drill speed and feed played a part, as did the number of times that a drill should be used. Drill wear was critical, and coupons used in the panel allowed the manufacturer to check the quality of the hole by drilling the coupon the first time the drill was used and then the last time.

Drills get hot, and as they remove material from the multilayer hole structure they are cutting through copper, glass reinforcement and resin; usually some form of epoxy. The heat of the drill causes some of the epoxy resin to be transferred to some of the copper at the ring intended to be connection to the barrel of the plated through hole. The term "Resin Smear" became a key element of the drilling process. How much was there? And how could it be removed before the plating was added to the inside of the hole? Some companies claimed that their drilling profiles did not produce any resin smear. Others used some chemistry or plasma gas etching to remove the excess resin in the hole. The National Security Administration (NSA) insisted that the holes needed to be etch-backed with sulfuric acid in order to remove some resin and glass and provide a three corner lock of the plated through hole as shown in Figure 1. And the reliability debate raged on for ten years.

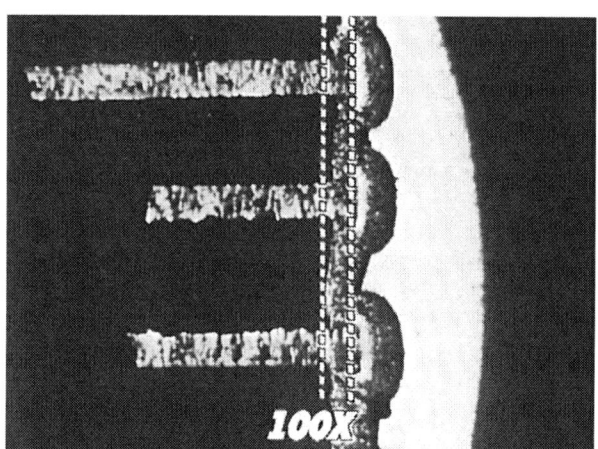

Figure 1 Plated-thru hole with etchback

The discussions on reliability took place in the multilayer committee of IPC. The only way to settle the issues was to perform a round robin test. The committee wrote a test plan, companies volunteered to build samples using and identifying their process, while others volunteered to stress the samples and prepare the final report. After Five Multilayer Round Robins what did the industry learn about hole de-smear and plating; etchback and non etchback? It appeared that if the process was good and there was no resin smear, or it was properly removed that the board circuitry would survive in the field. The effort was painful for the industry in order to establish plated thru hole quality requirements and how they related to reliability. Different plating cracks were identified and whether they promulgated or not. And techniques of using coupons on the panel were established as a measure of the quality of the plated through holes.

Figure 2 shows an example of a page from Multilayer Round Robin V which highlights the differences that were found in various samples. Not all the products submitted were acceptable and even then since the requirements for Commercial, Military and Hi-Reliability were different it became obvious that product would perform sufficiently in different environment due to the different stresses that were imposed by the environment. One needs to remember that the environment provides stress, but so does the processing of the product. As the industry learned over the years the products first sees the stresses imposed by the assembly process

conditions. Some of these include the repair or modifications. Second are the environmental conditions imposed by the use the of the product such as a board mounted to the engine block of an automobile, and finally the normal component and circuit wear-out due to the on/off, hot/cold differentials. Much of this was learned as the industry did other round robins on small hole, or stress sequence evaluations.

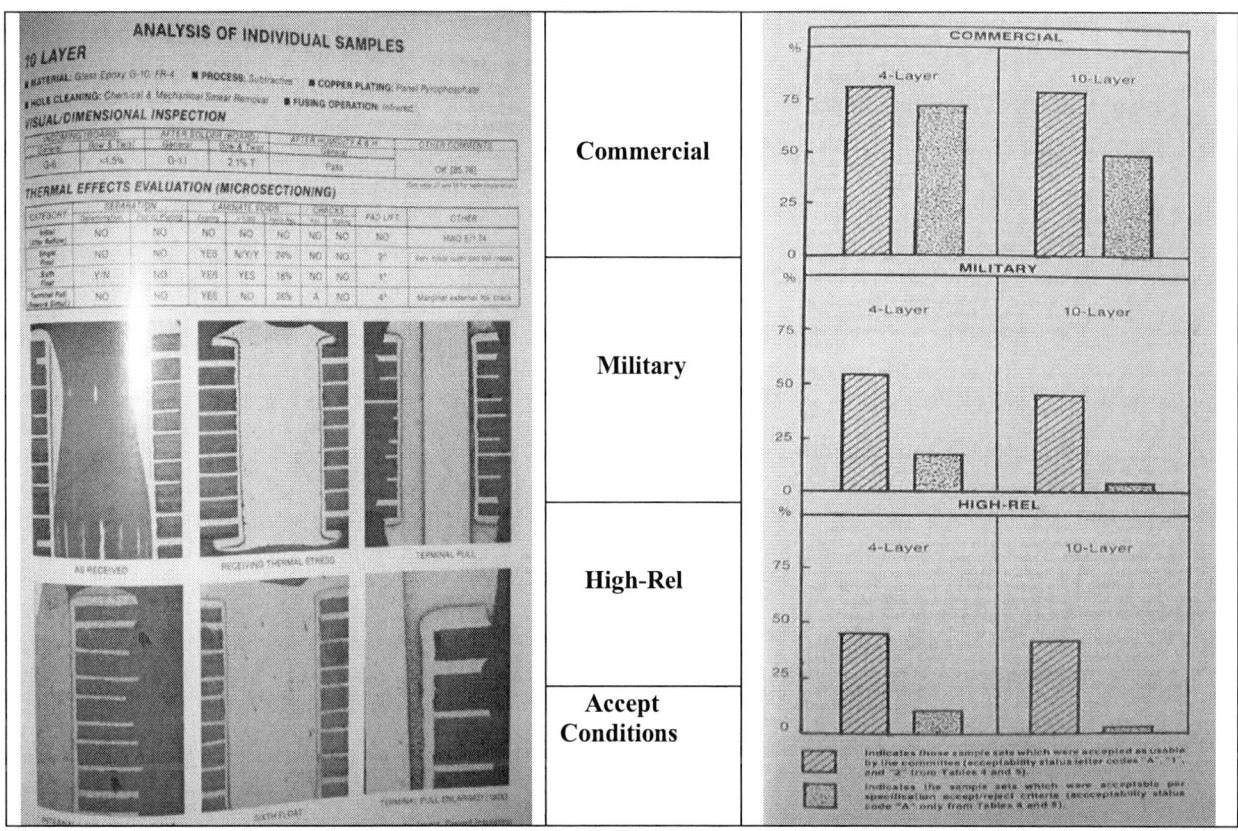

Figure 2 Sample pages from Multilayer RR V for Ten and Four layer Board Samples

It sometimes takes a long time for the industry to agree on a particular aspect when something unusual takes place. Just as all the work of the Multilayer Round robins was coming to closure during the coupon preparation someone noticed some dark spots near the inside of the plating of the hole. Identified as a sulfination void these dark spots were nicknamed "Resin Recession". It was explained that when the sample was floated in hot solder some of the resin which was not completely cured receded from the hole wall. When asked how much could be tolerated the military said none; while the industry said 100%. After 2 hours of discussion a compromise was reached that 40% resin recession was acceptable; while 41% and above was a reject. For the next five year the industry threw away panels where the coupons exhibited more the 40% resin recession.

At the end of the multilayer round robins it was decided to see what happened to the holes with examples of the black dots. Samples were solder-floated once, twice, three, four five and six times with no extra damage and all circuits still

connected. Finally the "thermal zone" was established in the standards as shown in Figure 3. This zone is used to examine the plating integrity and no longer evaluates the resin. The industry continues to be paranoid about the reliability of the product as well as the quality used to make sure that the reliability can be achieved.

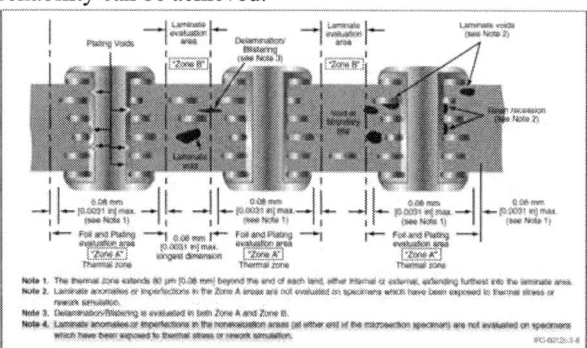

Figure 3 Coupon Examination to Determine Plated Through-hole Quality

SURFACE MOUNTING

With the move to surface mounting the industry faced new challenges on reliability. During that era many wondered if attaching parts to the surface of the conductive pattern would survive the stresses that an assembly would see. Not only was there concern about the solder acting as the only mechanical securing mechanism, there was also the possibility that the land pattern on the printed circuit board would tear away. Many tests were in existence that had been developed to assure copper foil adhesion. Some of these were peel strength while others dealt with hole plating pull-away. Never-the-less it was important to determine if the solder joint itself had sufficient strength to sustain the vibrational elements or bending moment of the mounting substrate that would stress the solder joint mechanical strength.

In the late 1980s four engineers from AT&T published a paper entitled "Surface Mount Solder Attachment Reliability Figures of Merit – "Design for Reliability Tools". This paper was presented as a part of the SMART conferences in January of 1989. By that time the engineers at AT&T had perfected the equations for determining the methodology for a good DFM model. They had proven the concepts by testing varies AT&T products to determine the possible failure mechanism based on the number of thermal cycles that the product would experience. Their concept consisted of four basic Figure of Merit (DF) equations starting with different components, whether they were leaded or leadless, and their size and expansion rate. The results were factored into the second equation which considered the mounting substrate and its expansion rate in the X & Y axis. Equation number 3 considered the upper and lower limits of the thermal characteristics to which the design would be exposed. The final equation into which the first three were factored considered the life cycle expectancy of the assembly. Essentially the concept considered:

- Determine the FM (comp) for all components in the design (these should be in the component catalog)
- Determine the FM(assy) for all components in the design based on the choice of substrate
- Determine the FM(env) for all components in the design based on the system thermal environment and component power dissipation
- Determine the FM(use) for all components in the design based on the expected product cycle life and the allowable failure probability
- Determine the FM(use) for the design from the lowest FM(use) for any of the components in the design provided the allowable failure probabilities were chosen to reflect the number of components and component mix in the design.

The final step in all the equations is the analysis that makes the determination as to whether the design meets reliability requirements. If the FM(use) was equal to 1 or greater the design was adequate; if the result of the four equations was 0.7 or less the design did not meet reliability requirements. Anything in between required some rethinking or contacting a reliability expert. The ideas of the AT&T work were discussed further by the SMT Accelerated Reliability Test Task Group and the focused on providing the industry with all the tools needed in order to establish good product that was reliable and by November 1992 they published the IPC-SM-785 "Guidelines for Accelerated Reliability Testing of Surface Mount Solder Attachments.

The work of this committee did more than just publish a great standard they also provided the industry with a snap shoot of nine end-use environments as shown in Table 1. The concepts identified the minimum and maximum temperature as well as the average range between the upper and lower limits in any use case; the number of hours between on/off that the product would operate each day and the number of cycles each year. Typical years of service and approximate failure risk allowable by the end-use customer were also identified.

The standard not only describes the characteristics of the nine use environments a recommendation is made for each condition as to how to determine the accelerated testing need to prove that the product will survive the expected end life conditions. The information from Table 1 is combined with some hypothetical example use conditions and used in some of the equations on solder fatigue a determination is made with some equivalent mean cyclic lives for the accelerated test conditions for both leadless and leaded surface mount solder attachments. These equivalent test cycles are determined for the range of Years of Service and the Acceptable Failure Risks in Table 1, but are the expected mean cycles to failure for the test conditions.

Table 1 End-use Environments Related to Solder Joint Wear-out and Fatigue

Worst-case use environment							
Use category	Tmin °C	Tmax °C	ΔT°C	t_D hrs	Cycles/ year	Typical years of service	Approx. accept. failure risk %
1) Consumer	0	+60	35	12	365	1-3	1
2) Computers	+15	+60	20	2	1 460	5	0.1
3) Telecom	-40	+85	35	12	365	7-20	0.01
4) Commercial aircraft	-55	+95	20	12	365	20	0,001
5) Industrial & automotive Passenger Compartment	-55	+95	20 &40 &60 &80	12 12 12 12	185 100 60 20	10	0.1
6)Military Ground & ship	-55	+95	40 &60	12 12	100 265	10	0.1
7) Space leo geo	-55	+95	3 to 100	1 12	8 760 365	5-30	0.001
8)Military avionics a b c	-55	+95	40 60 80 &20	2 2 2 1	365 365 365 365	10	0.01
9)Automotive under hood	-55	+125	60 &100 &140	1 1 2	1 000 300 40	5	0.1

The results show that for the more benign use conditions (Use Categories 1 through 6), the test regimes provide high acceleration factors; for the more severe use conditions, the test accelerations diminish or disappear entirely. This is the result of a number of reasons, such as severe thermal use conditions, long service lives, and low tolerances for failure acting singly or in concert. This reflects the fact that for these conditions the reliability of the products cannot be experimentally verified; reliability assurance has to depend solely on analytical reliability modeling.

Several additional technical papers have been published with the idea expressed that the strength of the solder joint and assembly was better than anticipated. No revisions have been made, or are contemplated as of this date by IPC, and the models that have been created by the IPC-SM-785 can still be used as a design guide for determining the methodology for testing the product to determine life cycle conditions.

When some of the industry was forced to move to new solder alloys that did not contain "Lead" the rules changed somewhat. New models were developed for the lead-free solders and some were very compatible with the existing work done by the SMT Accelerated Reliability Test Task Group, yet the new components being fostered on the industry require a new analysis. This is especially important with those components that are leadless and referred to as bottom termination components BTCs as shown in Figure 4. Some assemblers have forgotten that you needed solder volume under the part.

product, the industry is now made up of a variety of partnerships with the biggest concern being that of Supply Chain Management. With that situation comes the fact that the designs have become much more complex. The traditional multilayer board has been replaced by HDI products that may consist of many steps in the manufacturing process to laminate a re-laminate the completed structure. Examples of descriptions abound such as 3-6-3 or 2-8-2 each describing a 12 circuit layer board however the manufacturing sequence and the occurrences of vias that run from layer to layer may be very different. See Figure 5 for various examples of HDI constructions.

Figure 4 Bottom Termination Component examples

THE NEW RELIABILITY CHALLENGES

The world has changed to the point where most manufacturing is outsourced. Where at one time the electronic industry was made up of vertically integrated OEMs who designed, built, assembled and tested the

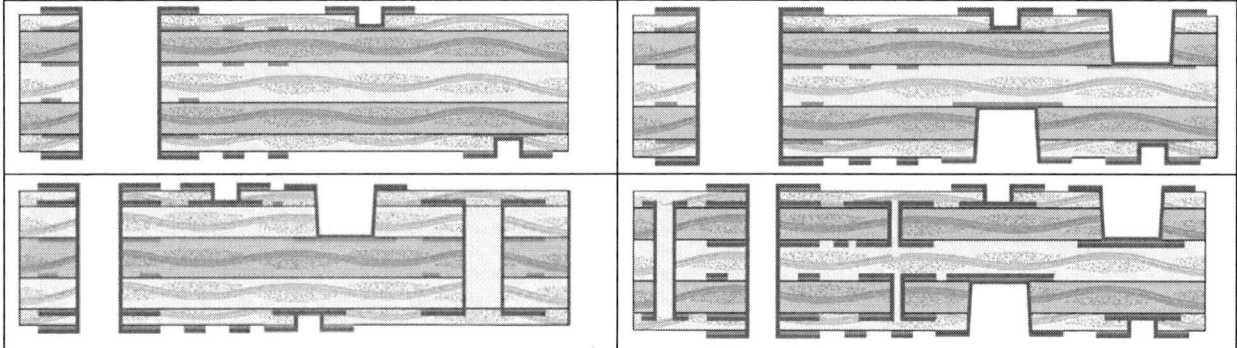

Figure 5 Examples of 4 different HDI products each requiring different process steps

With most of the complex printed board being outsourced the OEMs are struggling in order to determine the manner in which to convey the design concepts and the manufacturing requirements. In many instances the product will be described in the documentation package. This may consist of hard copy or electronic media. The manufacturing instructions are usually conveyed as Notes and in many instances reference industry specifications and the requirements that the customer wants delivered as part of his final product. These are the established requirements, but they do not guarantee the reliability of the product. The details establish the quality level required which is also predicated on the end-product use conditions.

Various performance classes have been developed by the industry and they are intended to convey the quality needed by the customer. The three classes established by the IPC are:

- *Class 1 General Electronic Products* Includes consumer products, some computer and computer peripherals, as well as general military hardware suitable for applications where cosmetic

imperfections are not important and the major requirement is function of the completed printed board or printed board assembly.

- *Class 2 Dedicated Service Electronic Products* Includes communications equipment, sophisticated business machines, instruments and military equipment where high performance and extended life is required, and for which uninterrupted service is desired but is not critical. Certain cosmetic imperfections are allowed.

- *Class 3 High Reliability Electronic Products* Includes the equipment for commercial and military products where continued performance or performance on demand is critical. Equipment downtime cannot be tolerated, and must function when required such as for life support items, or critical weapons systems. Printed boards and printed board assemblies in this class are suitable for applications where high levels of assurance are required and service is essential.

The classes set the requirements for the quality of the final product; however they are not reliability descriptions. It is understood that without the quality the product can never achieve the reliability goals. Thus one should consider:

- **Quality** is the ability to produce the product in the manner specified by the customer in the documentation package provided, including any test or legal requirements. The concept is one of meeting the requirements. The OEMs define the printed board quality requirements necessary to meet their reliability needs.

- **Reliability** is the ability to function as expected under the expected operating conditions for an expected time period without exceeding expected failure levels. It is **Proof of Performance.** The End item reliability can only be determined by the OEM. PWB Fabricators often have little or no visibility to end item requirements.

The decisions being made for the arrangement of conductive layers in a Multilayer board are becoming more crucial every day. Not only is cost a player, but also availability of material in the thickness that various analyses would recommend. Add to these considerations the fact that traditionally board manufacturers have developed their own suite of preferred material; both thickness and suppliers. The issues are many on what a design requires electrically and what compromises can be tolerated in order to still get a **reliable working product**.

OEM products must survive 2 primary environments:
1) Product Assembly (Reflow/wave & rework)
2) Field Service (thermal cycles & shock/vibration)

Traditional PWB Quality Requirements are primarily measurements used for PWB fabrication process validation but have limited use for determining reliability. The modern reliability challenges are many. Components are much smaller than they were a few years ago. Components are placed more densely on the printed boards. The printed boards go through more severe reflow processes, and often multiple times. Lead-Free solder increases processing temperatures.

As a result printed board designs have changed where the board features are very small, with high aspect ratio vias (old designs seldom higher than 5:1, today can be 10:1) and there are many more vias per printed board than in the past. In addition laminate materials must be more robust as assembly temperatures are higher and the materials must have low z-axis expansion for greater via life.

Traditional printed board quality measurements are still necessary to collect the data necessary to keep processes in control. This is because features are smaller they are susceptible to process variation that might affect their reliability. It should be understood that checking 3 or even 12

holes per panel for quality does not provide a sufficient screen for defects. High volume thermal stress tests can sample many vias and pinpoint weaknesses that are used for process improvement.

Currently some common High Volume (large via count) Stress Tests are used by the industry as the measure of assessing reliability. These are identified as Interconnect Stress Test (IST) and Highly Accelerated Thermal Shock (Hats).

Interconnect Stress Test (IST) is an accelerated stress test method used to evaluate the integrity of the Printed Circuit Board (PCB) interconnect structure. IST creates a thermal cycle that stresses a specifically designed coupon, while simultaneously monitoring the electrical integrity of plated through holes (PTHs) and internal interconnects (Posts). It's a test method that measures the integrity of different areas of the same structure. IST tests the PTHs and the Posts at the same time.

Highly Accelerated Thermal Stress (HATS) was developed to emulate traditional air-to-air test methods, while significantly reducing the drawbacks of traditional methods. The test uses a single chamber in which high volume hot and cold air pass stationary samples. The high volume air flow provides rapid thermal transfer to the device under test and reduces the time for the specimens to reach temperature equilibrium. This greatly reduces the time required for each cycle, and the stationary samples are easily fixtured to a high-speed precision resistance sampling network

How do these tests relate to reliability is a tricky question that have caused a lot of soul searching by the OEMs. There is no easy answer. Some OEMs have been able to correlate IST or HATS test results to via life, others have not. Is an answer necessary? An answer is good to help establish OEM design rules but, If a PWB can be fabricated to survive, for example, 400 accelerated thermal cycles (IST or HATS) and proves reliable in OEM Product tests, then the 400 cycle limit can be used as a measure of process consistency that can assure PWB performance from lot to lot. It is a measure of the process capability as it relates to reliability failure modes and may not directly relate to final product reliability. It is a measure of the Printed board's structure Durability.

End-Use Matrix for Printed Boards
Building on the concept used to determine reliability for surface mount solder joints the end use environments are used to characterize the differences needed for checking reliability of different products. The printed board structures have been segmented into five categories of mounting products. These were:

- *Interposer:* from 5 to 50 mm L x W aspect ratio equals 1:1.5
- *Module:* from 10 to 75 mm L x W aspect ratio equals 1:3

173

- *Portable Board:* from 20 to 150 mm L x W aspect ratio equals 2:4
- *Product Board:* from 40 to 400 mm L x W aspect ratio equals 3:4
- *Backplane*: from 100 to 800 mm L x W aspect ratio equals 3:6

The Interposer is used as a part of the semiconductor package. The Module is a small board that can be used many times and is mounted on a Product Board or Backplane. The Portable Board was established to accommodate those hand held products that needed finer features and greater precision. The Product Board is the mounting substrate used in most pieces of table top equipment, while the Backplane is the foundation of many systems that use the concept of daughter boards being inserted in the backplane unit.

The IPC Technology roadmap for the year 2011 used emulators to represent the products that provided an insight into the technology drivers for design, boards, assembly, and production sources. There were 14 emulators that represented the five mounting structures. Figure 6 show an example of the size variations of some of the products that were described in the roadmap which indicated their present characteristics and where they would be in ten years.

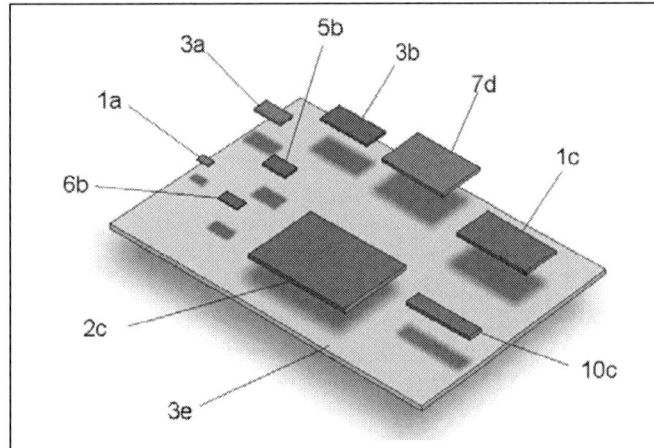

1a	Consumer Interposer
3a	Telecomm Interposer
3b	Telecomm Module
5b	Automotive Passenger Module
6b	Military ground /shipboard Module
1c	Consumer Portable Board
2c	Computer Tablet Portable Board
10c	Medical Portable Board
3d	Telecomm Product Board
5d	Industrial Product Board
7d	Space Shuttle Product Board
9d	Under Hood Product Board
3e	Telecom Backplane
8e	Military ATR Backplane

Figure 6 Emulator Board Size Comparisons

In order to establish some consistency between the end-use environments of the assemblies the same nine use conditions were identified as being representative of the industries uses of electronics. One slight change however became necessary since the industry has identified a specific need for electronics that are uses in medical equipment. As a result a tenth row was added to the matrix used to describe the reliability stress exposures that would be recommended for those products. The resulting matrix essentially consists of 50 cells that identify the needs of the 5 mounting platforms being used in the 10 end-use environments.

There is no doubt that just as in the 1970s the plated through hole was the weakest link, as the mounting structures have become more complex the stress simulation intended to establish the robustness of the interconnections needs to also match the environment. Cyclic stresses have always been used to establish the wear-out capability, so it is no surprise that they are used in today's methods of determining robustness. The differences between thermal extremes is important since it stresses the organic laminate material which when they expand add a stress to the interconnection barrel of the holes. Hole size and aspect ratio (hole length to hole diameter) also play a vital role in making the determination. The time that the stress is applied is a function of the use environment as are the extremes of temperature, humidity, or voltage surges that might be intended to simulate the field conditions.

Table 2 was developed to represent the testing of the mounting structure. It is only one example of what some OEMs feel is necessary in order to establish the robustness of the mounting structure. The OEM approach is to test the board material at 150°C, while the microvias, especially those that are stacked on one another, should be exposed to a cyclic condition of 190°C. The number of cycles is intended to establish the robustness of the product in each particular cell.

Table 2 Printed Board Construction to end-use Reliability Test Matrix

Product Application per end use					
End-use Environment	A-Interposer	B-Module	C-Portable	D-Product	E-Back Plane
1-Consumer	100 cycles @ 150	100 cycles @ 150	100 cycles @ 150	100 cycles @ 150	100 cycles @ 150
2-Computers and Peripherals	100 cycles @ 150	100 cycles @ 150	100 cycles @ 150	100 cycles @ 150	100 cycles @ 150
3-Telecomm	250 cycles @ 150	250 cycles @ 150	250 cycles @ 150	250 cycles @ 150	250 cycles @ 150
4-Commercial Aircraft	350 cycles @ 150	350 cycles @ 150	350 cycles @ 150	350 cycles @ 150	350 cycles @ 150
5-Industrial and Automotive Passenger Compartment	500 cycles @ 150	500 cycles @ 150	500 cycles @ 150	500 cycles @ 150	500 cycles @ 150
6-Military (ground and shipboard)	500 cycles @ 150	500 cycles @ 150	500 cycles @ 150	500 cycles @ 150	500 cycles @ 150
7-Space	1400 cycles @ 150	1400 cycles @ 150	1400 cycles @ 150	1400 cycles @ 150	1400 cycles @ 150
8-Military Aircraft	500 cycles @ 150	500 cycles @ 150	500 cycles @ 150	500 cycles @ 150	500 cycles @ 150
9-Automotive (under hood)	500 cycles @ 150	500 cycles @ 150	500 cycles @ 150	500 cycles @ 150	500 cycles @ 150
10- Bio Medical & Life support	500 cycles @ 150	500 cycles @ 150	500 cycles @ 150	500 cycles @ 150	500 cycles @ 150

Since the advent of using lead-free solder in the assembly operation many OEMs now require some form of assessment that the mounting structure will pass the exposure. Most printed board assemblies are exposed to the reflow oven profiles needed during the attachment of components to one or both sides of the substrate. Since the components have a variety of different features in their profile as well as the mounting terminations these exposures may vary to several iterations. Add to this that after testing some components may need to be replaced. The OEM wants to be sure that after all the assembly work is completed that the mounting structures' integrity is still intact.

Table 3 is an example of the 50 cell matrix used to identify the number of cycles that a mounting structure might be exposed to during the assembly operations. They are usually worse case conditions and are intended to emulate the laminate characteristics that identify their Td (Temperature at Decomposition). The Td was a new requirement imposed on laminate descriptions that were intended to be used with lead-free assembly process profiles. The IPC-4101C defined several new formulations that were identified as being lead-free capable and therefore provided the Td as well as other pertinent electrical and physical properties.

These thoughts were considered when a group of reliability committee members were asked to develop the ranges identified in Table 3.

Table 3 Printed Board Assembly Simulation to end-use Reliability Test Matrix

Product Application per end use					
End-use Environment	A-Interposer	B-Module	C-Portable	D-Product	E-Back Plane
1-Consumer	6X260°C	6X260°C	6X260°C	6X260°C	6X260°C
2-Computers and Peripherals	6X260°C	6X260°C	6X260°C	6X260°C	6X260°C
3-Telecomm	6X260°C	6X260°C	6X260°C	6X260°C	6X260°C
4-Commercial Aircraft	6X260°C	6X260°C	6X260°C	6X260°C	6X260°C
5-Industrial and Automotive Passenger Compartment	6X260°C	6X260°C	6X260°C	6X260°C	6X260°C
6-Military (ground and shipboard)	6X230°C	6X230°C	6X230°C	6X230°C	6X230°C
7-Space	6X230°C	6X230°C	6X230°C	6X230°C	6X230°C
8-Military Aircraft	6X230°C	6X230°C	6X230°C	6X230°C	6X230°C
9-Automotive (under hood)	6X260°C	6X260°C	6X260°C	6X260°C	6X260°C
10- Bio Medical & Life support	6X230°C	6X230°C	6X230°C	6X230°C	6X230°C

CONCLUSIONS

The aspects of quality and reliability are the concern of every member of the supply chain. Deviations from the indented requirements, as documented by the OEM, are never a reason for acceptance as they filter back to the concept of dissatisfied customers all along the supply chain; most of all the end-use customer. The industry and individual specifications identify the requirements for quality. These have been, and will continue to be, identified as:

- Visual Description
- Dimensional descriptions
- Interconnection Integrity (Microsection)
- Continuity/In-circuit test
- Customer Specific

With the need to establish a method of working with the members of the supply chain the OEMs need a new methodology to establish the fact that the new supplier, with slightly different materials and processes, can produce a product that is identical to that made during the prototype stages.

The term Process Robustness was coined by several OEMs to represent the test methods usually reserved for reliability evaluations now required of a new manufacturer. It is very similar to what the military once coined "First Article Inspection" and it may be for some contractual requirements that these issues will need to be revisited. The methods that would be used to establish "Durability" are:

- HATS test requirements
- IST test requirements
- Solder Float exposure
- Solder reflow simulation

These will become reliability requirements to ascertain product robustness.

ALTERNATIVES TO SOLDER IN INTERCONNECT, PACKAGING, AND ASSEMBLY

Herbert J. Neuhaus, Ph.D., and Charles E. Bauer, Ph.D.
TechLead Corporation
Portland, OR, USA
herb.neuhaus@techleadcorp.com

ABSTRACT

Solder plays a special role in the world of electronics manufacturing as evidenced by the disruptive nature of the lead-free movement. The intense search for attractive lead-free solders reveals the preeminent importance of solder to the industry. In fact, solder consumes so much attention that solder-less alternatives are often overlooked.

Material-based alternatives to solder include conductive adhesives and transient-phase compounds. Developments in nanotechnology spawned a virtual renaissance in conductive adhesives and other solder-less joining materials.

As a complement to the solder-less materials developments, embedded assemblies use conventional materials in novel ways to improve performance by cutting interconnect parasitics and increase reliability gains by eliminating wire-bonds and solder-bumps. Freescale, Imbera, GE, Verdant, and many others develop and employ diverse approaches to embedding active devices.

Particle Interconnect represents another solder alternative. While originally developed for automated test, particle interconnect holds considerable promise in a variety of applications including LED assembly and printed electronics.

This presentation surveys the landscape of alternatives to solder in interconnect, packaging, and assembly. Next, the presentation treats practical implementation challenges such as yield management strategies and supply chain restructuring. Finally, the presentation concludes with a discussion of scenarios in which solder alternatives offer highly compelling business and technical benefits.

Key words: lead-free, conductive adhesives, nano-scale fillers, nanotech fillers, TLPS, particle interconnect, embedded packaging.

INTRODUCTION

Solder embodies the chief method for attaching components to a printed wiring board (PWB) during the manufacturing of electronic assemblies. Long the primary choice for assembling electronics, eutectic tin-lead (SnPb) solder exhibits attractive reflow properties, low melting point, and ductility. Lead, however, suffers from increasing regulatory scrutiny due to its relatively high toxicity to human health and the environment. In 2001, the European Union (EU)

proposed the Waste Electronics and Electronic Equipment (WEEE) and the associated Restriction of Hazardous Substances (ROHS) directives that ban the use of lead in electronics devices sold in the EU beginning in July 2006.

The legislation and market trends leading toward the implementation of lead-free electronic assemblies raised several issues including the need to increase the thermal tolerances of electronic components. Lead-free solder alloys such as tin-silver-copper (SnAgCu) with a melting point of 217°C, require higher processing temperatures than traditional tin-lead (SnPb) alloys thereby reducing the process window and focusing on the need for rigorous control of the thermal process during soldering. Raising component thermal tolerances places a significant economic burden on electronics manufacturers since they face higher overall costs.

MATERIALS-BASED ALTERNATIVES
Conventional Conductive Adhesives

Electrically conductive adhesives (ECA) provide potential solder replacements in microelectronics assemblies. Two types of ECAs exist: isotropic conductive adhesive (ICA) and anisotropic conductive adhesive (ACA). Although ICAs and ACA employ different conduction mechanisms, both materials consist of a polymer matrix containing conductive fillers. ICAs conduct in all directions and typically contain conductive filler concentrations between 20 and 35% by volume. Hybrid applications and surface mount technology primarily utilize ICAs. In ACAs, electrical conduction occurs only in the direction of applied compression during curing. Typical ACA conductive filler volume loading ranges between 5 and 10%. ACA technology suits fine pitch technology especially flat panel display applications, flip chips and fine pitch surface mount devices [1].

Electrically conductive adhesives consist of a polymer binder that provides mechanical strength and conductive fillers, which offer electrical conduction. Polymers include both thermosets (such as epoxies, polyimides, silicones and acrylic adhesives) and thermoplastics. ECA conductive fillers consist of metallic materials such as gold, silver, copper, and nickel or nonmetallic materials such as carbon.

Compared to conventional solder interconnection technology, the advantages of conductive adhesives include:

- More environmental friendly than lead-based solder;

- Lower processing temperature requirements;
- Finer pitch capability (ACAs);
- Higher flexibility and greater fatigue resistance than solder;
- Simpler processing (no flux cleaning);
- Compatible with inexpensive and non-solderable substrates (e.g., glass).

Despite the advantages of ECA technology, lower electrical conductivity than solder, poor impact resistance, and long-term electrical and mechanical stability concerns limit widespread adoption of ECAs by the electronics industry.

Figure 1 illustrates a recent application of ACA in film form (ACF) for fine pitch chip on film (COF) assembly. In this scenario, the high probability of electrical shorts between adjacent bumps, due to the accumulation of conductive particles between the bump during the bonding process led to the development of a triple-layered ACF with functional layers on both sides of conventional ACF layer to improve interface adhesion and control bonding property for fine pitch application during thermo-compression bonding as shown in Figure 1 [2].

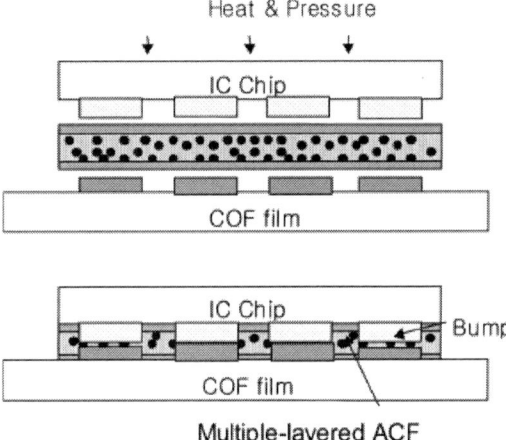

Figure 1. COF bonding process using triple-layered ACFs.

The joining of the driver ICs in tape-carrier packages (TCP) to LCD glass panels and other interconnection areas for flat panel display manufacturing comprise the most common use of ACFs. Figure 2 shows various packaging technologies using ACF for LCD modules; TCP, COG and COF bonding [3]

Nano Enhanced Conductive Adhesives
Developments in nanotechnology spawned a virtual renaissance in conductive adhesives and other solder-less joining materials. Nanotech enhances the performance of conductive adhesives via three distinct routes:

- Cost (reduced precious metal loading);
- Conductivity (improved filler packing and sintering);
- Reactivity (vast surface area).

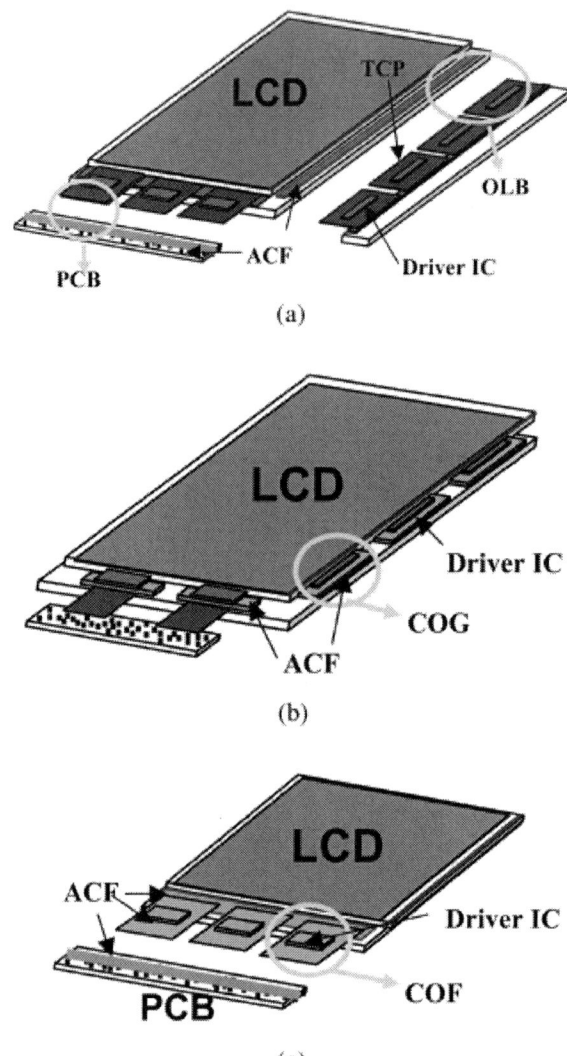

Figure 2. Various packaging technologies using ACF in LCD modules (a) TCP; Outer Lead Bonding (OLB) and PCB bonding, (b) COG bonding and (c) COF bonding.

Contact resistance between filler particles limits the conductivity of conventional conductive adhesives. The introduction of nano-scale fillers increases electrical conductivity by a combination of mechanisms, including more efficient filler packing and facile sintering of filler particles into high conductivity networks. The tendency for nanoparticles to agglomerate makes effective and uniform dispersion critical to practical nanoparticle conductive adhesives.

Researchers at Endicott Interconnect Technologies demonstrated the use of Cu, Ag, or low melting point alloy nanoparticle filled adhesives during lamination to enhance wiring density of organic laminate based electronic packages and circuit boards as shown in Figure 3.

178

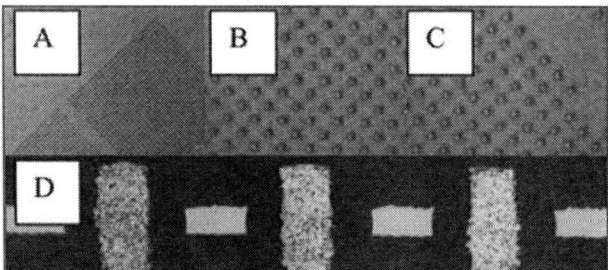

Figure 3. Optical photographs of adhesive filled joining core (A-C) top view, and (D) cross-section [4].

Frequently, silver flakes provide the conductive network in ICA adhesives. The overall resistance of the network consists of resistance of flakes and resistance of contacts between flakes. The addition of nano-scale particles enables the formation of additional bridges between flakes, which may increases the density of the conductive network connections and decreases overall resistance [5]. Figure 4 illustrates this concept.

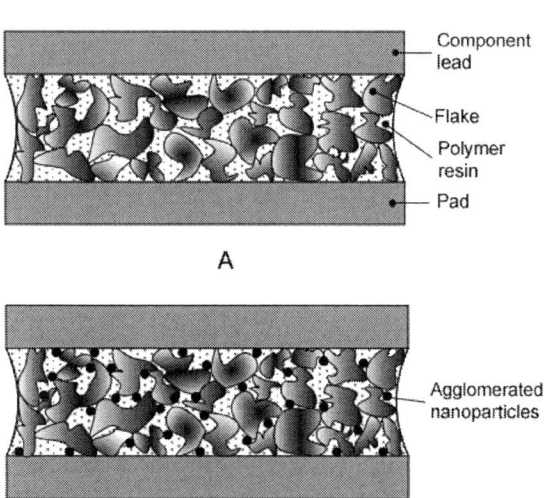

Figure 4. Adhesive joint without nanoparticles (A), and with added nanoparticles (B).

Transient-Phase Compounds

Transient liquid phase sintering conductive adhesives, with an interpenetrating polymer/metal network, mitigate some of the deficiencies of standard particle-filled conductive adhesives. A process known as transient liquid phase sintering (TLPS) forms a metal network in situ reinforced with the polymer matrix. Bulk and interface metallurgical electrical connections provide stable electrical and thermal conduction. The TLPS conductive adhesives utilize conventional surface mount technology dispensing and processing equipment. Electrical conductivity results also indicate values closer to those of traditional solder alloys. Reliability testing including humidity followed by air-to-air thermal shock (-55°C to +125°C) demonstrate that this type of adhesive performs substantially better than standard, passive filler loaded conductive adhesives.

Figure 5 schematically illustrates a transient phase compound developed by Ormet Circuits which begins as copper and alloy particles in a liquid organic formulation and sinters into an interpenetrating metal/polymer network. Figure 6 depicts an application of the material from Ormet Circuits as a die attach. [6]

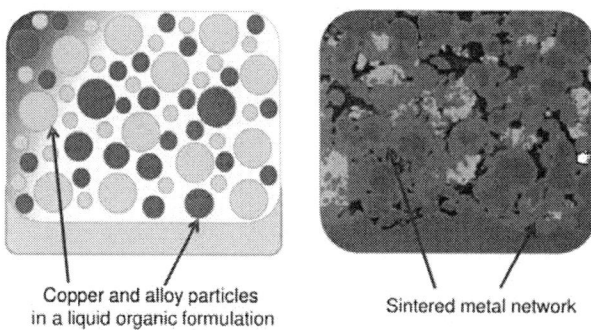

Figure 5. Ormet's transient liquid phase sintering material.

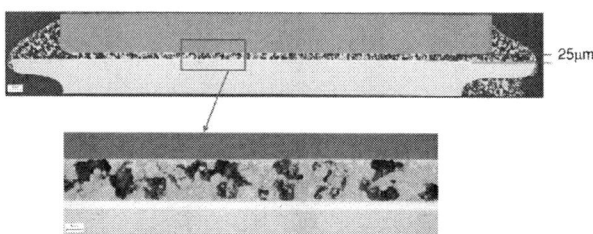

Figure 6. Application of Ormet material to die attach.

STRUCTURE-BASED ATERNATIVES

As a complement to the solder-less materials developments, embedded assemblies use conventional materials in novel ways to improve performance by cutting interconnect parasitics and increase reliability gains by eliminating wire-bonds and solder-bumps.

Freescale, Imbera, GE, Verdant, and many others develop and employ diverse approaches to embedding active devices.

Freescale – Redistributed Chip Package (RCP)

Freescale's RCP package employs thin-film build-up directly on devices embedded in molding compound, eliminating the need for wire-bonds and flip-chip bumping. Figure 7 depicts the RCP package.

RCP targets System-in-Package (SiP) for mobile electronics applications. [7] Electrical performance and the potential for miniaturization increase thanks to the lack of wire-bonds and flip-chip bumps. RCP and similar embedded chip package technologies disrupt the standard packaging food chain by moving bare die into interconnect fabrication. Freescale shipped limited products that utilize RCP starting in 2008.

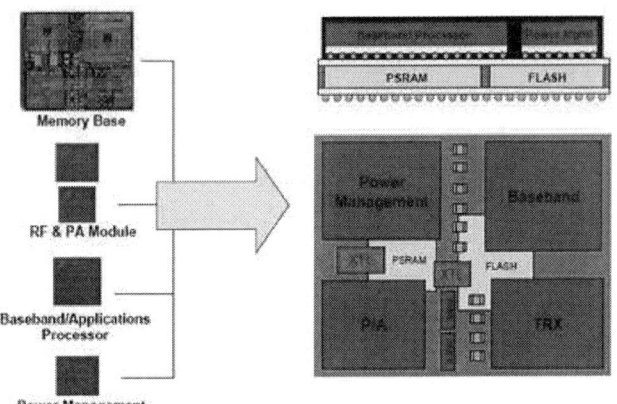

Figure 7. Freescale RCP.

Imbera – Integrated Module Board (IMB)

Imbera's Integrated Module Board (IMB) integrates active and passive devices in organic boards by laminating the bare die within conventional printed circuit board layups. After pressing, micro-vias and through-holes are drilled and plated. Chip connections include Cu/Cu or Cu/Au with no intermetallic compounds. Figure 8 shows the Imbera package.

Targeted applications include SiPs and System-in-Board for advanced consumer electronics. Based in conventional printed circuit board fabrication technology, IMB demonstrate moderate to low electrical performance and size reduction, but expects to offer strong cost advantages and good infrastructure compatibility.

GE – Embedded Chip Build-Up (ECBU)

The GE Embedded Chip Build-Up (ECBU) technology utilizes bare and pre-metallized flexible dielectrics to form thin-film build-up directly on devices without wire-bonds and flip-chip bumping. Figure 9 depicts the ECBU package [8].

GE promoted the ECBU technology in applications including microprocessors, video processors and ASICs with demanding interconnect and thermal requirements. The lack of wire-bonds and flip-chip bumps permits excellent electrical performance and very dense wiring capability. ECBU's placement of bare die in the interconnect structure requires some rearrangement of the existing packaging infrastructure and supply chain. Technical evaluations of ECBU currently underway at leading microprocessor and graphics processor suppliers indicate significant performance benefits.

Verdant Electronics – Occam Process

The Verdant Occam process, illustrated in Figure 10, positions pre-tested, burned-in components on an adhesive layer of a temporary or permanent. After encapsulating the components, the adhesive layer is then removed over the component leads mechanically or by laser ablation. Finally, plating the holes and forming traces with a conductive, copper connection provides an interconnection structure.

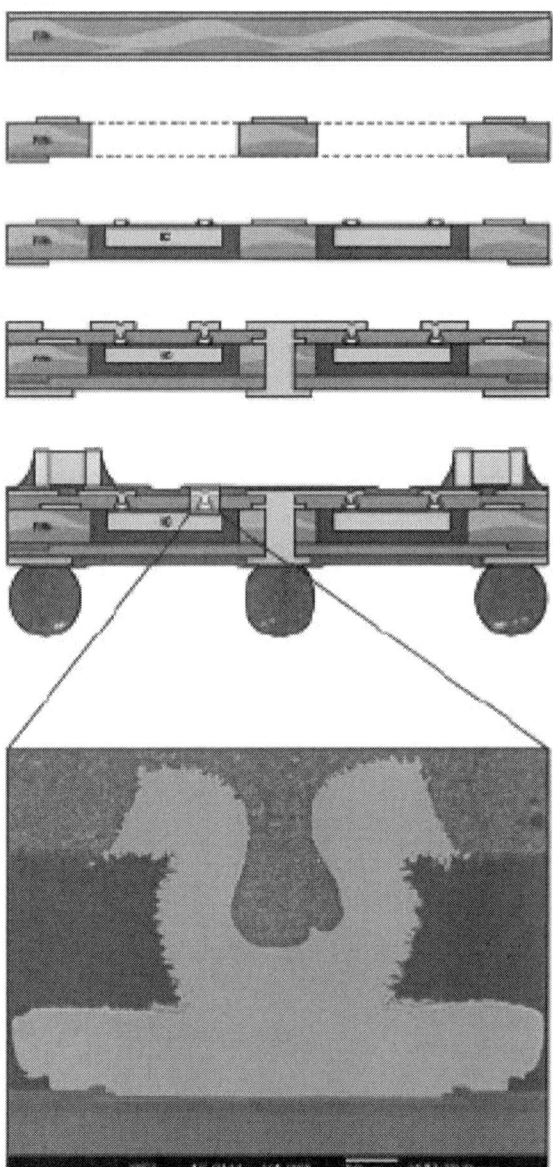

Figure 8. Imbera IMB.

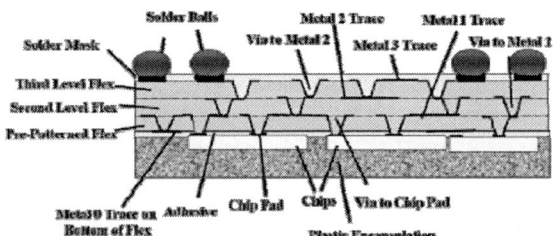

Figure 9. GE ECBU.

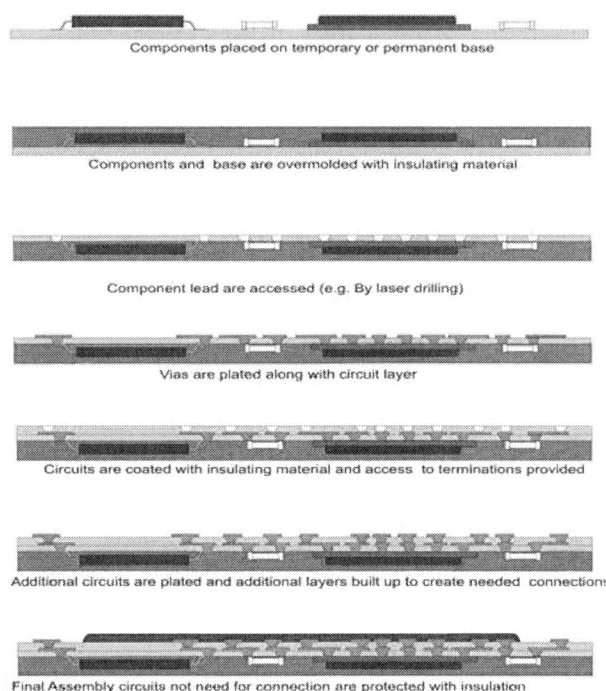

Figure 10. The Verdant Electronics Occam Process.

PROCESS-BASED ALTERNATIVES
Particle Interconnect

Particle interconnect represents another solder alternative. Originally developed for automated test, particle interconnect offers low contact resistance with low damage.

Particle Interconnect provides low contact resistance contact by means of hard and irregularly shaped particles on the bond pad. This particle enhanced surface easily pierces into the mating substrate even with the presence of a nonconductive oxides layer and adhesive on the mating substrate surface. Figure 11 shows a cross-section micrograph of such a piercing connection.

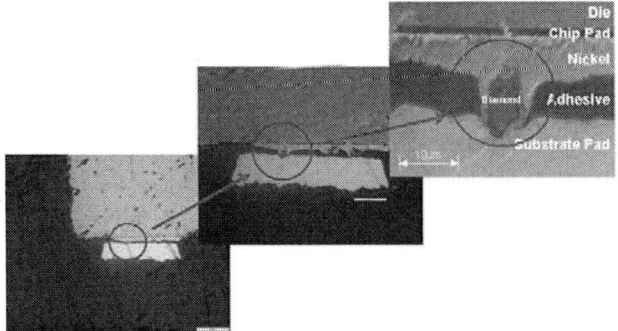

Figure 11. Particle enhanced contact between a Si chip bond pad and a pad on a PWB [9].

The schematic drawing shown in Figure 12 illustrates the principles of an electroless version of the Particle Interconnect process. Particle Interconnect begins surface preparation of Al bond pads on the wafer (cleaning and zincating), followed by a modified electroless nickel-

particle co-deposition. A second electroless nickel-plating is followed by a finally an immersion gold treatment.

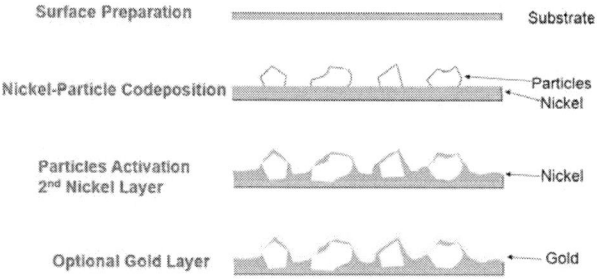

Figure 12. One of several plating-based processes for depositing particle interconnect

The first electroless nickel (EN) plating step utilizes a modified composite electroless nickel plating method to co-deposit nickel and particles onto Al bond pads by mixing hard particles with the EN solution. After the co-deposition, a particle surface activation step ensures adhesion of the metal to exposed particle surfaces during the second nickel plating. The second conventional electroless nickel plating step casts a layer of metal over deposited particles. Figure 13 shows a micrograph of a completed Particle Interconnect surface on Al bond pad.

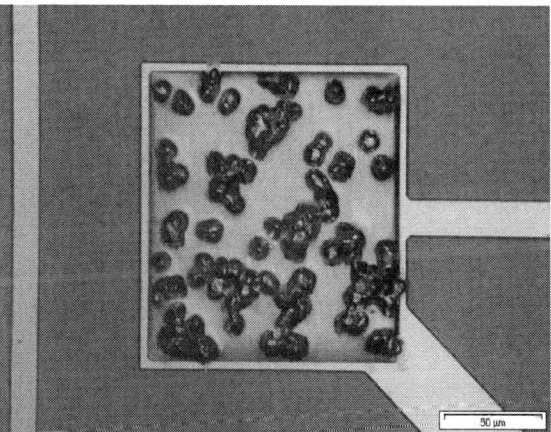

Figure 13. Particle Interconnect on an Al bond pad [9].

Particle Interconnect holds considerable promise as a solder replacement in a variety of applications including LED assembly and printed electronics.

Solder often attaches LED devices to package substrates. However, solder can short LED junctions by wicking up the sides of the device. Evaluations of Particle Interconnect together with a non-conductive adhesive indicate that the Ni-coated diamond particles provide enhanced electrical and thermal conductivity without danger of junction shorting. Figure 14 shows a LED with Particle Interconnect on the bottom face of the device.

Similarly, Particle Interconnect may prove beneficial in thermally sensitive applications such as printed, organic electronics.

181

Figure 14. Particle Interconnect on the bottom of an LED die provides superior thermal and electrical conductivity without danger of shorting the p-n junction.

SUMMARY

The disruptive nature of the effort to replace lead-based solders reveals the central role played by solder in electronics assembly. Lead-free solders require higher temperatures and increase the thermal tolerance requirements of electronic components. As a result, the industry has become increasingly open to solder-less alternatives.

TechLead has identified three broad families of solder alternatives: materials-based, process-based, and structure-based. Each family enjoys renewed interest and new applications.

As with many disruptive technologies, the need for some supply chain restructuring limits adoption. However, TechLead forecasts that demand for performance and reliability will overcome adoption barriers.

REFERENCES

1. K. Gilleo, Assembly with Conductive Adhesives, *Soldering and Surface Mount Technology*, No. 19, 12-17, 1995.
2. M. J. Yim, J. S. Hwang, J. G. Kim, J. Y. Ahn, H. J. Kim, W. S. Kwon, and K. W. Paik, J. Electron. Mater. 33, 76 (2004).
3. M. J. Yim, and K. W. Paik, Electronic Materials Letters, Vol. 2, No. 3 (2006), pp. 183-194.
4. V. R. Markovich, R. N. Das, M. Rowlands and J. Lauffer, Fabrication and Electrical Performance of Z-Axis Interconnections: An Application of Nano-Micro-Filled Conducting Adhesives.
5. Li, Q., ZHANG, J. Effects of Nano Fillers on the Conductivity and Reliability of Isotropic Conductive Adhesives (ICAs). Key Engineering Materials Vols., 353-358, 2007, pp 2789-2882.
6. http://www.ormetcircuits.com.
7. "Circuit device with at least partial packaging and method for forming" US Pat. 6838776 - Filed Apr 18, 2003 - Freescale Semiconductor, Inc. and "Circuit device with at least partial packaging, exposed active surface" US Pat. 6921975 - Filed Apr 18, 2003 - Freescale Semiconductor, Inc.
8. R. Fillion, C. Bauer, "High performance, high power, high I/O COF packaging," 15th European Microelectronics and Packaging Conference & Exhibition, Brugge, Belgium, June 2005.
9. H. J. Neuhaus and M. E. Wernle, "Advances in Materials for Low Cost Flip-Chip." Advancing Microelectronics, 2000, p. 12.

2013 iNEMI TECHNOLOGY ROADMAP OVERVIEW

Bill Bader and Chuck Richardson
International Electronics Manufacturing Initiative (iNEMI)
Herndon, VA, USA
bill.bader@inemi.org, chuck.richardson@inemi.org

ABSTRACT

iNEMI will release its 2013 Technology Roadmap to industry on April 4, 2013. Every two years iNEMI maps future manufacturing technology needs of the global electronics industry for the next ten years. It discusses the major business and technology issues, paradigm shifts, emerging technologies and markets, technology gaps, and identified needs in each of 6 product segments: aerospace/defense, automotive, consumer/portable, medical, netcom, and office/large business systems as well as forecasting the technical evolution associated with about 20 different areas that support these product area needs. The resulting gaps identified between the product needs and technical capabilities form the basis for iNEMI projects and potential research programs within and outside iNEMI. This paper discusses the iNEMI roadmap process and highlights 1 product and 3 technology areas: Medical Products, MEMS (Micro Electromechanical Systems), Optoelectronics and Packaging Technologies.

INTRODUCTION

The iNEMI roadmap charts future opportunities and challenges for the electronics manufacturing industry. Our widely utilized roadmaps influence Research & Development (R&D) investments and technology deployment around the world.

Updated every two years, the roadmap sets the direction for technology development and deployment by predicting future packaging, component and infrastructure challenges and describing critical technical and business elements required to support industry growth. This information provides the basis for collaborative or individual company projects undertaken by our members. The projects deliver solutions to identified gaps that allow the industry to continue on its fast paced speed.

ROADMAP PROCESS

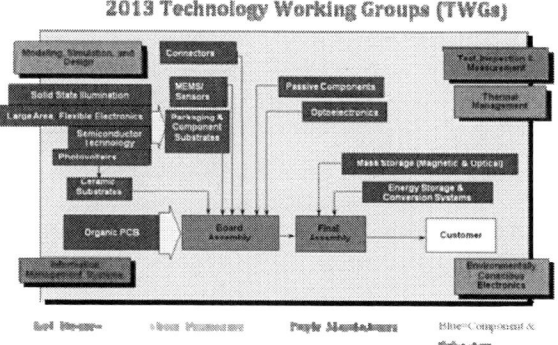

Figure 1. Twenty TWGs

The 2013 Roadmaps were developed by 20 Technology Working Groups, in response to inputs from representatives of OEMs in 6 Product Emulator Groups. These are; Automotive, Medical, Consumer / Portable, Office / Large Business Systems, Network /Datacom/ Telecom, Aerospace / Defense.

The leading electronic systems' manufacturers are basing their strategic planning for new products on the assumption that the electronics infrastructure will develop and implement the technology to meet these key drivers.

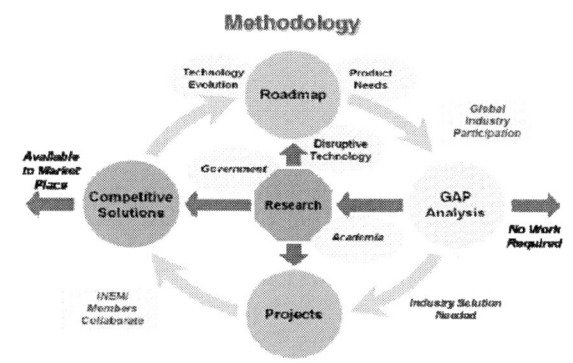

Figure 2. iNEMI Methodology

SITUATION ANALYSIS

Business/Technology

Driven by Performance and Size Requirements

The electronics industry has done better than most in seeing solid growing demand in many of the market segments as consumers continue to demand creative new products that can improve productivity, facilitate ubiquitous communication and entertain.

This demand for thin multifunctional products has led to increased pressure on alternative high density packaging technologies with High-density 3D packaging becoming the major technology challenge.

SiP (System in Package) is a Technology driver for small components, packaging, assembly processes and for high density substrates.

New sensors and MEMs are expected to see exponential growth driven by portable products with motion gesture sensors expanding use of 2D-axis & 3D-axis gyroscopes. This segment is maturing, encouraging industry collaboration.

Performance requirements such as increased bandwidth and lower power are driving the 3D ICs designed with through silicon vias.

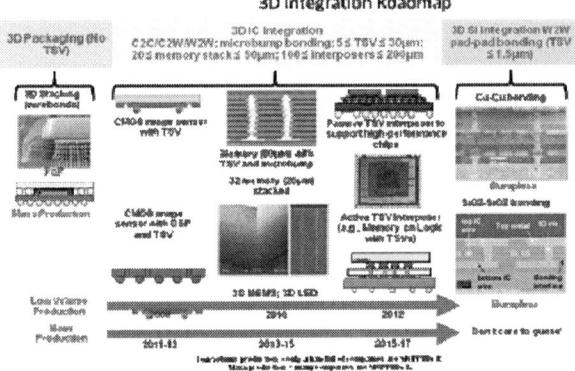

Figure 3: 3D Integration Roadmap

Rapid consumer product lifecycles quickly drive premium pricing towards commodity levels with only the most creative products enjoying a relatively longer period of healthy margins at the OEM level.

Market

The boundaries among computers, communications and entertainment products have blurred. Flat panel displays are the norm for virtually all applications with touch screen becoming more prevalent in a number of product categories. Wireless products continue to proliferate. Home and office functionality is being added to automotive products. The needs of the telecommunication and data communication infrastructures are converging. With the move to all digital communications and storage we see the convergence of a number of markets:

• Medical-Consumer

• Automotive-Entertainment

• Communication-Entertainment

• Computing-Entertainment

• Computing-Security

Technology

Multi-core processors are now the norm for most computing applications. A consequence of the expected demise of the traditional scaling of semiconductors is the increased need for improved cooling and operating junction temperature reduction.

The consumer's demand for thin multifunctional products has led to increased pressure on alternative high density packaging technologies. High-density three-dimensional (3D) packaging of complete functional blocks has become the major challenge in the industry. Other critical needs identified for 2013 are:

• New MEMS driven by Automotive, Medical, Games and Cell Phone applications

• Thermal Management for Portable Products

• Development of viable rework process for Pb-free soldering

• Cooling Solutions for Portable Electronics (3D-TSV)

• Reliability Evaluation and functional testing of MEMS

• Testing of Energy Managed modules

• Functional Testing of Complex SIPs

• Low Temperature Processing

Strategic Concerns

The 2013 iNEMI Roadmap has identified several strategic concerns listed here:

Restructuring from vertically integrated OEMs to multi-firm supply chains resulted in a disparity in R&D Needs vs. available resources.

Critical needs for R&D are struggling to be met due to the middle part of the Supply Chain being least capable of providing resources.

Industry collaboration is needed from all available sources such as University R&D centers, Industry consortia, and "ad-hoc" cross-company R&D teams.

The mechanisms for cooperation throughout the supply chain must be strengthened.

Cooperation among OEMs, ODMs, EMS firms and component suppliers is needed to focus on the right technology and to find a way to deploy it in a timely manner; collaboration is iNEMI's strength so we play an important role.

Paradigm Shifts

Several potential paradigm shifts have been identified in the roadmap for 2013. The first is a need for continuous introduction of complex multifunctional products to address converging markets which favors modular components or

SiP (2-D & 3-D). Reasons for this are modular components increase flexibility and shortened design cycles.

SiP/MCP FORECAST

Product/Package Type Volume (Bn Units)	2011	2016 Forecast	Leading Suppliers/Players
Stacked Die in Package	8.3	10.9	ASE, SPIL, Amkor, STATS ChipPAC, Samsung, Micron, Hynix, Toshiba, SanDisk
Stacked Package-on-Package (PoP/PiP)	0.7	1.5	Amkor, STATS ChipPAC, ASE, SPIL, TI, Samsung, Renesas, Sony, Panasonic
PA Centric RF Module	3.7	3.9	RFMD, Skyworks, Anadigics, Renesas, TriQuint, Avago
Connectivity Module (Bluetooth/WLAN)	0.5	0.6	Murata, Taiyo Yuden, ACSIP, ALPS
Graphics/GPU or ASIC MCP	0.1	0.2	Intel, IBM, Fujitsu
Leadframe Module (Power/Other)	3.0	5.0	NXP, STMicro, TI, Freescale, Toshiba, NEC, Infineon, Renesas, IR, ON Semi
TOTAL	14.8	22.1	

Table 1: Prismark Forecast for SiP/MCP

Cloud connected digital devices have the potential to enable major disruptions across the industry by causing major transitions in business models e.g. "rent vs. buy" for software (monthly usage fee model), new power distribution systems for data centers, huge data centers operating more like utilities (selling data services), slowing local compute and storage growth (as data moves to the cloud).

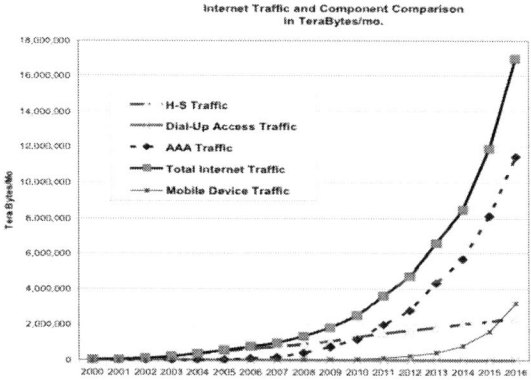

Figure 4: Forecast of Internet Traffic Growth

[Traffic by data content vs. year, both actual and forecast.
H-S= High Speed
Traffic; AAA=Advanced Architecture Traffic
Source: International Gatekeepers Inc. report "North American Network Traffic Forecast April, 2011"]

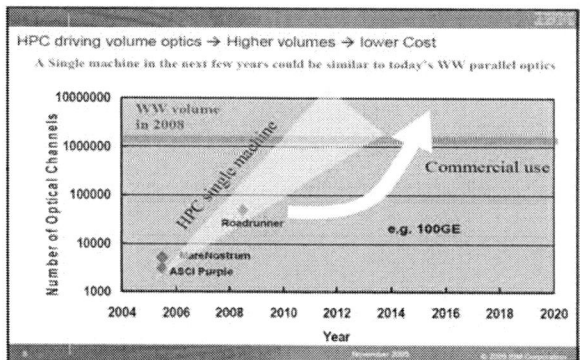

Figure 5: Potential Impact of HPC on Optical Interconnect Usage

Rapid evolution and new challenges in energy consuming products such as SSL (Solid State Lighting), Automotive and High Performance Computing will identify new unforeseen challenges.

Sensors everywhere – MEMS and wireless traffic!

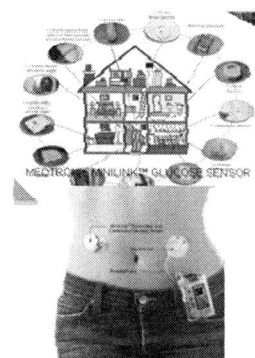

Figure 6: Example of a Body Area Network

Completing the 2013 iNEMI Roadmap
The 2013 iNEMI Roadmap Development Cycle is wrapping up! Global workshops were held in San Diego, CA 5/29/12; Berlin, Germany 6/12/12 and Hong Kong, China 6/14/12. Complete integration of chapters and editing is planned for 12/15/12 and will be made available to iNEMI Members by 12/30/12.

The 2013 iNEMI Roadmap will be introduced to the industry via public webinars on April 4, 2013. The 2013 iNEMI Roadmap will be available to the industry to order at www.inemi.org (watch web site for status). Individual roadmap chapters will also available for purchase as a PDF document at www.inemi.org.

ACKNOWLEDGEMENTS
The authors gratefully acknowledge the contributions of nearly 600 direct participants and hundreds more indirect participants in this cycle's roadmap. Many participants are members of the 14 collaborating organizations that contributed excerpts from their own roadmaps and helped to recruit participants from their membership.

3D IC INTEGRATION TECHNOLOGY DEVELOPMENT IN CHINA

Wei Koh, Ph.D.
Pacrim Technology
Irvine, CA, USA
kohmail@gmail.com

ABSTRACT

China's semiconductor foundry and microelectronic packaging industries are embracing the move to join 3D IC integration technology development with ample funding and rapid pace. An overview of the recent progress on the efforts in 3D IC integration technology development by the leading domestic companies and research institutes is provided here.

Because China still lacks in infrastructure for advanced and modern front end of the line (FEOL), backend of the line (BEOL), and middle-end of the line (MEOL) process capabilities for 300mm wafers, development efforts on 3D IC integration have many limitations to begin with. The efforts adopted by some leading research institutes and back-end packaging assembly and test companies; however, appear to be quite ingenuous and pragmatic, by selecting more easier "cutting-in" research projects and processes that require less initial capital investment and infrastructural establishments.

The pattern of latest development efforts can be divided into two major areas: 1) TSV materials, processing, and interconnection; 2) low-cost interposer and 2.5D integration assembly application in wafer level CSP packages for MEMS and sensors using small-sized wafers. The short term perspectives and longer term growth opportunities for China's indigenous development efforts on 3D IC integration is summarized in conclusion.

Key words: TSV, 3D IC Integration, Interposer

INTRODUCTION
China Semiconductor Industry

Before we discuss the red-hot topic of 3D IC integration technology, it is very helpful to review some background information and update on the status of China's semiconductor and back-end packaging assembly industry; how the industry's growth had been planned and cultivated to arrive at the present condition. Having established such bases of knowledge, it will then be much easier to understand and follow the approaches that the semiconductor and microelectronics assembly industries in China are taking to "catch up" with the 3D IC and TSV integration technologies.

In recent years (2010/2011), China purchased over 40% of world's semiconductor ICs worth more than $100 billion for use in electronic products and systems. Yet, ironically, less than five percent of these ICs were made locally by domestic manufacturers. There is, therefore, a large IC consumption/production gap that amounted to $87 billion in 2010 for China [1]. In fact, as of mid-2012, none of the top 20 of world's semiconductors suppliers (including foundries) is Chinese. For the record, the top 10 semiconductor suppliers for 1H2012 according to IC Insights [2] are shown in Table 1. Four countries contributed to the first nine suppliers: USA, Korea, Japan, and the Republic of China (Taiwan). China's top four semiconductor foundries--SMIC, Grace Semiconductor, HHNEC, and HLMC are not even on the list for the next 10 rankings. Furthermore, according to a recent forecast by SEMI on the 2013 semiconductor materials global market share [3], shown in Figure 1, China's share (as part of the "Rest of Asia" group of 12%) will be substantially behind its neighbors such as the Republic of China, Japan, and Korea. Per DigiTimes Research, the total China IC foundry industry output is estimated to be $3.29 billion for 2012 [4].

Table 1. Top Semiconductor Suppliers in 1H2012

Company	Country	Rank
Intel	USA	1
Samsung	Korea	2
TSMC	ROC (Taiwan)	3
TI	USA	4
Qualcomm	USA	5
Toshiba	Japan	6
Renesas	Japan	7
SK Hynix	Korea	8
Micron	USA	9
ST Microelectronics	Netherlands	10

Source: IC Insights McLean Report, Aug. 2012

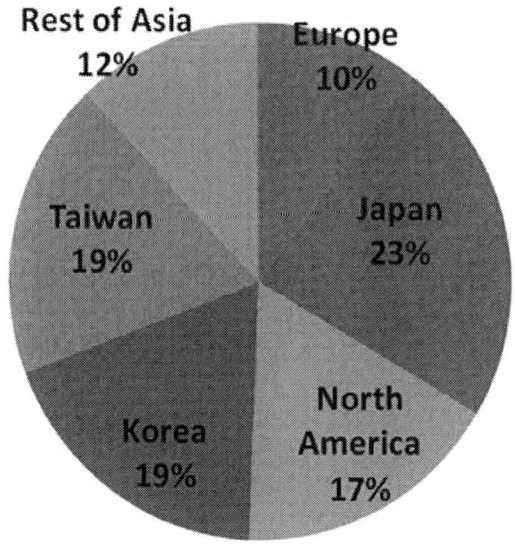

Figure 1 Global Market Share of Fab Materials (SEMI)

Because of this huge IC consumption/production deficit suffered by China and the relatively small scale of China's semiconductor suppliers, for over a decade, the Chinese government and its domestic semiconductor industry have been trying very hard to increase the capability in IC design and production capacity for silicon wafer fabs in order to remedy this situation. The most recent 12-5 Plan (12th Five-year plan for 2011-2015) has eight focal industries for major upgrade and funding. These industries are:

1. IT and electronics manufacturing
2. Steel
3. Automotive
4. Biotechnology
5. Machine manufacturing
6. Rare earth materials
7. Concrete materials
8. Aluminum industry

For high-end manufacturing, semiconductor and IT are of course selected as the key focal industries for funding to encourage growth and expansion of indigenous, homegrown semiconductor wafer fabs and back-end assembly industries. However, for many years, the barriers that have been hindering China's advance in semiconductor technology know-how and growth still existed:

- IC design capability—China still lags behind the Western and Japanese/Korean players in designing sophisticated, advanced ICs based on leading-edge process nodes such as 30nm and below. Such highly proprietary IP knowledge and experience cannot be gained overnight. Foreign players typically have prohibited the transfer or sales of advanced, leading-edge design and processing technology to China.
- Insufficient availability of talents of experienced engineers and technical personnel to carry out leading-edge research and development; domestic companies

often rely on technology collaboration with foreign players.

To remedy this situation, China government has implemented the "863 Project" on high-value, high-end advanced technologies that fund projects to research institutes and the industry to focus on research and development of next-generation technology platforms such as Internet of Things and Cloud Computing. Local companies are also more aggressive in requesting foreign partners to share or transfer their advanced IP and know-how in joint-venture investments. When the 3D IC integration using TSV technology came along in recent years, China sees it as a perfect opportunity and platform to further advance its semiconductor and backend packaging assembly industries.

China Semiconductor Foundries
Silicon wafer based CMOS technology form the largest segment in China semiconductor industry. The larger fabs and foundries, including SMIC, Grace Semi, HHNEC, and HeJian all have capabilities for processing 300mm wafers. However, there are also many smaller fabs focusing on other types of semiconductor, such as MEMS, III-V semiconductors, LED (light emitting diode), sensors and optoelectronics. In 2011, China already had 68 LED wafer fab companies. These types of fabs may use non-silicon wafers such as GAN and GaAs wafers. For example, the company CSMC still processes 150mm wafers for MEMS. Total investment spending in China for 2012, as shown in Table 2 [5], is catching up with EU and Japan.

Table 2. Global Semiconductor Industry Spending (K$)

	2011	2012	
Americas	8944	6158	
China	3126	2934	
EU/Mideast	3767	3041	
Japan	5525	3825	
Korea	7400	10255	
SE Asia	2471	1792	
Taiwan	7997	7048	
Total	39230	35053	

(SEMI "World Fab Forecast" Nov 2011) [4]

CHINA 3D IC TECHNOLOGY STATUS
The approaches taken by many of the domestic institutes and companies to pursue the 3D IC integration are similar: aiming at efforts that require less initial extensive capital investments and lower technical barriers. With plenty of funding, many companies started such approaches to initiate TSV and 3D stacking/wafer level packaging technology without extensive scope or broad objectives. Thus, while in other regions of the world, modern advanced 3D IC integration using TSV are aimed at applications for high performance memory on logic, high performance CPU and GPU and other ASIC chips as the "mainstream" focus, the approaches taken by China are

somewhat different, with beginning emphases on low cost TSV and 3D stacking applied to MEMS, sensors, and camera modules.

3D IC Research Topics
At the August, 2012 ICEPT in Quilin, China's premier packaging conference, no less than forty papers dealt with 3D IC integration and through silicon via (TSV). Considering that TSV and 3D stacking is still in an embryonic development stage for China's backend manufacturing industry, such pace of progress is indeed quite impressive. By reviewing the relevant reports given in this conference, a quick glimpse of the development efforts undertaken by the leading institutes may be revealed. Table 3 lists the four topics of interest: TSV process and materials; wafer thinning, handing, and bonding; silicon and glass interposer for 2.5D integration; and 3D packaging and applications. For each technical field, the contributing institutes and companies are listed.

Table 3. Summary of Topics and Contributors

Research Topics	Technical Field	Institute/Company
TSV	Wet etching Copper plating Stress analysis Modeling/test	SIMIT-CAS Peking Univ, Huazhong Univ. Tsinghua Univ. IME-CAS, Fudan Xidian Univ. BUT
Wafer Handling	CMP process Carrier-less Low temperature bond Glass bonding	Tsinghua Univ. Huazhong Univ. SIMIT-CAS Peking Univ.
Interposer	Glass TGV Micro bumping Silicon interposer	IME-CAS Shanghai JiaoTong BUT
Packaging	Wafer level package LED MEMS Sensors Optical	JCAP Co., Fudan Univ., Peking U. Shanghai Univ. Guilin Univ. Peking Univ. IME-CAS

(**Note:** SIMIT=Shanghai Institute of Microsystem and Information Technology, CAS=Chinese Academy of Sciences. IME=Institute of Microelectronics, BUT= Beijing University of Technology, JCAP= Jiangyin Changdian Advanced Packaging Co.)

It can be said that the major 3D IC integration technology development efforts are spear-headed by CAS--the Chinese Academy of Sciences and some of the major State Key Laboratories. Within CAS, the two largest research organizations are the Institute of Microelectronics (IME) in Beijing and the Shanghai Institute of Microsystems and Information Technology (SIMIT). Top notch universities such as Tsinghua University, Peking University, Shanghai JiaoTong University, Fudan University, and Huazhong University all have research projects relating to 3D TSV, interposers, and 3D WLP packaging applications. In the following section, detailed research reports and results for some of the studies are presented.

TSV Materials and Process Development
Many leading research institutes and semiconductor foundries are devoting their development efforts on TSV. Some sample research projects are described below.

Void-Free Bottom-up Via Filling Process
Peking University researchers have studies focused on TSV copper fill plating materials and process optimization using plating solutions provided from a local company, Shanghai Sinyang Corp. In one paper [6], they compared bath solutions with different levels of additives to examine the effects of additives including suppressors, accelerators, and levelers. A numerical simulation model was employed to describe the "absorption" (note: should be adsorption) of the suppressor and accelerator behavior during the plating process. The experimental result of filled TSV cross-section using one sample bath solution is shown in Figure 2; and that using an optimized bath is given in Figure 3.

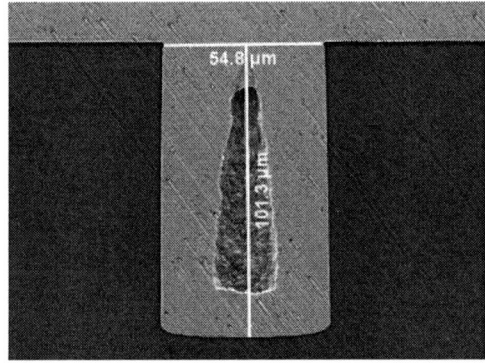

Figure 2. TSV Fill with Solution A

Figure 3. TSV Fill with Solution B

Low-cost TSV for MEMS Application
A rising star in MEMS and optoelectronics research, Huazhong University (Wuhan) published quite a few papers on TSV, particularly low cost process for use in 4-in wafers used in MEMS packaging. In one example [7], TSV having 60 μm size diameters are made by DRIE in 4-

inch wafers that are 370 μm thick, as shown in Figure 4. After bottom-up copper plating (Figure 5) showing cross-section), the wafer was sealed to a Pyrex7740 glass plate by anodic bonding; no CMP process was employed prior to the bonding to save cost. The authors claim that as long as the protruded caps on the top of TSV are less than 100 μm in height, the bonding results remained satisfactory.

Figure 4. DRIE etched TSV

Figure 5. TSV after Bottom-up Copper Fill

Thermal Modeling of TSV in Package
Another Xidian University research paper [8] compared thermal distribution inside a package with TSV interposer underneath a power chip and compared that for a wire bonded package. There is a slight lowering in the temperature distribution. The study also found that the density of TSV (i.e., smaller pitches) would influence the thermal distribution. Figures 6 and 7 show the modeling analysis results for the TSV package and a conventional wire bond package, respectively.

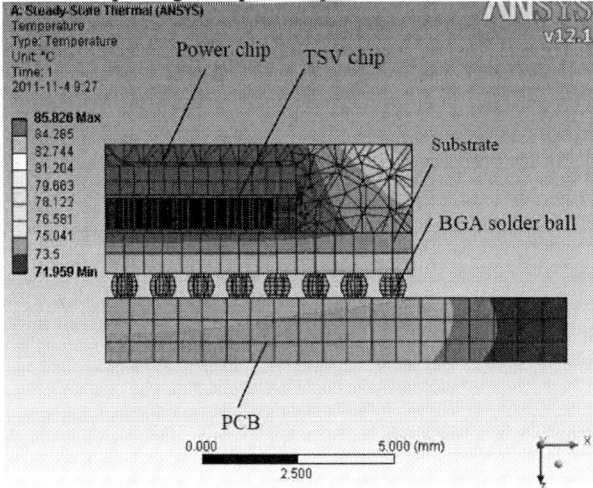

Figure 6. Temperature Distribution in a TSV substrate Package

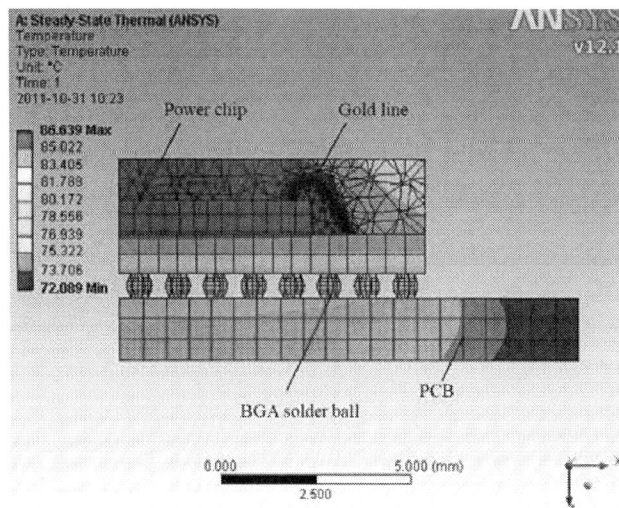

Figure 7. Temperature Distribution in a Wire Bond Package

Wafer Thinning-CMP Process
Tsinghua University presented a compressive review of the CMP process (Figure 8) and slurry application [9]. Using DOE studies of different slurry compositions, they found that higher peroxide concentration slurries should be used for wafers with thicker surface copper, due to accelerated CMP rates. The optimized slurry also resulted in very uniform wafer thickness, as shown in Figure 9, four different areas of a 50 um thinned wafer all measured to have a thickness of 48um, with the surface uniformity of less than 16nm RMS obtained.

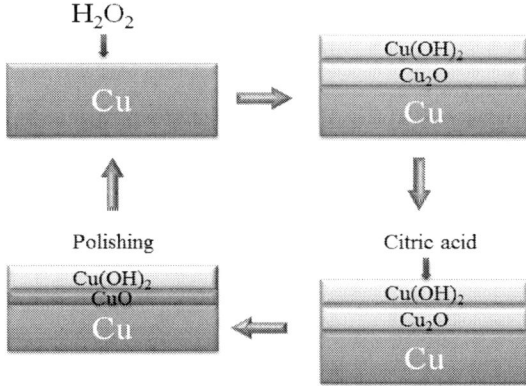

Figure 8. Peroxide based Slurry for CMP Process

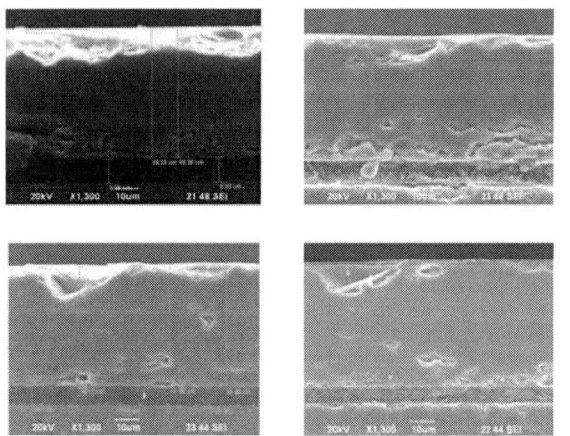

Figure 9. Thickness Measurement of Four Different areas in a 50um thinned Wafer

Low Temperature Wafer to Wafer Bonding

For applications in MEMS, this Peking University investigation [10] used two 4-in. silicon wafers with sputter coated Sn/Al surface, each 500 nm thick. The bonding conditions employed low temperature (280°C), low pressure (0.25MPa), and short duration (3 minutes) in vacuum. After bonding, diced chips (10 mm x 5 mm) are examined for bond integrity and strength. Shear strengths between 3.1 and 5.7MPa are found. Figure 10 illustrates SEM images of fractured bond metal on the wafer surface after shearing.

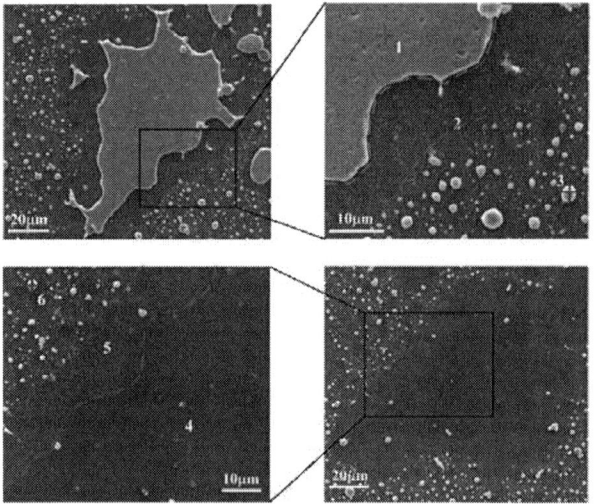

Figure 10. SEM Images of Fractured Metal Surfaces between top and bottom wafer after Shearing

INTERPOSER AND 2.5D INTEGRATION

In the package structure shown in Figure 11 by SIMIT-CAS, a silicon TSV interposer is bonded to a Si chip using indium micrbumps [11]. While fabricating the high-aspect ratio interposer with TSV, a supporting Si carrier must first be bonded to the bottom side of the interposer to form the bottom seed layer for subsequent bottom-up copper filling of the TSV. An Au-to-Au wafer level diffusion bonding process is applied as illustrated in Figure 12. With a 300°C

bonding temperature, the two separate gold surface coatings formed a join layer after solid state diffusion and remain electrically conductive. When the bonding temperature is increased to 400°C, however, some Si and SiO2 may diffuse into the gold layer, forming an Au/Si eutectic alloy layer that is unsuitable to use as the conductive seed layer for copper electroplating in the TSV.

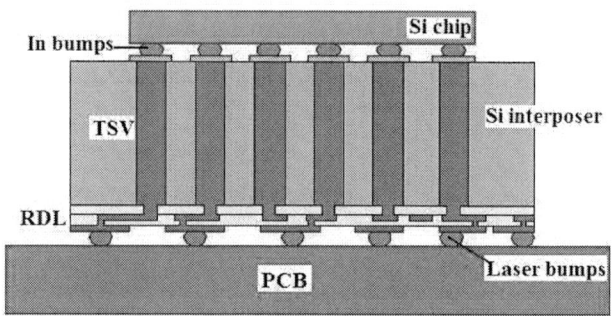

Figure 11. SIMIT-CAS 3D TSV Interposer

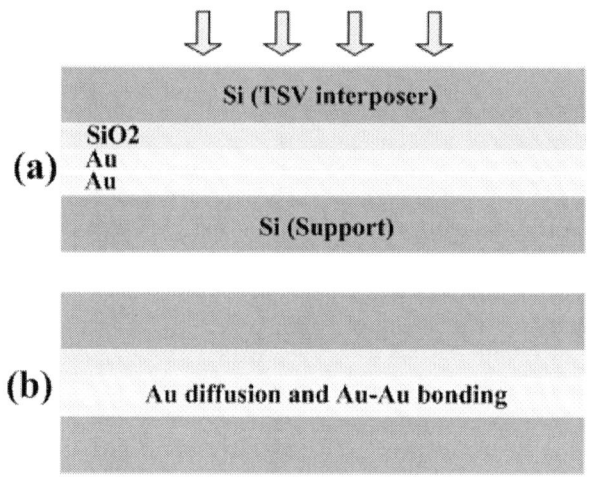

Figure 12. Au-to-Au Diffusion Bonding

Huawei 2.5D Packaging

Recently, Huawei announced an ambitious program with Altera to make 2.5 D interposer for integrating FPGA and wide I/O memory for networking applications [12]. Jiangyin Changdian Advanced Packaging Co. (JCAP) in Jiangsu province, who may be a partner of Huawei, announced an agreement recently signed with Singapore A*Star IME on jointly developing through silicon interposer (TSI) technology [13]. Figure 13 is an IME slide showing its 2.5D approach.

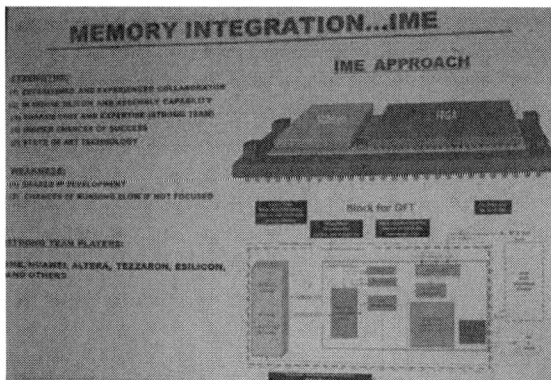

Figure 13. IME TSI 2.5D Package Concept

FOUNDRY EFFORTS

SMIC is also active in 3D IC, it formed a TSV technology department and its SVP, Dr ShiuhWuu Lee, presented a talk called "SMIC's Perspectives, Current Activities and High level plan on 2.5D/3D IC" at the 2012 SEMICON China 3DIC Technology Forum in March, 2012.

The upcoming SEMICON China 2013 (March 19, 2013, Shanghai) will have a 3D IC technology forum session, topics of interest include 3D IC design and manufacturing, EDA design tools, and IC manufacturing technology for telecom and wireless applications.

CONCLUSIONS

A multi-pronged progress in 3D IC integration research has been made by several leading research institutes and companies such as Huawei and JCAP within just a few years. Many applications are aimed at using 4 and 6-in size wafers, where wafer thinning, bonding and TSV fabrication are relatively easier compared to that for 300mm wafers. The near term fruit and resulting strength in China's efforts will be in practical, low-cost applications for camera modules, sensors, and MEMS. Some companies (Huawei, SMIC) are also planning to move into more complex interposers and 2.5D integration applications. In China, there are still plenty of market opportunities and growth potentials for mobile and computing devices to adopt 3D IC integration technologies for eventual high volume manufacturing in the coming years.

REFERENCES

1. "Continued Growth: China's Impact on the Semiconductor Industry 2011 Update" PWC, Nov. 2011.
2. IC Insights Strategic Reviews Database, McClean Report, August, 2012.
3. Dan Tracy, Japan 2012 Semi Market, Oct 3, 2012 SEMI.
4. N. Chai, "2013 Greater China IC Foundry Industry Forecast," DIGITIMES Research, Oct. 2012.
5. SEMI "World Fab Forecast" Nov. 2011.
6. Y. Zhu, et al, "Effect of Additives on Copper Electroplating Profile for TSV Filling," ICEPT-HDP 2012, Guilin, China, pp. 56-59.

7. C. Xu, et al, "Void Free Filling of TSV Vias by Bottom up Copper Electroplating for Wafer Level MEMS Vacuum Packaging," Ibid., pp. 64-67.
8. W. Tian et al, "TSV Modeling and Thermal Analysis Based on 3D Package," Ibid. pp. 546-548.
9. Z. Liu et al, "Copper Chemical Mechanical Polishing and Wafer Thinning with Temporary Bonding for Through Silicon Via Interconnect," Ibid. pp. 488-493.
10. Z. Zhu et al, "Low Temperature Al based Wafer Bonding using Sn as Intermediate Layer," ibid, pp. 127-129.
11. X. Chen et al, "TSV Interposer with Au-Au Diffusion Bonding Technology for Wafer Level Fabrication," Ibid. pp. 926-929.
12. R. Merritt, "Huawei, Altera mix FPGA, memory in 2.5D Device," EETimes, Nov. 14, 2012.
13. "A*Star Institute of Microelectronics and Huawei Announce Joint Effort to Develop 2.5D/3D Through-Silicon Interposer Technology," Press Release, August 17, 2012 by A*Star IME, Singapore.

Author Index

2013 Pan Pacific Symposium

Anselm, Ph.D., Martin	Tools and Techniques for Material Assessment in Advanced Technologies
Anselm, Ph.D., Martin	A Mechanistically Justified Model for Life of SnAgCu Solder Joints in Thermal Cycling
Aschenbrenner, Ph.D., R.	Panel Level Packaging – A Manufacturing Solution for Cost Effective Systems
Bader, Bill	2013 iNEMI Technology Roadmap Overview
Bailey, Ph.D., Chris	Impact of Lead-Free Components and Technology Scaling for High Reliability Applications
Bauer, Ph.D., Charles E.	Alternatives to Solder in Interconnect Packaging and Assembly
Becker, K-F	Panel Level Packaging – A Manufacturing Solution for Cost Effective Systems
Bergman, Dieter W.	The Quest for Reliability Standards
Berndt, Hartmut	Requirements on a Class "0" EPA – Basics Standards ESD Equipments and Measurements
Blazej, Ph.D., Daniel	Aerosol Jet® Printing of Conductive Epoxy for 3D Packaging
Borgesen, P.	A Mechanistically Justified Model for Life of SnAgCu Solder Joints in Thermal Cycling
Borkes, Tom	A New Manufacturing Model for Successfully Competing in High Labor Rate Markets
Bowen, Terry	Silicon V-Groove Alignment Bench for Optical Component Assembly
Braun, T.	Panel Level Packaging – A Manufacturing Solution for Cost Effective Systems
Bruderer, A.	Dielectrics for Embedding Active and Passive Components
Chang, Jackson	The Challenges of LGA Server Socket Trends
Chen, Chih	Microstructure Control of Uni-Directional Growth of η-Cu6Sn5 in Microbumps on (111) Oriented and Nanotwinned Cu
Chen, Chih	Side Wall Wetting Induced Void Formation Due To Small Solder Volume in Microbumps of Ni/SnAg/Ni Upon Reflow
Chen, Delphic	Microstructure Control of Uni-Directional Growth of η-Cu6Sn5 in Microbumps on (111) Oriented and Nanotwinned Cu
Chen, Ming-Kun	Design and Fabrication of Ultra-Thin Flexible Substrate

Chen, Qiao	Low-cost and High Performance Silicon Interposers and Packages (LSIP) – A New Georgia Tech PRC Industry Consortium
Cho, SH	Dielectrics for Embedding Active and Passive Components
Choi, Yongwon	3D-TSV Vertical Interconnection Using Cu/SnAg Double Bumps and Non-Conductive Films (NCFs)
Christenson, Ph.D., Kurt K.	Aerosol Jet® Printing of Conductive Epoxy for 3D Packaging
Cuthbert, Keith	Tamper Proof Tamper Evident Encryption Technology
Farajzadeh, A.	3D Integration A Thermal-Electrical-Mechanical-Reliability Study
Fisher, Michael J.	Tamper Proof Tamper Evident Encryption Technology
Forsythe, Thomas M.	Cleaning High Reliability Assemblies with Tight Gaps a Detailed Analysis
Frémont, H.	3D Integration A Thermal-Electrical-Mechanical-Reliability Study
Fu, Shen-Li	Design and Fabrication of Ultra-Thin Flexible Substrate
Galster, N.	Dielectrics for Embedding Active and Passive Components
Gattuso, Andrew	The Challenges of LGA Server Socket Trends
Giroux, Donald	Aerosol Jet® Printing of Conductive Epoxy for 3D Packaging
Huang, Yu-Jung	Design and Fabrication of Ultra-Thin Flexible Substrate
Hung, Michael	The Challenges of LGA Server Socket Trends
Isaacs, Phil	Tamper Proof Tamper Evident Encryption Technology
Jones, Wayne	Tools and Techniques for Material Assessment in Advanced Technologies
Kludt, J.	3D Integration A Thermal-Electrical-Mechanical-Reliability Study
Koh, Ph.D., Wei	3D IC Integration Technology Development in China
Komatsu, Teruo	A Nano Silver Replacement for High Lead Solders in Semiconductor Junctions
Kress, J.	Dielectrics for Embedding Active and Passive Components
Kumar, Gokul	Low-cost and High Performance Silicon Interposers and Packages (LSIP) – A New Georgia Tech PRC Industry Consortium
Kuo, Jui-Chao	Microstructure Control of Uni-Directional Growth of η-Cu6Sn5 in Microbumps on (111) Oriented and Nanotwinned Cu
Liang, Y.C.	Side Wall Wetting Induced Void Formation Due To Small Solder Volume in Microbumps of Ni/SnAg/Ni Upon Reflow
Liao, Bono	The Challenges of LGA Server Socket Trends
Lin, Han-Wen	Microstructure Control of Uni-Directional Growth of η-Cu6Sn5 in Microbumps on (111) Oriented and Nanotwinned Cu
Lin, Nick	The Challenges of LGA Server Socket Trends

Lin, Yi-Lung	Design and Fabrication of Ultra-Thin Flexible Substrate
Liu, Chen-Min	Microstructure Control of Uni-Directional Growth of η-Cu6Sn5 in Microbumps on (111) Oriented and Nanotwinned Cu
Lu, Hao	Low-cost and High Performance Silicon Interposers and Packages (LSIP) – A New Georgia Tech PRC Industry Consortium
Lu, Jia-Ling	Microstructure Control of Uni-Directional Growth of η-Cu6Sn5 in Microbumps on (111) Oriented and Nanotwinned Cu
McHugh, Bob	The Challenges of LGA Server Socket Trends
Mohammed, Ilyas	BVA: Solution for Next Generation Very Fine-Pitch Package-on-Package (PoP) Applications
Morris, Jr., Thomas	Tamper Proof Tamper Evident Encryption Technology
Neuhaus, Ph.D., Herbert J.	Alternatives to Solder in Interconnect Packaging and Assembly
Nishimura, Tetsuro	A Nano Silver Replacement for High Lead Solders in Semiconductor Junctions
Ostmann, A.	Panel Level Packaging – A Manufacturing Solution for Cost Effective Systems
Paik, Kyung-Wook	3D-TSV Vertical Interconnection Using Cu/SnAg Double Bumps and Non-Conductive Films (NCFs)
Park, R.	Dielectrics for Embedding Active and Passive Components
Qasaimeh, A.	A Mechanistically Justified Model for Life of SnAgCu Solder Joints in Thermal Cycling
Quinones, Ph.D., Horatio	Jetting Fine Lines for High Viscosity Fluids onto 2D and 3D Electronic Packages
Renn, Ph.D., Michael J.	Aerosol Jet® Printing of Conductive Epoxy for 3D Packaging
Richardson, Chuck	2014 iNEMI Technology Roadmap Overview
Schlobohm, J.	3D Integration A Thermal-Electrical-Mechanical-Reliability Study
Semmens, Janet E.	Acoustic Micro Imaging Analysis Methods for 3D Packages
Shea, Chrys	Technical Communication: Strategies for Success
Shin, Jiwon	3D-TSV Vertical Interconnection Using Cu/SnAg Double Bumps and Non-Conductive Films (NCFs)
Solberg, Vern	BVA: Solution for Next Generation Very Fine-Pitch Package-on-Package (PoP) Applications
Sundaram, Ph.D., Venkatesh	Low-cost and High Performance Silicon Interposers and Packages (LSIP) – A New Georgia Tech PRC Industry Consortium
Sweatman, Keith	A Nano Silver Replacement for High Lead Solders in Semiconductor Junctions
Tomokage, Ph.D., Hajime	Three Dimensional Integration Research Focusing on Device Embedded Substrate

Tu, King-ning	Microstructure Control of Uni-Directional Growth of η-Cu6Sn5 in Microbumps on (111) Oriented and Nanotwinned Cu
Tu, King-ning	Side Wall Wetting Induced Void Formation Due To Small Solder Volume in Microbumps of Ni/SnAg/Ni Upon Reflow
Tummala, Ph.D., Rao R.	Low-cost and High Performance Silicon Interposers and Packages (LSIP) – A New Georgia Tech PRC Industry Consortium
Weide-Zaage, K.	3D Integration A Thermal-Electrical-Mechanical-Reliability Study
Wren, Wesley F.	Computed Tomography on Electronic Components-Better Ways to Do Failure Analysis Plus 4D CT The New Frontier
Yang, L.	A Mechanistically Justified Model for Life of SnAgCu Solder Joints in Thermal Cycling
Yin, L.	A Mechanistically Justified Model for Life of SnAgCu Solder Joints in Thermal Cycling